International Conference on The Impact of Digital Microelectronics and Microprocessors on Particle Physics

International Conference on The Impact of Digital Microelectronics and Microprocessors on Particle Physics

Microprocessor Laboratory, Trieste
28-30 March 1988

Editors
M Budinich
E Castelli
A Colavita

World Scientific
Singapore • New Jersey • London • Hong Kong

Published by

World Scientific Publishing Co. Pte. Ltd.
P O Box 128, Farrer Road, Singapore 9128

USA office: World Scientific Publishing Co., Inc.
687 Hartwell Street, Teaneck, NJ 07666, USA

UK office: World Scientific Publishing Co. Pte. Ltd.
P O Box 379, London N12 7JS, England

INTERNATIONAL CONFERENCE ON THE IMPACT OF DIGITAL MICROELECTRONICS AND MICROPROCESSORS ON PARTICLE PHYSICS

ISBN. 9971-50-742-0

Printed in Singapore by Utopia Press.

Foreword

It is completely pointless to discuss the merits of using microprocessors in all fields of physics and particularly so in High Energy Physics (HEP) experiments. We can safely say that HEP experiments can no longer be done without them. The same is true for most fields within experimental - and now also theoretical - physics.

The International Centre for Theoretical Physics (ICTP) perceived this change at the beginning of the decade, hence it decided to organize a programme to spread the knowledge of this technology mainly for the benefit of Third World scientists. ICTP requested the help of CERN, wich at that time was preparing a similar activity for its own internal use. The first course was held in Trieste in 1981 and since then approximately one thousand scientists have passed through the various colleges held in Trieste and outside (Sri Lanka, Portugal, Colombia, Mexico, China, Argentina).

As the sophistication and quantity of the course equipment increased, the mere complexity required a permanent space and staff to take care of the logistics involved. At the same time, the need of a close connection between the training activity and the up-to-date research in the microelectronics field was recognized.

As a consequence, the Microprocessor Laboratory was created in 1985 as a joint venture between the International Centre for Theoretical Physics (ICTP) and the Italian National Institute for Nuclear Physics (INFN), supported by the United Nations University. The Laboratory has a dual role as it is meant to expand the knowledge of microprocessor and digital electronics in the Third World and on the other hand to develop specialized instrumentation for the needs of frontier physics experiments, in particular HEP experiments.

As part of the activities of the Laboratory, its Scientific Committee decided to organize a Conference on the uses of microprocessors in high energy experiments and closely related applications. During the meeting the advances of on-line and off-line applications were reviewed allowing lively discussions to take place.

All the sessions had invited and contributed papers. Most of the contributed papers were presented as posters due to the lack of time for session presentations. Scientists from all parts of the world contributed and attended the meeting, making it a truly international conference.

There is no need to summarize the results and news presented during the conference; this is left to the reader who will find in the following pages the "state of the art" as it was at that time.

The Conference took place in the Main Building of ICTP, which also provided for most of the physical needs (lecture rooms, lodging, etc.) of the meeting, as well as the secretarial staff for such a large undertaking.

It is a pleasure for us to thank here all the people who helped in the organization of this meeting and in particular the conference secretary, Ms. Ave Lusenti, for her excellent work.

M. Budinich

E. Castelli

A. Colavita

Trieste, July 1988

Scientific Advisory Committee

Barrelet E.	University of Paris VI
Cabibbo N.	University of Roma-Tor Vergata and INFN
Cittolin S.	CERN
Grayer G.	Rutherford Laboratory
Halatsis C.	University of Thessaloniki
Hertzberger L. O.	University of Amsterdam
Innocenti P. G.	CERN
Kunz P.	SLAC
Nash T.	FNAL
Notz D.	DESY
van Dam A.	Brown University
von Ruden W.	CERN
Williams D. O.	CERN
Zanella P.	CERN

Organizing Committee

Bertocchi L.	ICTP, Trieste
Budinich M.	University of Trieste
Castelli E.	University of Trieste and INFN
Centro S.	University of Padova
Colavita A.	ICTP and INFN, Trieste
Gavillet Ph.	CERN
Verkerk C.	CERN
Zanello L.	University of Roma-La Sapienza

Conference programme

Monday March 28 1988

TIME		SPEAKER	TITLE
8:00			Registrations.
Session I		Chairman: S.Centro	
9:00		A.Salam	Welcome address.
9:10	Invited Talk	I.Gaines	Multi-Processor Developments in the USA For the Future HEP Experiments and Accelerators
9:50	Invited Talk	S.Falciano	Trigger for LEP Experiments
10:30	Contribution	S.Quinton	The Delphi First and Second Level Trigger
10:50		COFFEE BREAK	
Session II		Chairman: P.G.Innocenti	
11:20	Invited Talk	G.Heath	Trigger Problems for Future Accelerators
12:00	Contribution	J.J.Thaler	Design Automation for SSC Data Acquisition
12:20	Contribution	T.Glanzman	Fastbus Based Software Trigger for Mark II at SLC
12:40	Contribution	R.Tripiccione	From APE to APE-100: present and future of the APE project
13:00		LUNCH	
Session III		Chairman: W. von Rüden	
14:30	Invited Talk	G.M.McPherson	Special Purpose Processor for High Energy Physics
15:10	Contribution	J.Lecoq	The Second Level Trigger of L3 Experiment
15:30	Contribution	Ch.Ender	Multiprocessor Event Filtering at the Heidelberg/Darmstadt Crystal Ball
15:50	Contribution	V.Mertens	EVI: a High Speed Interface between Fastbus and VAX-BI
16:10		COFFEE BREAK	
Session IV		Chairman: D.O.Williams	
16:45	Invited Talk	P.G.Innocenti	Microprocessors in LEP accelerator control
17:25	Contribution	D.Cutts	Data Acquisition Hardware for the D0 MicroVAX Farm
17:45	Contribution	B.Jost	The Aleph Event Processor
18:05	Contribution	P.Giannetti	A Multiuser Data Acquisition Based on the Macintosh Computer
18:25	Contribution	P.Le Dû	The LEP OPAL Event Selection System
18:45		END OF FIRST DAY	
		WELCOME PARTY	

Tuesday March 29 1988

TIME		SPEAKER	TITLE
Session V		Chairman: R.Mount	
9:00	Invited Talk	A.Marchioro	Multiprocessor Systems in Fastbus
9.40	Contribution	D.Crosetto	Use of Digital Signal Processors (DSP) in High Energy Physics Experiments
10.00	Contribution	N.Bingefors	SIROCCO IV: Front End Readout Processor for the Delphi Microvertex
10:20	Contribution	M.Pitke	Architecture of a Parallel Processor for Image Processing Application
10:40		COFFEE BREAK	
Session VI		Chairman: G.Grayer	
11:00	Poster session		Please stay near your poster to give informations
11:40	Contribution	J.Perrier	Charged Particle Trigger for the L3 Detector
12:00	Contribution	S.Jaroslawski	A Fast Track Trigger Processor for OPAL Experiment
12:20	Contribution	E.Eisenhandler	The New UA1 First Level Trigger Processor
12:40		LUNCH	
Session VII		Chairman: P.Gavillet	
14:30	Invited Talk	G.Parisi	Neutral Networks
15:10	Invited Talk	S.Conetti	On-line Application of the ACP
15:50	Contribution	B.Denby	Neural Networks & Cellular Automata in Experimental High Energy Physics
16:10		COFFEE BREAK	
Session VIII		Chairman: L. Zanello	
16:45	Invited Talk	T.Toffoli	Cellular Automata
17:25	Invited Talk	E.Petrolo	Use of Optical Data Transmission in HEP
17:45	Contribution	P.Finocchiaro	The Data Acquisition System for the Crystal Ball at LNS
18:05	Contribution	R.McLaren	The CERN Host Interface Optical Interconnect
18.25		END OF SECOND DAY	

Wednesday March 30 1988

TIME		SPEAKER	TITLE
Session IX		Chairman: I.Gaines	
9:00	Invited Talk	H.Heijne	Data Processing for Particle Physics in a Silicon Detector Environment
9:40	Contribution	M.J.French	An OPAL 32 BIT Coincidence Array Integrated Circuit
10:00	Contribution	J.P.Vanuxem	Application of Bipolar Cell Array Technology to the development of a Time Digitizer
10:20	Contribution	G.Darbo	The Contiguity Processor. A SIMD Architecture for a Second Level Track Trigger
10:40		COFFEE BREAK	
Session X		Chairman: D.Notz	
11:00	Invited Talk	D.May	Transputers & Occam
11:40	Contribution	J.Carter	Recent Experience with Transputer Based Processor Farm
12:00	Contribution	J.C.Vermeulen	A Transputer Based Second-Level Calorimeter Trigger System for the ZEUS Experiment
12:20	Contribution	T.Nash	A Site Oriented Supercomputer for Theoretical Physics
12:40		LUNCH	
Session XI		Chairman: B.Dobinson	
14:30	Invited Talk	H.Spieler	Integrated Microsystems as Driving Force in Future Detector Design
15:10	Contribution	L.Ristori	VLSI Structures for Track Finding
15:30	Contribution	K.Einsweiler	The UA2 Data Acquisition System
15:50	Contribution	S.Stapnes	The XOP Trigger Processor Integrated into a High Energy Physics Experiment
16:10		COFFEE BREAK	
Session XII		Chairman: E.Castelli	
16:30	Contribution	J.Hoek	SU(3) Lattice Gauge Theory Calculations on T800 Transputer Arrays
16:50	Invited Talk	T.Nash	Concluding Remarks
17:30		END OF THE CONFERENCE	

TABLE OF CONTENTS

Multi-Processor Developments in the United States for Future High Energy Physics Experiments and Accelerators *

Irwin Gaines

For The Advanced Computer Program
H. Areti, R. Atac, J. Biel, A. Cook, J. Deppe, M. Edel, M. Fischler, R. Hance, D. Husby, T. Nash, T. Pham, and T. Zmuda
Fermi National Accelerator Laboratory+
Batavia, Illinois 60510 USA

ABSTRACT

The use of multi-processors for analysis and high-level triggering in High Energy Physics experiments, pioneered by the early emulator systems, has reached maturity, in particular with the multiple microprocessor systems in use at Fermilab. It is widely acknowledged that such systems will fulfill the major portion of the computing needs of future large experiments.

Recent developments at Fermilab's Advanced Computer Program will make such systems even more powerful, cost-effective, and easier to use than they are at present. The next generation of microprocessors, already available, will provide CPU power of about one VAX 780 equivalent/$300, while supporting most VMS FORTRAN extensions and large (>8MB) amounts of memory. Low cost high density mass storage devices (based on video tape cartridge technology) will allow parallel I/O to remove potential I/O bottlenecks in systems of over 1000 VAX equivalent processors. New interconnection schemes and system software will allow more flexible topologies and extremely high data bandwidth, especially for on-line systems. This talk will summarize the work of the Advanced Computer Program and the rest of the U.S. in this field.

1. INTRODUCTION

High-Energy physicists have always wanted more computer power than they could afford to buy in the commercial marketplace. However, the natural parallelism inherent in the HEP computing problem (running the identical program on many millions of different events) suggests a simple parallel processing solution. The pioneering emulator work by Kunz *et al.* at SLAC demonstrated the feasibility of using multiprocessor systems to provide cost-effective computing. Indeed, one of the earliest nine processor 168E systems is still doing work today running Monte Carlo programs for the LASS detector, almost a decade after it was first commissioned.

More recently, the advent of powerful 32-bit microprocessors has allowed the development of even more convenient and cost-effective parallel processing systems. Such systems are now an acknowledged important component of computing in high-energy physics.

---

*Talk given by Irwin Gaines at the Adriatico Conference on the "Impact of Digital Microelectronics and Microprocessors on Particle Physics," International Centre for Theoretical Physics, Trieste, Italy, March 28-30, 1988.

+Fermilab is operated by Universities Research Association, Inc., under contract with the U.S. Department of Energy.

They have made crucial contributions to data analysis in a number of current generation experiments, and will be indispensable for the large experiments of the SSC era.

In particular, the systems developed by the Advanced Computer Program (ACP) at Fermilab are rapidly becoming a standard, with over 30 installations in universities and laboratories worldwide. The first generation ACP systems, which provide CPU power at a cost of less than $2500 per VAX 780 equivalent, were brought on-line two years ago.[1] The initial 110 processor system has been heavily used at Fermilab for physics event reconstruction during this period, while a number of additional systems for both off-line and on-line use are currently being installed.

In this paper I will discuss the new developments which will lead to ACP systems of close to an order of magnitude greater cost effectiveness in the next year. This second generation ACP project will also allow much higher bandwidth for both I/O and interprocessor communication, and will have software tools allowing almost any UNIX or VMS based processor to be used as a node in a multiprocessor ACP system. I will also describe several other U.S. processor projects, concentrating on off-line aspects, since on-line use will be covered in Conetti's talk at this conference.

2. NON-ACP PROJECTS

Although there are no plans for future generations of emulators, both the early 168E and current 3081E systems are in use for data analysis and Monte Carlo. 3081E production systems are running at SLAC, Santa Cruz, Harvard and Oklahoma.[2] They are also being used on-line in a FASTBUS environment for triggering the MARK II experiment at SLC.

The most promising commercial multiprocessor system is scheduled to be installed in May at the University of Florida for use by the CLEO collaboration. It will consist of 32 MicroVAX 3200's (without the keyboard, monitor and graphics boards), 6 full MicroVAX 3200 workstations, and 5 MicroVAX 3600's connected by Ethernet, with 8 MBytes of memory on each processor and 5 RA82 disks (a total of 3 GBytes). The system will be run as a loosely coupled multiprocessor, with VMS acting as a file server for the diskless 3200's, although ACP-like master/slave host/node software may also be written for the system. This system will provide over 100 VAX 780 equivalents of processing power in a convenient package, at a cost (based on a research agreement with Digital) of under $4000 per VAX 780 equivalent. Costs on the open market will likely be somewhat higher. Disadvantages of this system are the limited bandwidth provided by Ethernet and the relatively high cost, particularly compared with second generation ACP-like systems. It is nevertheless an important indication that a leading computer company recognized and is beginning to respond to the availability of low-cost high-power multiprocessors. Sun Microsystems is also apparently committed to low-cost farmlike architectures using their SPARC CPU.

An unusual and powerful approach to multiprocessing is the data driven hardware processor designed at Nevis Laboratory by Bruce Knapp and Bill Sippach. The processor consists of 380 boards of 40 different types and is programmed by the way the individual modules are cabled together and downloaded with constants. When configured to do track finding it can reconstruct 100,000 events per second, while the VAX 780 takes 0.1-1.0 sec. to reconstruct an 8 track event. This enormous increase in processing power comes from a combination of pipelining and parallelism (ordinarily all 380 boards are processing simultaneously), fast cycle times (one operation on each module every 25 nsec), and because the modules carry out powerful operations that are specialized for track finding. The processor was used during the summer of 1987 to process 10^9 events from 8000 data tapes from BNL E766, a heavy quark spectroscopy experiment, and will be used for on-line processing in E690 at Fermilab.

An interesting and ambitious new approach to on-line multiprocessor systems that has been proposed by the Computing and Electronics/Electrical Departments at Fermilab goes by the name of FUSE, for FASTBUS Uniform System Elements. The proposal is aimed at architectures for greatly increased data acquisition throughput for next generation fixed target and collider experiments. The proposal specifies a number of building block functional units, which can then be combined in a variety of ways to build FASTBUS modules optimized for different applications.

Presently identified system elements include:

1) FASTBUS Master Interface, which provides a high-speed FASTBUS crate interface allowing efficient execution of sequential sets of FASTBUS operations for flexible readout of data from multiple slaves;
2) FASTBUS Auxiliary Port Interface, which provides a second data port onto the module through the FASTBUS auxiliary connector to any of a number of buses, including Lecroy ECLine, RS485, SCSI, and ACP Branchbus.
3) Module Control Element, which provides centralized control of the entire module, including connection to a serial port and a local area network; and
4) Event Processing Units, which provide application specific processing for data formatting and compression and event monitoring tasks as part of the readout system.

A typical FASTBUS module will include several processing elements, which can be digital signal processors, floating point engines, or general purpose processors depending on the application. These uniform system elements can then be combined into readout controllers, event builders, and front-end processors for future high-performance data acquisition systems.

3. SECOND GENERATION ACP SYSTEMS - NEW CPUS

Future multiprocessor systems will clearly benefit from the increasingly powerful processors now becoming available. Besides the "trivial" speedups by using newer versions of the processors already in use (for example, 25 MHz 68030s replacing 16 MHz 68020s) there are entirely new families of chips that can be used. In particular, the popular Reduced Instruction Set Computer (RISC) architecture has led to several new processors. Also heartening is the trend that the new processors often have software (including FORTRAN and COMPILERS and UNIX operating systems) available for them even before the hardware is available.

New processors that will offer at least a factor of three, and in some cases a factor of ten, more performance than the first generation ACP processors include:

1) The R2000 RISC chip from MIPS
2) The Fairchild, now Intergraph Clipper chip set
3) The AM29000 RISC chip from AMD
4) The SPARC RISC chip from SUN
5) The T800 transputer from INMOS
6) The 88000 RISC chip from Motorola
7) The 32532 processor from National Semiconductor

Extravagant claims are made for all of these processors by their manufacturers, suggesting potential performance of up to 17 million instructions per second (MIPS). Such claims need to be taken with a grain of salt, as a single instruction does not do the same thing on different processors. For high energy physics use, a natural standard is to take the VAX 11/780 as representing 1 MIPS in performance, and to measure the new processors only in comparison to a VAX on high energy physics codes written in high level languages, thus evaluating both the hardware and the compilers. It matters not how many instructions per second the CPU can execute if the instructions are not useful in FORTRAN or C and if the compilers fail to provide sufficient optimization.

The ACP has performed such physics benchmarks on several of the new chips. Based on these results, we have chosen to design a VME CPU board based on the MIPS R2000 CPU. The standard ACP FORTRAN benchmark suite (consisting of three programs: a small Monte Carlo/track reconstruction package; an actual fixed target tracking code running on experimental data; and a floating point intensive theoretical calculation) was run on a 16 MHz MIPS system. For the three benchmarks, performances were 7.9, 6.4, and 7.4 times that of a VAX 780, and in each case more than a factor of 10 greater than the present generation ACP 68020 boards.

The ACP MIPS processor module (see the block diagram in Figure 1) consists of:

1) a 16 MHz MIPS R2000 CPU;
2) a 16 MHz MIPS R2010 floating point unit;
3) Four 16 MHz Write Buffers;
4) a 32 KByte instruction cache;
5) a 32 KByte data cache;
6) 8 MBytes of parity checking main memory, made up of 1 Mbit (100 nsec nibble mode access) DRAMs, expandable to 16 MBytes;
7) interval timers that can interrupt the CPU; and
8) a full function VME Master/Slave interface supporting 20 MBytes/sec block transfers.
9) possibly a VSB interface if board layout permits.

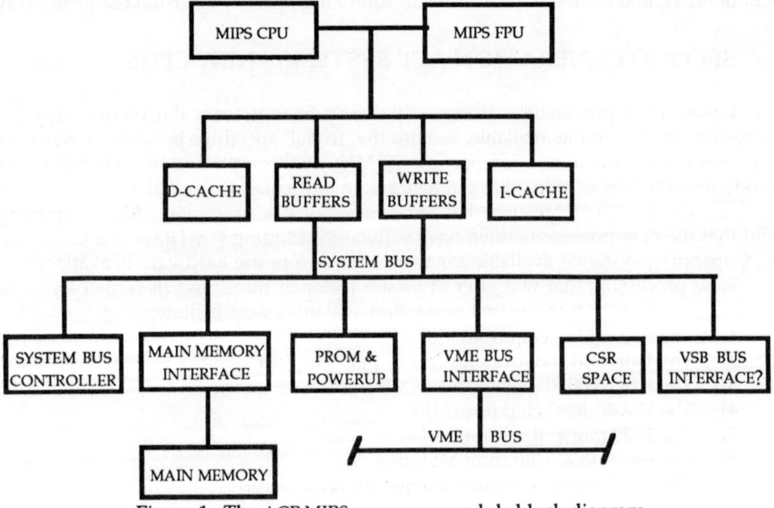

Figure 1. The ACP MIPS processor module block diagram.

Since the MIPS CPU chip has on-chip memory management, the module will be able to run the full UNIX operating system, booting either from a VME disk drive or using the Network File System (NFS) over the Branchbus.

This module will provide high-level language processing power with a cost effectiveness of roughly $300/VAX 780 equivalent, based on the physics benchmarks described and the features listed above. The FORTRAN compiler for these chips is the best we have encountered for a microprocessor, supporting many VMS extensions and comparing favorably with the VMS compiler in convenience and sophistication. Full UNIX program development tools are available. This processor will form the cornerstone of the second generation ACP systems. As discussed below, other VME processors running UNIX can also be used in the system, giving an opportunity for single board computer manufacturers to compete directly at the board level. Whenever a system is assembled, the most cost-effective processors currently available can be used.

4. SECOND GENERATION ACP SYSTEMS - NEW INTERCONNECT TOPOLOGIES

The ACP Branchbus was developed to deal with the problem of linking several high performance local buses to a host and allowing high speed block transfers.[3] No commercial alternative was available. The original Branchbus is a 32-bit bus connecting a single master (QBus, Unibus or FASTBUS) to multiple crates (VME), supporting block transfers at up to 20 MBytes/sec. Three improvements to the Branchbus will allow higher performance and more complex interconnection schemes in future systems.

First, the Branchbus now supports multiple masters. A distributed arbitration scheme similar to that used on the SCSI bus allows up to 16 masters to share the bus. Existing masters can be used in multi-master systems by replacing their existing Branchbus Interface Daughter Board with a new Multi Master Branchbus Interface Daughter Board which handles the arbitration transparently to the user.

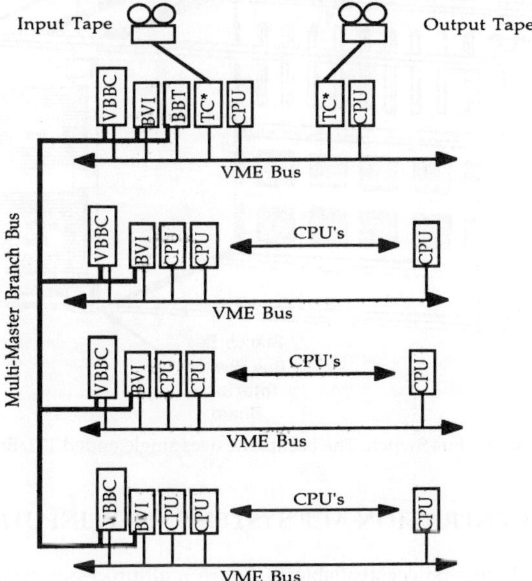

Figure 2. A Second Generation Off-Line ACP Multiprocessor System using the VBBC.

The VBBC (VME Branchbus Controller) allows any VME master, in particular any node in an ACP multiprocessor system, to act as a Branchbus master and read or write to any

Branchbus address. This allows any processor in the system to communicate with any other processor without host intervention, allowing the more elegant system architectures described below. (Figure 2 shows a block diagram of a second generation off-line ACP system using the VBBC.) The VBBC is a VME slave and Branchbus master. It is a shared resource in the VME crate, allocated on a first-come first-served basis by a test-and-set bit in its control register. Once programmed with the Branchbus control words and address, the Branchbus cycles occur transparently to the VME master which simply reads or writes data to the VBBC. The design of the VBBC is complete, and first prototypes will be available in early summer.

The ACP Branchbus Switch allows full crossbar interconnection of up to 16 Branchbuses (or more using multiple switches). With this switch, any Branchbus master device can connect to any slave in the entire switch connected system. All channels of the switch can be active simultaneously. For example, eight of the Branchbuses could be connected to the other eight, all transferring data simultaneously giving an aggregate bandwidth of 8x20 MBytes/sec or 160 MBytes/sec (in addition to any local bus activity on any of the VME crates in the system). The Switch is based on the TI 74AS8840 16x16 four bit crossbar chip. The Switch is a backplane incorporating 14 of these chips. Modules may be plugged into the 6Ux280mm Eurocard Switch Crate (see Figure 3) much as with VME. However, instead of the signals being connected in a bus structure, each slot in the crate is a crossbar switch point. Use of the Switch will allow future multiprocessor systems to obtain as much bandwidth for interprocessor communication as is required, up to 160 MBytes/sec per Switch. The first two Switch crates are built and working.

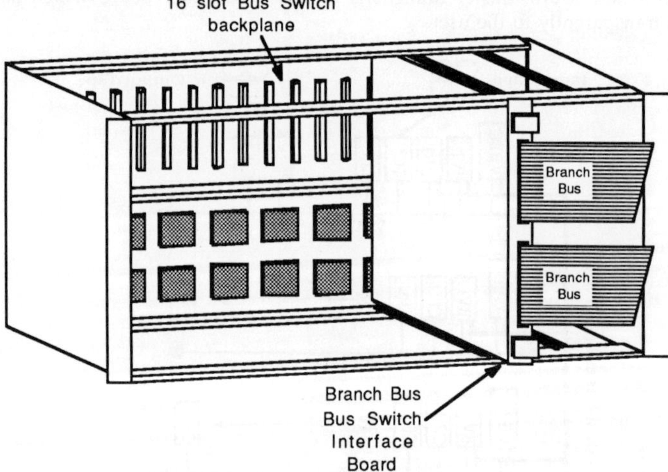

Figure 3. The ACP Branchbus Switch. The backplane uses single ended TTL Branchbus protocol.

5. SECOND GENERATION ACP SYSTEMS - NEW INPUT/OUTPUT

The increased CPU power available in future multiprocessor systems will require concomitant increases in input/output capabilities beyond the present standard of 6250 BPI magnetic tape. There will be a demand both for significantly higher density mass storage devices (to accommodate the ever-increasing amounts of data planned for future experiments, one Fermilab experiment is talking about tens of billions of events) as well as increases in I/O bandwidth.

Three developments will help confront the I/O problem. First is an increasing reliance on I/O devices that interface directly to the multiprocessor system bus. Unibus and QBus tape drives are being replaced by VME tape interfaces, such as the Ciprico TM3000. Such VME interfaces have a bandwidth potential of the 20-30 MBytes/sec allowed by VME compared to the roughly 1 MByte/sec possible with minicomputer buses. A fundamental constraint on future high rate I/O is thus removed.

Second, new I/O devices are replacing magnetic tape. Optical disks, video tape cartridges (both VHS and 8mm formats) and digital audio tapehold out promise of much more compact (and cost effective) mass storage than conventional tapes. Current devices allow bandwidths comparable to those achieved with mag tapes, with further improvements expected in the next year. For example, the ACP group is studying the 8mm video tape device manufactured by Exabyte Corporation. Currently available devices store 2 GBytes (as much as 12 conventional mag tapes) on a standard $10 tape cartridge. The drives cost less than $3000 and can deliver data at 250 KBytes/sec to VME (through a SCSI bus interface) or to QBus. Both density and speed are expected to double in the next year. No standard has yet emerged in this rapidly developing field, and the long-term reliability of these devices has not yet been established. However, it is clear that future systems can count on considerably better I/O performance than available in current systems.

Thirdly, this availability of cheap high-capacity mass storage devices which interface directly to the multiprocessor bus allows the possibility of parallel I/O. The multiprocessor system will have available to it many devices all reading and writing simultaneously, allowing the total I/O bandwidth to be increased to whatever level is required. In addition, some of the video tape cartridge devices will allow cheap and simple mechanical loading devices to avoid waiting for human operators to mount tapes. We are planning such "juke boxes" with optical bar code labels to insure mounting of the correct set of tapes from a large data sample. With the system architectures discussed below, these three developments will insure that the voracious appetite of the new processors for data will be satisfied.

6. SECOND GENERATION ACP SYSTEMS - SYSTEM ARCHITECTURE

The next generation ACP system will use MIPS (or competing) CPUs with at least an order of magnitude more processing power than the first generation 68020's. This increase in CPU power makes it imperative to provide a system architecture that removes the potential bottlenecks preventing the current systems from being scaled up by an order of magnitude. Such bottlenecks include the I/O bandwidth, interprocessor communication bandwidth, and CPU power available in the host processor. Moreover, we would like the software for the next generation system to provide greater flexibility with less complexity.

The second generation ACP system will meet these goals through a redesign of the system software. It will allow existing applications to run virtually unchanged, yet will provide a variety of powerful new features to allow users to realize the full potential of the new processors. (The new processors can also be used in first generation ACP systems alongside existing 68020 processors when the applications do not require greater performance than provided by the original systems which are limited by the MicroVAX host.)

In the second generation system *any node* can assume the functions previously exercised only by the MicroVAX host processor. In particular, any node in the system can do `send`, `get`, `broadcast` and `accumulate` type operations to or from an available individual node (or set of nodes) in a given class or rank. As before, the system software will automatically find an available node, freeing the user from the details of the hardware architecture. Also, any node in the system can do I/O, reading or writing data tapes and accessing disk files.

An example of a configuration for a reconstruction problem with multiple input tapes is shown in Figure 4. Note that this is a software configuration; the actual hardware connection of the nodes is over Branchbus via VBBCs and Bus Switches (if necessary) and is transparent to the programmer. Nodes in Rank 1 read events from data tapes and pass them along to either Class 2 or Class 3 nodes (Rank 2), which process events of different trigger types. Nodes in Rank 3 collect events from any nodes in Rank 2, either Class 2 or Class 3, for output to tape. The user may choose to push (send) or pull (get) data between the classes in either direction.

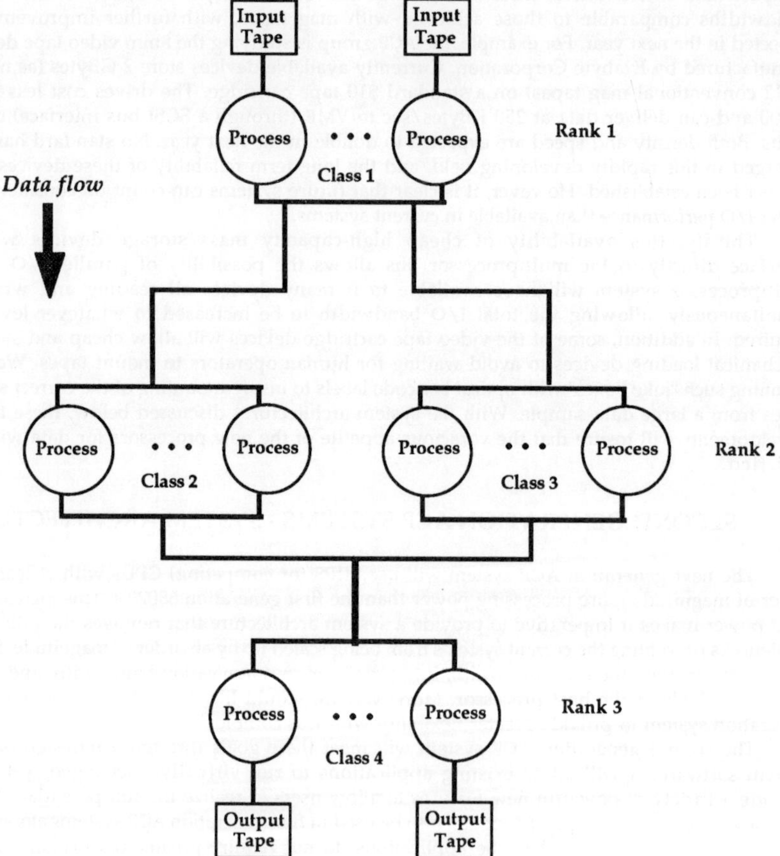

Figure 4. A powerful second generation Off-line ACP Multiprocessor Configuration.

The important new features of the second generation systems include:

1) The system will no longer be configured as a single master host with many node slaves. Any node in the system can assume some of the host functions. (This eliminates host CPU power as a bottleneck, as well as naturally allowing the parallel I/O described in 2.)

2) Input and output can be carried out by many nodes simultaneously. This capability is enhanced by using low cost cartridge tape devices. (This eliminates I/O bandwidth as a bottleneck.)
3) Any node in the system can communicate directly with any other node without intervention by a host. (This eliminates internode communication bottlenecks, since the communication can now proceed in parallel without the system software overheads present in first generation systems.)
4) The nodes can be (logically) configured in ranks, with data flowing smoothly from each rank to the next, doing input at the first rank, output at the last, and processing in all intervening ranks. This provides a natural match to the structure of the HEP computing problems as well as providing easier software development since the program in each rank will serve only a single function.
5) Any CPU running either VMS or UNIX and having a Branch Bus or Ethernet connection can be used as a node in the multiprocessor system, providing for great system flexibility as well as the ability to use existing hardware and software where appropriate.
6) Program development is done using the full set of UNIX (or VMS) compilers, linkers and debuggers for the class of nodes on which the process will run. The programs for each class of nodes can be developed independently.

Thus, the second generation system architecture provides a set of building blocks allowing a particular system to be matched to the set of applications it will run. Enough I/O devices and Branchbus interconnects should be provided so there are no bandwidth limitations. Enough standard nodes are added for the desired CPU power and any special purpose nodes (such as workstations for graphics) are also supplied. As each job is run on the system, nodes will be assigned to run particular user processes (input, output, or event processing) as appropriate. Not only the traditional compute bound event reconstruction tasks but also more I/O intensive data analysis jobs will find a home on these systems.

7. CONCLUSIONS

The use of multiprocessor systems for experimental event reconstruction is essentially a solved problem. Existing systems demonstrate that effective use can be made of systems with order 100 individual processors and up to 100 VAX 780 equivalents in total processing power. Improvements in system architectures and power of the processors in future systems will extend this to over 1000 VAX equivalents in a system.

8. REFERENCES

1] Gaines, I., Areti, H., Atac., R., Biel, J., Cook, A., Fischler, M., Hance, R., Husby, D., Nash, T., and Zmuda, T., "The ACP Multiprocessor System at Fermilab," Comp. Phys. Comm. 45 323-329 (1987); Biel, J., Areti, H., Atac., R., Cook, A., Fischler, M., Gaines, I.,Hance, R., Husby, D., Nash, T., and Zmuda, T., "Software for the ACP Multiprocessor System," Comp. Phys. Comm. 45 331-3379 (1987).
2] Kunz, P. F., Nucl. Instrum. Methods 135 435-440 (1976); Kunz, P.F., Gravina, M., Oxoby, G., Trang, Q., Fucci, A., Jacobs, D., Martin, B., and Storr, K., "The 3081/E Processor," Proc. Three Day In-Depth Rev. on the Impact of Specialized Processors in Elementary Part. Phys., Padova, 1983, 83-100 (1983).
3] Hance, R., Areti, H., Atac, R., Biel, J., Cook, A., Fischler, M., Gaines, I., Husby, D., Nash., T., and Zmuda, T., "The ACP Branchbus and Real Time Applications of the ACP Multiprocessor System" FERMILAB-Conf-87/76 (1987).

TRIGGERS FOR LEP EXPERIMENTS

Speranza Falciano

Istituto Nazionale di Fisica Nucleare, Sezione di Roma
Piazzale A.Moro, 2 I-00185 Roma
ITALY

ABSTRACT

An overview of the triggers developed for the LEP experiments is given. The requirements of those systems, dictated by the experimental framework, are discussed. The various trigger architectures, made by several filtering stages, are compared and examples of technical implementations are reported.

1. THE SCENARIO

Trigger systems at LEP are almost under completion. Their properties have been determined by the machine parameters, the physics programme, the detection techniques used by the experiments and the technology available at the time of the project. In the following we briefly present this scenario and then we move to the presentation of the trigger systems giving implementation examples and comparing the various architectures.

1.1 The LEP Design

LEP, the Large Electron-Positron Collider, is the largest and latest machine built at CERN. It has a circumference of 27 km and accelerates particles up to 50 GeV energy per beam. The expected luminosity is about 10^{-31} cm^{-2} s^{-1}. Eight interaction regions are foreseen but only four will be equipped with detectors. Therefore the machine will operate with 4 bunches per beam and the bunch crossing will occur every 22 μs at the interaction points. LEP is expected to turn on late in 1989.

1.2 The Physics Context

At the beginning of the LEP operations the physics programme will be concentrated at the Z^0 peak where a lot of topics can be addressed. Here, with the design luminosity and a total cross section of ~40 nb, the expected event rate from collision is about 1 Hz. Hopefully the programme will move soon to the search for Higgs bosons, top quark, super-symmetric particles and in general for new phenomena.

1.3 The LEP Detectors

Four experiments have been approved for the first phase of LEP

(100 GeV center of mass energy) : ALEPH, DELPHI, L3 and OPAL. Each of them has specific aims and therefore the detectors are characterized by peculiar and complementary technological features. ALEPH and OPAL are considered as classical experiments because they largely conform to known techniques. DELPHI and L3 incorporate detection techniques never used before on a so large scale to provide accuracy on particle identification (DELPHI) and muons and low energy photons (L3). The four detectors are made by the same components which may be summarized as follows :

* *Vertex Detector,* close to the beam pipe, measures the tracks of short-lived secondary particles (Example: The DELPHI Microvertex Detector).
* *Tracking Chamber,* to measure the tracks of the emerging charged particles (Example: The OPAL JET Chamber and the ALEPH TPC).
* *Electromagnetic Calorimeter,* to measure the electromagnetic energy (Example: the L3 BGO).
* *Hadron Calorimeter,* to measure hadronic energy.
* *Muon Chambers,* to detect penetrating muons.

The detector also includes a magnet to provide the field for momentum measurements.

2. THE TRIGGER SYSTEMS

The huge size and the fine granularity of the LEP apparata leads to the handling of more than 100 000 electronic channels which will produce Megabytes of data. Fast data handling instrumentation and filtering systems are necessary to reduce and record the data in between bunch crossings (22 μs). Therefore the main task of the trigger and data acquisition systems (DAQ) is to select and collect events of physics interest removing the background efficiently with a minimum of dead time.

To satisfy the conflicting requirements of power and speed the LEP triggers are made by three to four filtering stages: each stage decreases the event rate by a percentage which varies from one experiment to another and allows for an increasing processing time. The resulting multi-level structure may be summarized as follows:

* *First filtering stage*
  Implemented by hardware processors taking decisions in a few microseconds.
* *Second filtering stage*
  Performed by fast but programmable processors computing trigger algorithms within tens of milliseconds.
* *Third filtering/tagging stage*
  Pure software trigger running on sophisticated processors taking decisions within hundreds of milliseconds to a few seconds and eventually performing event tagging.

LEP is a new machine and variable running conditions and unexpected physics must be taken into account. Furthermore several sub-detectors with a large number of electronic channels need to be operated independently: these constraints can only be satisfied by a flexible and modular trigger and DAQ system, that is partitioning the read-out and the DAQ into several branches each corresponding to a sub-detector or part of it with local event building and data moving facilities. The convergence of data from the full detector is achieved at a central place near by the main DAQ computer. A typical DAQ architecture is shown in Fig.1.

Triggers are embedded in this resulting tree-like structure. The fast ones are implemented within the corresponding sub-detector partition allowing for parallel trigger calculations. The most complex filtering algorithms apply to the full data from the whole detector and this must happen just before the event is transfered to the main computer that is at the root of the tree.

Because of the different properties of the four detectors, the technical implementations, the response time, the selectivity power of the triggers and the detector elements partecipating to the various trigger components may vary a lot among the four experiments [1,2,3,4].

For physics requirements the trigger has to identify all the possible experimental signatures which may occurr in the apparatus as those shown in Fig.2. To perform this task the triggers designed for the four experiments have the following properties:

* They are sensitive to tracks and/or energy deposits.
* They are not necessarily associated with a particular physics process, but one can build *Physics Triggers* like *Dimuon trigger, Single Photon trigger, Jet trigger, etc.*
* They are redundant : most of the conventional events will fulfill more than one trigger condition (*Overlapping Triggers*) assuring reliability and high efficiency very important for a precise study of the e^+e^- physics.
* They are detector oriented yielding high flexibility for any change of detectors and physics.

Comparing the trigger categories implemented by ALEPH, DELPHI, L3 and OPAL we end up with 4 trigger types built around the single sub-detectors:

* *Energy trigger* (Electromagnetic and Hadron Calorimetry)
* *Charged Particle trigger* (Tracking Chamber)
* *Muon trigger* (Muon Counters)
* *Luminosity trigger* (Forward Detector).

While they contribute to the first level decision in an independent way, correlations among them to better classify the event are achieved at the next levels where programmable devices are used and a longer processing time is allowed.

In the following we shall show the principles and the construction characteristics of the various filtering stages used to achieve the final decision of recording or not the event on a storage medium.

3. THE FIRST FILTERING STAGE

This stage includes the fast triggers working at the beam crossing frequency. The aim is to reduce the 50 kHz input rate evaluating in a few microseconds energy deposits and track segments pointing to the interaction vertex. Fast decisions are based on analog sums, pick-up signals and low precision digitized data elaborated by hardwired processors made by look-up tables, hit arrays and coincidence matrices.

As shown in Tab.1, ALEPH and DELPHI differ from OPAL and L3 as the fast decision is taken in two steps. In the ALEPH experiment to control the gating rate of the TPC (<500 Hz) a pre-trigger is delivered in 5 μs from the time of beam cross-over. Then the second step validates or rejects the Level-1 track trigger using the drift time informations from the TPC. This Level-2 trigger comes after 50 μs from the beam crossing and if the event must be rejected, data are taken at the third bunch crossing introducing some dead-time in the system.

Here we want to give an example of a fast trigger at LEP and mention the principles and the technical implementation of the L3 Level-1 *Energy trigger.*

In this experiment, for each sub-detector, the trigger data are encoded and read-out separately. In particular, the individual channels of the Electromagnetic Calorimeter (BGO) and two layers of the Hadron Calorimeter (HCAL) are grouped respectively into 32x16, 16x11 and 16x13 (ϕ,θ) intervals (896 channels in total) each delivering analog sums converted by a fast FERA ADC system into 10-11 bit digitized data.

The trigger processor may be seen as built up of two parts : the first performs the analog to digital conversions and organizes the data flow, the second elaborates the information supplied by the previous step (see Fig.3). It is built with ~300 CAMAC modules but CAMAC is only used for initial parameters downloading (thresholds, strobe patterns, trigger conditions, etc.) and diagnostic purposes. The data will flow on front panel ECL differential busses with 16 bits of data plus a Data Ready strobe used to validate and synchronize the data.

The basic modules of the processor are:

* *ALU (Arithmetic and Logic Unit)* to perform arithmetic and logic operations, data accumulation.
* *MLU (Memory Look-up Unit)* for Boolean logic, multiplications, divisions, loop counting.
* *DS (Data Stack),* 256 words deep used as FIFO memory.
* *ED (ECLine Driver),* source of 256 words.

The Energy Trigger will calculate and apply thresholds on:

* *Total Energy* (BGO + HCAL Calorimeters). The total energy deposit must exceed the 20% of the center of mass energy.
* *Partial Energy* (Large Angle, BGO only). At the background free region ($\theta \sim 90^0$) a lower energy threshold is allowed.
* *Longitudinal Energy Balance.* This is particularly suitable for events with missing energy.
* *Ratio of BGO Energy to Total Energy.* Contributes to background rejection.
* *Total Energy in the Small Angle BGO Monitor.* Used for monitoring purposes.
* *Energy Cluster.* It is formed by the TEC (Vertex Chamber), BGO, Scintillators and HCAL.
* *Single Photon.* The identification is performed by BGO and TEC: the sum of two adjacent rows and columns of the BGO (ϕ,θ) matrix must exceed the 80% of the total energy. Of course no tracks in the corresponding TEC segment must be found.

* *Overflow Trigger.* Indicates problems in the hardware processor : in this case the event is accepted and analyzed by the next level trigger.

The processor works with a clock of 60 ns and takes the decision within 16.8 μs. The remaining time before the next bunch crossing allows for a fast clear of the front-end electronics if the event is rejected.

4. THE SECOND FILTERING STAGE

This further level of trigger runs asynchronously to the previous one as multi-event buffers are largely used at the output of the fast triggers and close to the front-end electronics to derandomize the rates. This filtering process works on trigger and/or detector data performing trigger algorithms more refined than the previous stage by correlating informations from different sub-detectors.

To accomplish this task fast but programmable devices are used. At this level the advantage of a software based trigger is its enormous flexibility : if a tuning of the filter algorithms according to the running conditions is required, a simple program downloading operation changes the trigger criteria.

An example is the Level-3 trigger of the DELPHI experiment whose architecture is shown in Fig.4. It is software based and runs on a number of parallel microprocessors sending results to a Trigger Supervisor.

The parallel computations are performed locally at each sub-detector partition on the so called *local T3* FASTBUS processors and benefit of the read-out architecture with embedded MC68000 based *Local Supervisors.* At this stage, where only data from individual sub-detectors or part of them are treated, the input data, the complexity of the computation and the output results may vary a lot. For instance, for a complex sub-detector the Level-2 trigger provides pointers to the data of possible interest (e.g. (ϕ,θ) pointers to track candidates) in order to speed up the T3 calculations which are mainly devoted to validate and improve the decisions of the fast triggers.

Then the final trigger software runs on the so called *T3 Supervisor* , a MC68000 based FASTBUS Master : here the whole event is available in form of raw or pre-processed data and correlations of results from different partitions are possible at this time (combination of track segments and matching of tracks and energy deposits in the calorimeters). At this level the determination of the event origin is crucial for the rejection of the background mainly due to beam-gas interactions, cosmics and electronic noise.

The DELPHI Level-3 trigger has ~ 20 ms available to take the decision and reduce the event rate from < 20 Hz to 5 Hz (see Tab.2).

Another example is the OPAL Level-2 trigger implemented as a *VME Filter Matrix.* It differs from similar triggers of the other experiments for the expected input rate which is very low, 4 Hz. The tasks assigned to the Matrix are listed below :

* *Event building,* that is assembling a complete event from the various detector branches.
* *Fake trigger rejection* if the input rate exceeds 4 Hz.

* *Event selection and flagging* to signal the event to the next trigger processors.
* *Special calibrations and Monitoring.*

As for the previous example, the event rejection is based on the recomputation of the Level-1 trigger algorithms (hardwired) with higher accuracy as all the data with final digitization are available and correlations of different sub-detectors are possible.

From the hardware point of view the VME Filter Matrix is made by a series of FIF (Filter) Processors housed in the so called Filter Processor Crate. The FIFs are MC68030 processors with 68881 Floating Point Coprocessor running at 25 MHz speed and 1 to 4 Mbytes of SRAMs as local memory. A Manager Interface Processor (MIF), sitting in the same crate, supervises the whole system and synchronizes the detector read-out. Each FIF is attached via the vertical VSB bus to a dual-port 4 Mbyte Memory card which may supply a complete event to the attached processor. Up to 12 memory cards are housed in the VME Memory Crate where a Detector Matrix Interface Processor (DIF, MC68020 based) takes care of the read-out from the various branches. A schematics of this trigger is reported in Fig.5.

In the worst case of 20 Hz input rate, 200-300 ms are available to the Filter Matrix to analyze an event without introducing any dead-time in the system.

5. THE LAST FILTERING / TAGGING STAGE

The task of this last level trigger is to perform online event reconstruction by running sophisticated algorithms similar to those used by the offline software. Here, for the first time, all the detector data with final digitization are available. This essentially means a final validation of the previous triggers, data formatting, monitoring of the full detector, tagging of events of particular interest to ease the offline analysis.

To achieve this task powerful machines running high level languages as the mainframe have to be used and interfacing to the DAQ bus must be provided. The event processing time varies from hundreds of milliseconds to a few seconds (see Tab.3) according to the input rate, the CPU power and the number of processors in the farm.

The event processors adopted by the four experiments are the following :

* *DEC ACP* system (Attached CPUs) on BI bus (ALEPH)
* *370 Emulator* farm interfaced to VME (OPAL)
* *3081 Emulator* farm interfaced to FASTBUS (DELPHI and L3)

The ALEPH scheme is shown in Fig.6. The Attached CPUs (ACPs) are DIGITAL MicroVax II CPUs on BI with 1 Mbyte of local memory. Up to 4 BI adapters can be implemented on the VAX 8700 each supporting up to 10 ACPs. On each bus one ACP acts as a master and allocates events (~100 kbytes of raw data) to the other slave processors. The ALEPH solution allows for software compatibility between the online and offline environments.

DELPHI and L3 use a farm of 3081E emulators located near by the main DAQ computer and embedded into the FASTBUS DAQ architecture. A 3081E is equivalent to 1 IBM 370/168 that is 4 Vaxes 780.

By a specially developed interface the 3081E becomes a FASTBUS slave geographycally addressable on Cable Segment. The interface implements 2 FASTBUS ports completly independent and symmetric, one used for data input, another for data output : I/O operations may be done by one port at the time while the Control and Status Register (CSR#0), common to both ports, can be simultaneously accessed by them.

In Fig.7 we report the DELPHI Level-4 trigger scheme. A complete event is assembled by an Event-Builder from several source memories and sent to a free emulator via the input segment at a speed of 16 Mbytes/s. If the event is accepted, the CHI FASTBUS Interface or another FASTBUS master will take care of the data transfer from the emulator memory to the data logger.

The same phylosophy is adopted by the OPAL experiment for its Level-3 trigger which is based on a 370E farm integrated into the VME DAQ system.

6. CONCLUSIONS

A lot of work has been carried out by the 4 LEP experiments for the implementation of the Trigger and DAQ systems which must be fully operational by the summer 1989 when the first e^+e^- collisions are expected. The large amount of data to be handled requires systems able to face with analog signals from the detector up to the full event reconstruction in order to reduce the event rate to an acceptable level. Several advanced techniques have been used for solving the problems but also well experienced solutions. Even in this last case the huge amount of involved electronics and processors makes the system very complex and at the same time a big software effort is required to provide initialization, control and debugging facilities.

AKNOWLEDGEMENTS

I would like to thank all the people of the ALEPH, DELPHI, OPAL and L3 Collaborations who provided me with the material to prepare this talk.

REFERENCES

[1] ALEPH Collaboration, CERN/LEPC 84-8.
[2] DELPHI Collaboration, CERN/LEPC 84-3.
[3] L3 Collaboration, CERN/LEPC 84-5.
[4] OPAL Collaboration, CERN/LEPC 84-4, 84-2

Table 1: Summary of the fast triggers

	Trigger level	Input rate	Output rate	Decision time
ALEPH	Level_1 Level_2	50 KHz < 500 KHz	< 500 Hz < 20 Hz	5 μs 50 μs
DELPHI	Level_1 Level_2	50 KHz < 1 KHz	< 1 KHz < 20 Hz	3 μs 35 μs
L3	Level_1	50 KHz	100 Hz	< 17 μs
OPAL	Level_1	50 KHz	4 Hz	< 16 μs

Table 2: Summary of the intermediate level triggers

	Trigger level	Input rate	Output rate	Decision time
ALEPH	-	-	-	-
DELPHI	Level_3	< 20 Hz	5 Hz	$\sim$ 20 ms
L3	Level_2	100 Hz	10 Hz	< 8 ms
OPAL	Level_2	4 Hz	4 Hz	< 1 s

Table 3: Summary of the last level triggers

	Trigger level	Input rate	Output rate	Decision time
ALEPH	Level_3	< 20 Hz	$\leq$ 1 Hz	500 ms
DELPHI	Level_4	5 Hz	$\leq$ 1 Hz	500 ms
L3	Level_3	10 Hz	$\leq$ 1 Hz	$\leq$ 300 ms
OPAL	Level_3	4 Hz	$\leq$ 1 Hz	> 2 s

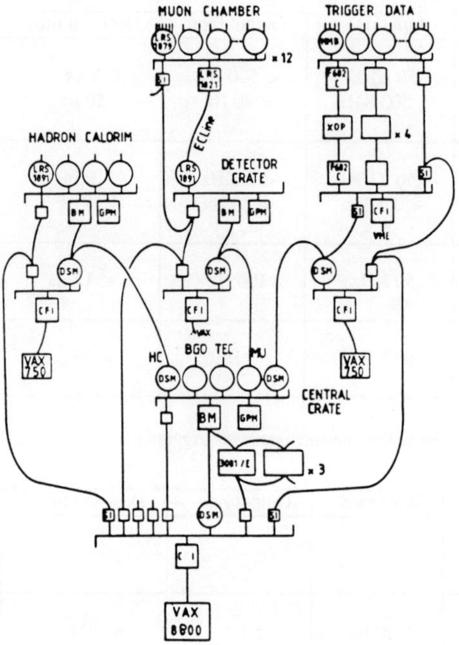

Fig.1 : The L3 DAQ Architecture

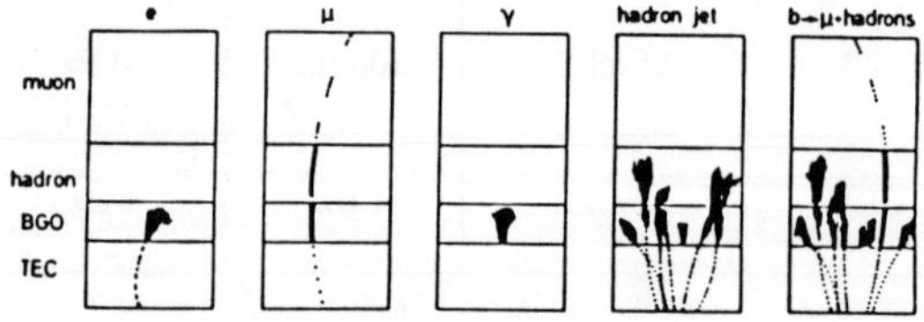

Fig.2 : View of events

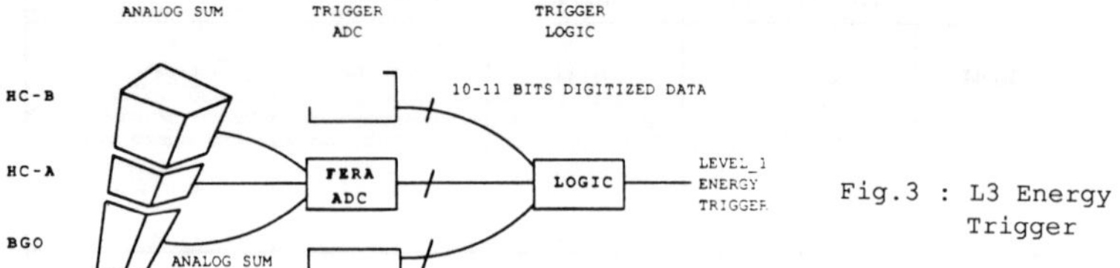

Fig.3 : L3 Energy Trigger

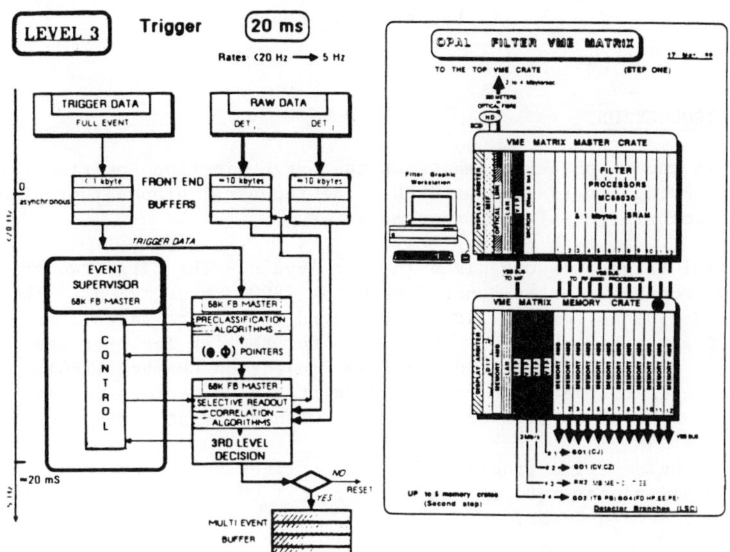

Fig. 4 : The DELPHI Level_3 Trigger

Fig. 5 : The OPAL VME Filter Matrix

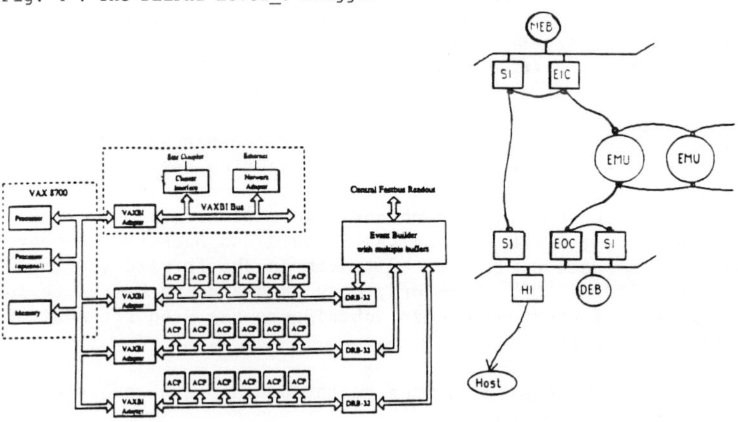

Fig. 6 : ALEPH Event Processors

Fig. 7 : The DELPHI Level_4 Trigger

THE FIRST AND SECOND LEVEL TRIGGER OF DELPHI

S. Quinton, L. Cerrito, Ph. Charpentier, K.E.Johansson,
L. Lanceri, R. Lucock,

1. INTRODUCTION

Delphi is one of the experiments at the Large Electron Position (LEP) collider being constructed at CERN. The technical proposal provides a general description of the experiment[1).

The Delphi trigger is organised in four levels. The first two triggers are synchronous with the Bunch Cross-Over (BCO), every 22 μs. The trigger decisions for the first and second levels are taken 3 and 38 μs respectively after the BCO. The timing for the 1st level trigger minimises the space charge in the Time Projection Chamber (TPC). For the 2nd level trigger data from long drift time detectors is used causing one bunch cross-over to be lost when an event passes level 1.

To limit the deadtime fraction for each trigger to the order of 2% the first level trigger rate should not exceed 1 kHz and the second level should not exceed 20 Hz.

2. PARTITIONS AND SUBTRIGGERS

The data acquisition and trigger system is organised as 17 detector or subdetector partitions implemented in more than 140 Fastbus crates. Trigger data is generated locally at the partition level and some trigger criteria are implemented here, e.g. for the requirement for a minimum energy deposition for the energy trigger.

The local trigger data is combined by seven independent subtriggers. Five of them are used to select e^+e^- interaction candidates: Track, Electromagnetic Energy, Hadronic Energy, Muon, and Total Energy. For example the track subtrigger combines information from the Inner and Outer detectors, the TPC and the the Forward and Backward chambers. Two bit outputs indicate 0, 1, 2 or >2 tracks in the forward (20° - 40°), very forward (10° - 20°), backward, very backward and barrel regions. The Small Angle Tagger and Cosmic subtriggers are used for monitoring and calibration data.

The final trigger decision is made on the basis of the subtrigger data by the central trigger supervisor

3. TRIGGER SYSTEM ARCHITECTURE

The architecture of the 1st and 2nd level trigger is shown in Figure 1 using the barrel track trigger as an example. The partitions usually run under the control of the Trigger Supervisor (TS) but they can also run in stand alone mode under control of a Local Trigger Supervisor (LTS).

The control functions are performed centrally by the TS control module, Zeus, communicating via the Trigger Control Lines (TCLs) to the LTS control modules Pandoras. Pandoras provide local control for the front end electronics and the Front End Freeing Controllers (FEFCs) which read out the front end data on a T2-YES signal.

The central trigger decision is taken by the TS decision module, Pythia, which receives the data prepared by the individual subtriggers through the Trigger Data Lines (TDL's). There is a separate set of Pythia modules and TDLs for the 1st and 2nd level triggers.

The Zeus and Pythia modules are implemented in Fastbus and installed in the Olympus crate together with the Trigger Supervisor Master (TSM) and a Segment Interconnect to the Event Supervisor crate. The TSM provides the FEFC and front end buffer functions for the TS and subtrigger decision boxes.

Pandora is also a Fastbus module as are the subtrigger decision modules and some of the LTS decision modules.

3.1 TS Control Module, Zeus[2)]

Zeus generates and distributes the timing sequence for the trigger and synchronous readout phase using the RF and Bunch Cross-Over (BCO) signals from LEP, or by generating them internally during LEP shut down. Zeus passes on to the Pandora modules the trigger decision T1-YES, T1-NO, T2-YES or T2-NO received from Pythia, the signal COS-EN if the Cosmic trigger is enabled and the signal COSMIC to indicate an enabled Cosmic trigger. Cosmic Triggers can be accepted in the period between T1-NO and the next BCO. T1-NO and T2-NO signals cause resets of the front end electronics through the Pandoras. T2-YES starts a readout cycle, front end ready, (FE-RDY,), being returned when this operation is complete. Zeus will detect and flag FE-RDY timeouts and assert the corresponding TIMEOUT signal to freeze the Pandora state for diagnosis. A global (GBL) signal is asserted by each Pandora to indicates its presence in the system.

3.2 LTS Control Module Pandora[2)]

The Pandora communicates with Zeus as described above and with the front end electronics through the Local Trigger Control Lines (LTCLs). A number of programmable pulses and clocks are provided for timing in the front end electronics as is a front end trigger enable signal (TRIG-EN). Pandora can run in a number of trigger modes. Global modes are under control of Zeus with real or dummy 1st and 2nd level decisions. Local modes can use Zeus timing with local real or dummy trigger decisions or local timing can be used when calibrating or triggering on Cosmics.

3.3 TS Decision Module, Pythia[3)]

Data is passed to Pythia through the TDLs accompanied by a strobe and warning flag. Fatal errors are indicated by the absence of a strobe, flagged by Pythia as a timeout. The warning flag indicates a non fatal error such as part of a subtrigger not functioning. The trigger data is used to form the address of a look up table architecture shown in Figure 2. Each level A table has 12 inputs and 16 outputs. The level A outputs are OR'ed together, each output using a mutually exclusive bit field in the 16 bit level B input word. Similarly the level B outputs are grouped to form the level C input. Although the tables can be used in a completely general way they will probably be assigned to detectors and detector areas as shown in the diagram. The final 4 bit level C output will be equipped with scalars before the final OR into a yes/no decision.

At each stage in the Pythia module, Fastbus accessible registers will capture intermediate values for monitoring Pythia operation and for providing information on the derivation of the trigger. It will also be possible to force data into these points from Fastbus to enable computer verification of the modules. The implementation of similar test schemes in the other trigger decision boxes is left to the individual designers. However a Fastbus accessible I/O register at the output of each subtrigger is mandatory.

4. CURRENT STATUS

A prototype Pandora module has been designed, built and tested at the LPNHE, Paris VI, and the production version is being developed. About 30 of these modules will be required in the final system. The Zeus prototype is being designed by the INFN, Rome. The Pythia modules have been designed and built at the Rutherford Appleton Laboratory and are under test. It is expected to have a demonstration system running at CERN in June this year.

5. REFERENCES

1) Delphi Technical Proposal CERN/LEPC/83-3
2) Cerrito, L. et al The Delphi Trigger Control System, Zeus and Pandora. Delphi 86-106 DAS-45
3) Holmgren, S.O. et al The First and Second Level Trigger. Delphi 87-12 DAS 48

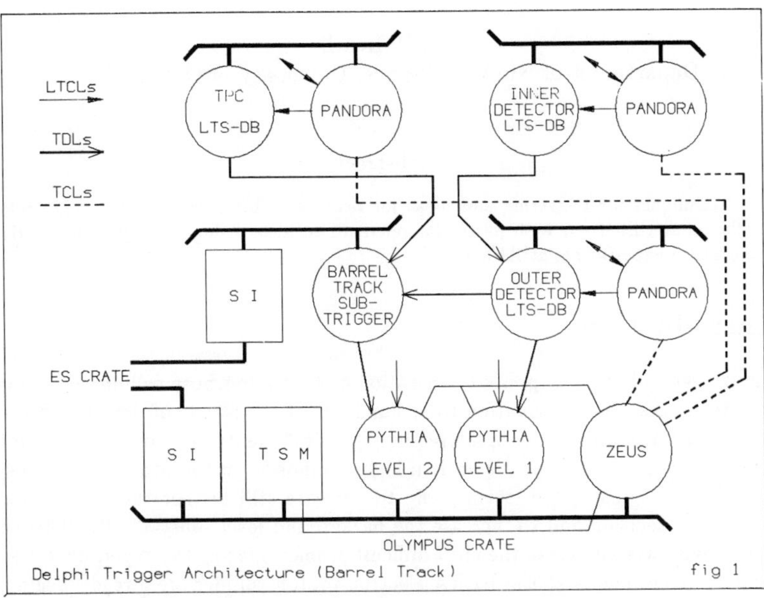

Delphi Trigger Architecture (Barrel Track) fig 1

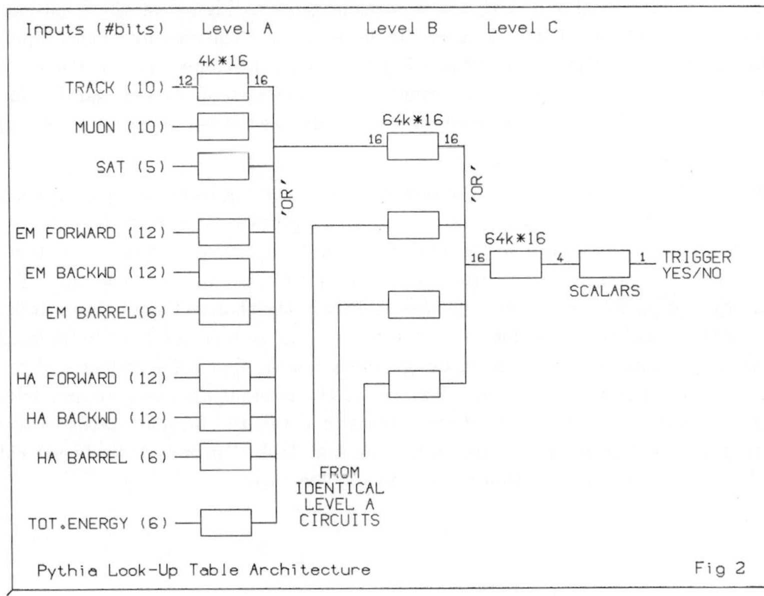

Pythia Look-Up Table Architecture Fig 2

Triggering at High Crossing Rate Colliders

G.P. Heath
Department of Nuclear Physics, University of Oxford, U.K.

Abstract

Techniques for triggering and ideas for readout architecture at future high-rate colliders are discussed with particular reference to ideas being developed in the H1 and ZEUS experiments at HERA.

1 Introduction

The trigger logic for the present generation of general-purpose collider experiments is required to process the information from hundreds of thousands of detector channels in order to extract a small number of interesting "physics" events from an enormous potential background. Future trends, particularly if high-luminosity proton-proton colliders such as SSC or LHC are constructed, seem likely to decrease still further the time available to make a trigger decision and to increase the background level substantially. These trends make the trigger system one of the most difficult problems facing the experiment designers. Fortunately, experiments at the HERA electron-proton collider are already beginning to face the problems associated with beam-crossing rates so high that the front-end data must be stored until a trigger decision can be reached. Schemes for such "pipelining" of both the data and the trigger processing have been developed by the H1 and ZEUS experiments. In this paper, therefore, some of their ideas are discussed and developed as examples of the kind of approach possible to triggering at the next generation of high-rate colliders.

Clearly it is impracticable to go in one step of processing from an input rate of many MHz to the few Hz which it is possible to write to a permanent storage medium. Several stages are therefore used to give the required complexity. The first of these stages uses fast, purpose designed hardware processors to perform a preliminary selection of events, with later stages in software to give more flexibility in the choice of algorithms. It is at the first stage that most of the problems unique to this kind of machine have to be faced, and this paper therefore concentrates on design ideas for the first level trigger and readout. In Section 2 we describe the general features of the readout architecture and triggering strategies at such a machine. We then discuss in detail the present status of ideas for triggering at HERA in Section 3, and in Section 4 we look at proposals for future colliders in the light of these ideas. Section 5 contains the conclusions.

2 Readout Architecture at a High Crossing Rate Collider

2.1 First Level Readout Pipeline

For a collider detector where the time required to form a first level trigger decision is longer than the interval between bunch crossings, the data from each crossing have to be delayed, or stored until a trigger decision arrives. A number of different techniques are available for providing this delay, the choice clearly depending to some extent on the type of detector. The most serious problems are posed by the calorimeter readout, and the solution adopted can have important consequences for the trigger design. If the first level decision time is short enough cable delay may be adequate, as long as the degradation in signal quality remains tolerable, or shaping may be used to lengthen the pulse duration. For longer decision times the data must be sampled continuously and the samples held in some form of electronic storage, either analogue or digital. This pipelined data storage acts in principle like a fixed length delay line – when the data for a particular crossing arrive at the end of the pipeline they must either be gated into some buffer by the arrival of a first level trigger signal, or are simply discarded. For any of these forms of delay, then, the timing of the first level trigger determines the slice of the pipeline to be read out.

The stage of transferring data from the pipeline to a local buffer can also influence the trigger design. Ideally one would like to do this without disabling the pipeline input, so that no dead time is incurred. This could in principle be achieved using a FIFO buffer, although sufficiently fast devices are not available today. In the HERA experiments, therefore, the pipelines have to be disabled on data transfer, and some means found to keep the total dead time loss at the expected first level trigger rates within acceptable bounds.

2.2 Pipelined Hardware Triggers

The first level of trigger processing is typically based on a combination of searches for energy deposits and tracks, and signatures for high-energy leptons and jets. In general as one goes to higher energies, calorimetric measurements assume greater importance relative to tracking. It should be noted however that at HERA triggers from tracking detectors will play an important part in discriminating against the dominant backgrounds of proton beam-gas and beam-wall interactions.

The trigger decision for each crossing will be formed by combining information on calorimeter energy and track multiplicity, together with electron and jet signals from calorimeter processors, muons from muon chambers (optionally supported by tracking information), and missing energy. The crucial point about the pipelined processors necessary at high-rate colliders is that the time taken for data from a particular crossing to flow through the trigger logic must be the same for all crossings. This means that for each component of the trigger, the time required to reach a decision must be independent of the activity in the detector in that crossing. It is this feature which imposes the most severe restrictions on the type of algorithms which can be used in these processors. Many of the types of algorithm which one would like to build into hardware, e.g. for isolated electron or jet finding in calorimeters, involve variable numbers of processing steps according to the degree of activity. To implement these in a truly pipelined processor it is necessary to build separate components to search for all allowed patterns in parallel. While this can lead to very fast processing, the requirements on cost and space quickly become prohibitive.

Pipelined triggers also require that the time resolution of the input signals should be good enough to associate each with a particular beam crossing. This can either be achieved using dedicated "trigger counters" such as MWPCs which have an inherently good timing resolution or via slower detectors such as liquid Argon calorimeters or drift chambers, provided the hardware is designed with the requirement to deliver a trigger signal in mind. Examples of both of these approaches from the H1 and ZEUS detectors respectively are discussed below.

2.3 Higher Level Architecture

The form of processing to be used for the higher levels of triggering will be determined largely by the expected output rate from the preceding levels. A further, non-pipelined stage of hardware processing based on the information used in the first level can be used to refine the calculations made there. Alternatively, if the input rate from the first level is low enough and the readout dead time not too large, all levels of trigger higher than the first can be implemented in software. In parallel with the processing for the trigger, the raw data from each event have to be compacted and formatted, and collected together for transfer to permanent storage. We can distinguish between intermediate level triggers based on distributed processing, with the data stored in buffers in the individual detector components, and high level filters working on the full event. Intermediate level triggering will again be based on fairly general requests for activity found in the detector, using more precise information from the components. The high level triggers use more detailed physics selection criteria and are typically performed by a "farm" of dedicated microprocessors as seen in most modern collider detector designs. A vital part of the design of such a multi-level trigger system is to optimise the amount of buffering at the input to each level to de-randomise the incoming event rate and thereby maximise the available computation time.

The task of data compaction at intermediate processing levels is a formidable one requiring considerable power. The use of continuous sampling readout as described in Section 2.1 generates enormous raw data rates (the ZEUS tracking system will produce 1 Tbyte/s of FADC digitisings), and many samples per channel may need to be read from the pipeline for each first level trigger to give a full description of the detector response. It is clearly highly desirable to distribute this processing load, in order to reduce the data rate radically before it reaches any data backplane. This can be achieved with great flexibility by having intelligent microprocessors to do zero-suppression and pulse shape parametrisation at the front-end electronics.

3 Triggering at HERA

3.1 HERA Detectors

The two HERA experiments, H1[1] and ZEUS[2], are scheduled to begin physics data-taking in mid-1990. In this Section we look in some detail at the trigger schemes for the two experiments. In both cases information from calorimetry, tracking and μ chamber systems will be used, with correlations between different detectors being important even at the first level of trigger. The principal sources of background particles in a HERA detector are predicted to be interactions of the proton beam upstream of the detector with residual gas molecules or with the beam pipe, collimators, etc. These interactions will produce showers in a forward calorimeter at rates in excess of 10kHz, compared with the highest rate ep processes which will be detected at around 100Hz. Tracking triggers with a good resolution on vertex position (in 3 dimensions) are therefore important.

The low-level trigger architectures of the two experiments show some differences in detail which can largely be traced to different choices of calorimeter. H1 have a liquid argon calorimeter with "long" signal collection times (up to $600ns$), and fine segmentation both laterally and in depth, giving a total of 65,000 readout channels. Shaping amplifiers are used to give output voltage signals with rise times of $2\mu s$, after which time a first level trigger decision must be available to gate sample-and-hold circuits for the readout. The ZEUS calorimeter has five times fewer channels and uses scintillator readout which will give faster signals. Here an analogue readout pipeline formed from a switched array of capacitors similar in concept to the SLAC AMU will be used, allowing longer processing times at the first level. In the ZEUS experiment a first level rate of 1kHz is aimed for which, with a time of around $60\mu s$ required for digitisation of the calorimeter signals gives a dead time of 6% and a mean second level processing time of $1ms$. For H1, however, the ADCs for the calorimeter are considerably more heavily multiplexed, so that an additional level of hardware trigger is required after the first, pipelined level to reduce the rate to around 100Hz. In ZEUS, possibilities for reducing the dead time further, either by rejecting some fraction of the first level rate before the digitisation is complete using fast hardware processors in the second level trigger, or with a stage of analogue buffering on the pipeline chip, are currently being investigated.

3.2 Hardware Trigger Processors in H1 and ZEUS

3.2.1 Calorimeter triggers. The raw input to the ZEUS calorimeter trigger is the summed energy (electromagnetic plus hadronic) from each of $\sim$ 2000 towers. These will be digitised by two 8-bit FADCs with a gain difference of a factor of 16, yielding a dynamic range of 4096:1. The pulse heights can then be combined, linearised, pedestal subtracted and checked against a threshold by table lookup. The trigger first forms the total energy, transverse energy and missing p_t sums. Given a tower sum of energy E_i located at polar angle θ_i and azimuthal angle ϕ_i, the three sums can be formed from E_i, $E_i \cos\theta_i$, $E_i \sin\theta_i \cos\phi_i$ and $E_i \sin\theta_i \sin\phi_i$, again calculated by table lookup at intervals of $24ns$. Each of the four quantities is then injected into an eleven stage ($2048=2^{11}$) summing network built from fast ECL parallel adders which produces the four sums in $180ns$. The sums from 16 sub-regions of the full calorimeter acceptance will also be provided. In addition to these global energy sums, ZEUS hopes to trigger on isolated electrons and possibly jets at the first level although, as discussed above, the formation of these triggers in a pipelined fashion is considerably more complicated. One possibility is to provide for some non-pipelined processing to be performed for crossings where the energy in a particular region is above some low threshold, with a limit on the time allowed for this processing after which the event will be accepted anyway. A hardware jet-finding processor for fast

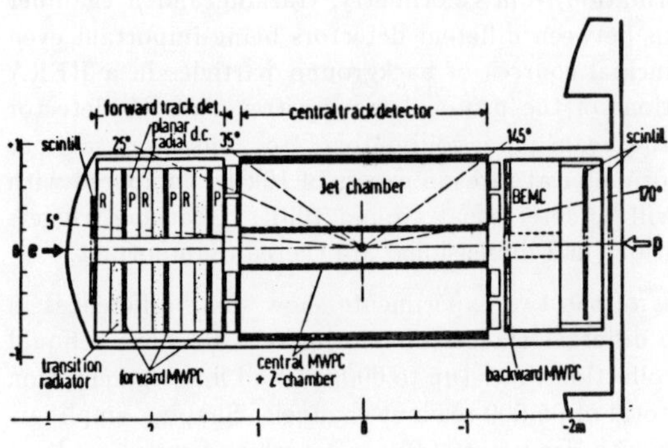

Figure 1: The H1 tracking system

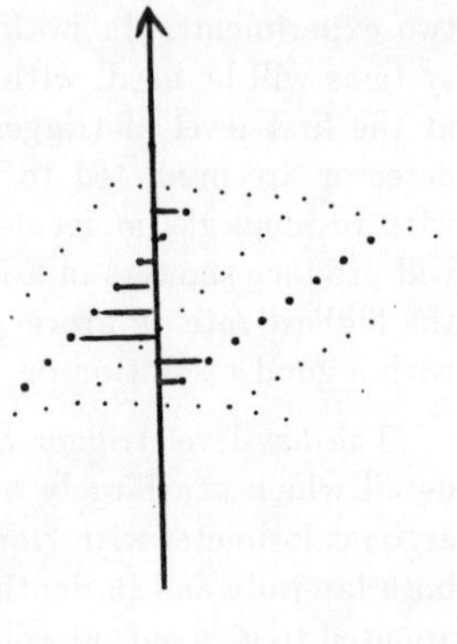

Figure 2: A drift cell of the ZEUS CTD

decision-making in the second level trigger is also under study.

The ZEUS calorimeter signals are uniquely identified with a beam crossing. For the H1 calorimeter the beam crossing associated with each signal has to be found by shaping the signal with a time constant of the order of 200-300ns, so that the shaped signal will pass through zero at a fixed time. Each trigger tower in the H1 calorimeter then produces two energy signals; a raw signal for use in combinational triggers where the other component has identified the beam crossing, and a signal gated by the zero-cross of the shaped pulse. Again both overall sums and regional sums will be available in the trigger.

3.2.2 **Tracking triggers.** Contrasting approaches between the two experiments are again evident in the area of tracking triggers. The tracking system for H1, shown if Figure 1, incorporates several sets of drift chambers with wires in orthogonal planes so that all track coordinates are measured with equal accuracy, together with MWPCs with fast response times specifically for triggering. These MWPCs will provide a charged multiplicity trigger with constraints on the z of the vertex, and individual track signals formed from hardware coincidence units, acting on sectors of the detector acceptance which can be matched up with the calorimeter trigger towers. Some drift chamber information is also expected to be used in the first level trigger but the main use will be at the second, non-pipelined trigger level.

The ZEUS tracking system is rather simpler than for H1, being made up of a single jet-type chamber in the central region, planar drift chambers at forward angles and a small vertex detector around the beam pipe. The trigger is expected to use mainly information from the Central Track Detector (CTD), where around 15% of the wires will be equipped with readout to measure the z-coordinates of tracks from the difference in arrival times of hits at the two ends of the wires. This information is provided principally for use in the trigger. The design of pipelined trigger processors for this type of chamber is complicated still further by the long drift time (maximum $\sim$ 500ns), which means that the hit information associated with a single track arrives at the processors distributed over 5 beam crossing intervals. Although the majority of the processing here is directed towards finding tracks pointing to the vertex in r-z, an important task is to pick the crossing with which a track is to be associated. The processor to do this relies on the fact that the drift cells are tilted with respect to the direction of a high-p_t track (to allow for the Lorentz angle of drift in the magnetic field) so that the arrival times of hits from such a track are distributed evenly over the full drift time, as illustrated in Figure 2.

3.3 Higher Level Processing

As for all large collider experiments, the data acquisition systems for H1 and ZEUS will include distributed systems of many hundreds of commercial microprocessors. These will be used for data compaction and formatting, control of data flow and general system monitoring functions as well as higher level trigger processing. Both experiments will use VME as the standard data bus, and for many of these functions the 68000 series of general-purpose microprocessors, together with associated support hardware (floating point coprocessors, DMA controllers) will be used. However, particularly in lower levels of the system where speed of operation is still of importance relative to full generality and ease of programming, some use will be made of Reduced Instruction Set Computing (RISC-type) processors. Digital Signal Processors (DSPs) featuring very fast integer arithmetic are now available from several manufacturers, and will be used at HERA for data compaction at front-end board or crate level. In ZEUS, the second level trigger system and Event Builder will be based around the INMOS Transputer, where the combination of a powerful microprocessor with memory and communications hardware on a chip should allow high-performance, flexible architectures to be built up relatively cheaply. A general-purpose VME Transputer board is currently under development for this system. For the final levels of software filter the two experiments are basing their developments around the Fermilab ACP project [3].

4 Future High Crossing Rate Colliders

4.1 Triggering at SSC and LHC

The severe problems of triggering at high energy hadron colliders with higher crossing rates than at HERA and several interactions per crossing have been studied at a number of recent workshops. Bunch spacings of $16ns$ for SSC and $25ns$ for LHC are proposed, with luminosities of $10^{33}cm^{-2}s^{-1}$ giving interaction rates of 10^8Hz. Here we examine present ideas on the structure of the triggering and data acquisition chain for an SSC or LHC detector. General areas where different approaches can be seen among the study groups are in the pipeline length and implementation, and the number and organisation of higher trigger levels. We look particularly at these areas in the light of HERA plans.

4.2 First Level Trigger and Readout Pipelines

Previous studies[4,5] have concluded that reductions to 10^4-10^5Hz can be achieved with triggers based on E_t and missing p_t sums, high-E_t electrons and jets and high-p_t muons.

The use of tracking information at the first level is generally not considered to be practicable, although if implemented successfully it would give benefits for the electron trigger, which is generally the noisiest component due to π^0 background. The processing of the calorimeter data foreseen at this level is thus similar in principle to HERA. The high probability of overlapping multiple interactions will be a problem for the trigger, and will seriously limit the usefulness of triggers based on combinations of signals, such as electron plus jet triggers. A veto on very large total energy may be necessary to cut out crossings with, say, 5 or more inelastic events, which are predicted to occur at around 10^5Hz.

The reduction factors attainable will clearly depend on the available processing time. This in turn will depend on the delay time in the readout, although the useful processing time will be significantly less than the total delay due to the time required for collecting trigger signals from each detector component into a central processor and for distributing the final decision. To allow for a reasonable degree of flexibility in the trigger processing, therefore, a relatively long signal delay time is desirable.

As argued in 2.1 above, longer delays will involve sampling and pipelining the data. However, the higher interaction rates at these machines mean that significant advances in technology will be necessary to achieve this without unacceptable dead time losses. Any dead time introduced by the readout of the pipelines is required to be much shorter than at HERA, probably well below $1\mu s$. If the inputs have to be disabled during readout, it becomes necessary to resynchronise the pipeline operation in different parts of the detector following each first level trigger. This will lead to an extra dead time, on top of the time to shift data out of the pipelines, of the order of the pipeline length. Strictly deadtimeless operation is thus a requirement for pipeline lengths of $1\mu s$ or longer. An attractive possibility under study for ZEUS is to build multiple parallel digital pipelines (to store FADC data from drift chambers) into a single chip. A trigger signal would stop just one of the pipelines, whose contents would then be accessible to a local processor. This chip then acts as both a deadtimeless pipeline and de-randomising buffer for the input to second level processing. The construction of analogue storage organised in a similar way, for input to a multiplexed ADC, would provide one route towards longer processing times. Alternative solutions for calorimetry require the development of fast, very high-resolution FADCs and digital pipelining.

4.3 Higher Level Triggers

For intermediate trigger levels at future colliders, it is expected that the trend towards high speed distributed processing seen at HERA will continue, with extensive use being made of VLSI technology. The introduction of tracking information should certainly be possible in the second level trigger. The postulated first level rate leads to an available processing time of $10\text{-}100\mu s$ so that a choice between the use of sophisticated hardware processors or high speed software triggers has to be made. One possible approach would be to develop custom-designed processor chips. However industry is also producing very fast processors of various sorts for use in the general area of signal processing, and it could well be that such products offer more cost effective solutions to the problems of pulse shape analysis and pattern matching encountered in intermediate level triggers.

Following the second level, and depending on the reduction factor expected, the data can either be transferred directly to a massive processor farm or another stage of distributed processing used. One proposed scheme[4] calls for a farm of 1000 processors of 5 MIPs power, with 25 parallel FASTBUS links carrying the input data rate of 1 Gbyte/s. However an extra level of filtering to reduce the data traffic into a processor farm might prove a more attractive option. Final event rates to tape are expected to be comparable with the 1-5Hz at present-day machines.

5 Conclusions

It is clear that the design of a triggering system for a general purpose experiment at SSC or LHC will be extremely challenging. At these energies the calorimeters are the most important detectors and their use is essential in all levels of the trigger. However, tracking detectors can also be very useful, particularly if they give good timing resolution and can be used in combination with the calorimeter to give electron triggers. We have seen that many of the ideas presently being developed for the HERA detectors could well be useful at SSC or LHC. Their usefulness however depends critically on developments in the electronics market. It is clear that widespread use of custom VLSI techniques in both the readout and trigger systems will be necessary for the next generation of experiments. However commercial developments over the next few years, in the areas of several hundred MHz FIFO's, fast high dynamic range FADCs or high-speed cheap RISC processors could dramatically change the feasibility of some of the ideas discussed in this paper. In any case, the difficulty of trigger design at these machines and the way in which the performance of each level of trigger is very tightly constrained by the available processing time and reduction rate required has strong implications of the overall design of the experiment. These must be taken into account at the earliest possible stage. Hardware, software and trigger design must be developed hand-in-hand if the experiments at the next generation of hadron colliders are to be successful.

References

[1] H1 Collaboration, Technical Proposal, March 1986 (unpublished)

[2] ZEUS Collaboration, Technical Proposal, March 1986 (unpublished)

[3] S. Conetti, these proceedings.

[4] Physics at the Suprconducting Supercollider, Snowmass 1986

[5] Proceedings of the Workshop on Physics at Future Accelerators, CERN 87-07(1987)

Design Automation for SSC Data Acquisition

Jon J. Thaler
Physics Department - University of Illinois
Urbana, Illinois USA

ABSTRACT

High data rates and complex event patterns will severely stress the data acquisition systems in future particle physics experiments. At the SSC event rates will be about 10^3 times greater than at presently existing hadron colliders. In order to build reliable data acquisition systems, more capable design tools than currently exist must be developed. These tools will also be applicable to other areas of experimental research.

1. INTRODUCTION

In this talk, I will discuss research at the University of Illinois on tools for the design of data acquisition and trigger systems in particle physics experiments. This research is an outgrowth of our involvement in the designs of the SLD and CDF triggers, and in the establishment of the Fastbus standard. This research has expanded beyond high energy physics; tools development now involves people from computer science and electrical engineering, and experimental applications now include biochemistry and atmospheric sciences. We also have collaborations with several computer, electronics, and software companies.

Our general approach is to use computer aided engineering (CAE) tools, developed for the design of printed circuit board electronics, to design large scale systems. This extension of CAE is currently an active field of research in the computer and electronics industry. Improved database technology, simulation methods, and information sharing protocols are needed.

2. EXPERIMENTAL ENVIRONMENT

The previous speaker discussed the experimental situation at future accelerators. I summarize here the main points that are relevant to the development of design tools.

2.1 LEP/SLC

At a luminosity of 10^{31}, there is about a 1 Hz event rate and 10^5 bytes per second from the detector. Thus, no filtering needs to be done beyond suppressing non-physics background. The bunch crossing times are many microseconds (or even milliseconds), so there is plenty of time to perform level-0 pattern recognition. Good events mostly contain two jets and the background is mostly cosmic rays or beam associated junk.

2.2 SppS/Tevatron

A luminosity of 10^{30} yields a 10^5 Hz event rate and 10^{10} bps. A 10^{-5} filter is required; its job is complicated by the fact that most uninteresting data is due to real beam-beam interactions. The beam collision times of several μs allow a level-0 trigger to proceed, and the per-channel data rate of 10^5 bps is still manageable.

2.3 SSC

The expected 10^{33} luminosity will give 10^8 events per second and a total data rate of about 10^{14} bps (10^8 per channel). These rates are several orders of magnitude greater than any open geometry experiment has ever had to handle. In addition, the proposed bunch crossing time is 33 ns, which not only leaves no time for a level-0 trigger, but also results in multiple simultaneous interactions. Thus, there are several aspects of the SSC environment which will complicate the data acquisition task:

- Overlapping events will make pattern recognition difficult, for some processes.
- Slewing of the output from various detector components will make it difficult to separate data from different events.
- The high data rate and short beam collision time require a pipelined data flow.
- The data rate per channel is too high for digital processing at the front end. Thus, a mixed analog/digital system is probably necessary.
- The data rate is too high for current bus technology. Parallel processing, with each stream autonomous, will be necessary. System coordination will be a complex task.
- The complexity of the data acquisition systems will make efficient monitoring of apparatus performance even more

important than it already is. Verifiability of proper operation must be one of the design criteria.

- Detector design optimization must include both data acquisition and offline analysis criteria. Both aspects of detector performance must be incorporated at the earliest design stages.

SSC era data acquisition systems will be designed by large experimental collaborations. Information exchange mechanisms are marginal at present and will become inadequate in the future. The design tools used in the future must establish and enforce adherence to detector specifications. Such tools do not now exist and must be developed as soon as possible. We are pursuing the possibility of adapting current CAE techniques, used for printed circuit design, to the problem.

3. PRESENT PC BOARD AND IC DESIGN METHODS (CAE)

The principal feature of a modern design system is the near automation of the entire design and manufacturing process. The designer enters a circuit schematic in the system, after which every step of the design process is tested by the CAE system. This includes connectivity checks, assignment of physical components to logical functions, timing and logic simulation using known component behavior, fault simulation, layout, and manufacture. Standard data interchange formats are being developed to allow different systems to communicate.

A CAE system offers several advantages to the user. The most obvious is the simplest; namely that by treating a circuit design in a manner similar to a computer program changes can be easily and reliably made, a design history can be kept, and documentation is naturally generated. In addition, useful parts of previous designs are easily incorporated into new projects.

There are more powerful advantages, also. A CAE system can simulate the behavior of a circuit, taking as input any set of signals that the designer cares to generate. The state of any set of signals in the circuit can be followed as a function of time, so that the proper logical design can be verified. In addition, gates can be made bad, so that the designer can learn what measurements are needed to detect faults. In this way a comprehensive set of test vectors can be generated for later diagnostic use.

CAE tools already are useful for system analysis. We have implemented a simulation of Fastbus which allows us, when designing a FB slave, to simulate the entire environment in which it will reside and verify its adherence to proper FB protocol. The slave control logic (SCL) has been implemented as a distinct piece of hardware which is merely plugged into any FB slave.

The practical advantages of the use of this CAE system also include large savings of time and money. We have designed several 300 IC FB boards for CDF and SLD. The time from initial schematic entry to final manufacture is typically 4 to 6 months, and *no prototypes are ever built.* The completed boards perform reliability, with little or no debugging necessary.

Experience with CAE design techniques has led us to identify features that will be needed in the future:

- The ability to simulate embedded CPUs, so that mixed hardware and software systems can be designed.
- Automatic test vector generation. At this time, faults must be put in by hand and the proper test vector identified.
- Cost and performance estimators. In complex systems, it would be useful to be able to compare alternative designs.
- Behavioral synthesis. Present CAE systems are able to optimize sets of gates ("logic synthesis"). This does not use any knowledge of the anticipated data environment. A higher level optimization ("behavioral synthesis"), akin to that used for computer operating systems, would use the simulator for optimization.
- Design verification. In order to compare hardware and software implementations of a given algorithm, it is necessary to be able to verify that the device is executing the proper algorithm. This requires a common language for the specification of the algorithm.

4. AREAS THAT NEED WORK

4.1 Databases

New database technology is required both to maintain the design information and to manage the experimental data. Experimental data volumes are already approaching 10^{13} bytes per year. Design information is smaller, but has more internal links. The effects of design changes throughout the design hierarchy must be tracked.

4.2 Truth Maintenance

In the context of system design, this means mostly the global communication of decisions. When a specification for one component is changed, that change must be broadcast. Designers may want to work on different versions of designs; this is not a simple procedure to automate. In data analysis, truth maintenance is related to data editors, by which analysis decisions are consistently applied to the entire data set.

4.3 Pattern Recognition

Real time pattern recognition will be very difficult at the SSC. It is important to develop efficient algorithms, and to compare their implementation in several technologies (analog and digital hardware and software) in order to be able to design data acquisition systems that can survive the SSC environment. The algorithms must be part of the detector design specification.

4.4 Data Flow

Huge data rates do not allow a simple bus architecture. It is necessary to understand how to synchronize parallel data processing so that an ultimate trigger decision can be made and the appropriate information saved. This is a much more difficult problem than the usual parallel computing environment, because data acquisition is interrupt driven and tends to be asynchronous.

5. OUR APPROACH TO THE PROBLEM

We are pursuing a three-fold approach to the development of new design tools. The first is a straightforward extension of our existing CAE tools to larger, system level, problems. For example, we have a graduate student simulating the CDF data acquisition system. The intent is to identify problems with our design tools by applying them to a solved problem. We have found that the language invented for PCB simulation, which was successfully applied to the Fastbus protocol, is inadequate for describing larger systems. This work has attracted interest from the CAE community.

The second approach is to work with CAE professionals to develop the new kinds of tools that are needed. We have formed a collaboration at Illinois with computer scientists and engineers who work on behavioral synthesis, databases, and system integration and have proposed to establish a center for the study of the issues pertaining to large scale data acquisition and data management problems.

The third approach is to identify other scientific disciplines which share similar problems of data acquisition, analysis, and management. We have established a collaboration with several.

6. OTHER POTENTIAL USERS OF SYSTEM LEVEL TOOLS

6.1 Genome analysis.

This is the field of biochemistry which studies the genetic sequences of various organisms. They have serious problems with automated film scanning, large databases, and truth maintenance.

6.2 Atmospheric Science

The complex data needed to study weather systems is well known. Our collaborators are cloud physicists who must try to integrate heterogeneous data sets into a coherent picture. They need new data acquisition, database, and truth maintenance techniques.

6.3 Autonomous Navigating Robots

How can a mobile Mars lander find interesting objects and avoid the potholes? Of all the other disciplines, this is the one with real time problems most similar to high energy physics. Although the data rates are somewhat lower, the pattern recognition is far more difficult.

A FASTBUS–BASED SOFTWARE TRIGGER FOR THE MARK II DETECTOR AT THE SLC*†

R. ALEKSAN,[a] D. BRIGGS,[a] T. GLANZMAN,[a] P. GROSSE–WIESMANN,[a] S. HOLMGREN,[b] S. KOMAMIYA,[a] M. SCHAAD,[b] J. TINSMAN[a]

1. INTRODUCTION

An intensive commissioning effort is currently under way to establish e^+e^- collisions at a center of mass energy of 93 GeV in the new SLAC linear collider or SLC. This machine was designed and built both to test the ideas of a new linear collider technology [1] and for the production of Z^0 bosons for physics analysis.[2,3] The initial physics run will be made with an upgraded Mark II detector, a veteran of two previous electron-positron colliders: SPEAR and PEP.

A new software trigger scheme has been developed to augment and enhance the existing charged and neutral triggers by providing sensitivity to new event topologies and some level of control over accelerator-induced backgrounds. Historically, the Mark II existed with two primary trigger components: a charged track finder based upon the central and vertex drift chambers and the time-of-flight counters; and an electromagnetic trigger based upon the total energy deposited in each of ten calorimeter modules. The trigger component of the new system is based upon the Mark II electromagnetic calorimetry but with significantly increased granularity and the inherent flexibility of software. Trigger processing also benefits from the relatively long period of time (up to 8.3 ms) between SLC beam crossings.

The production of long-lived neutral particles provides an example of an event topology which would not have triggered in the old system. By decaying beyond the first few drift chamber layers, such particles avoid the charged particle trigger, yet could produce clear signals in the calorimeters. Another example is the class of events containing a single photon as the only visible particle such as occur in the reaction $e^+e^- \rightarrow Z^0\gamma \rightarrow \nu\nu\gamma$. Sensitivity to this reaction is necessary to measure the number of neutrino generations. One goal of the new trigger is to achieve nearly 100% efficiency for single photons of energy above 750 MeV. Such a trigger necessarily depends upon a very low and well understood accelerator background for success. A minimum ionizing particle trigger will provide good efficiency for $Z^0 \rightarrow \mu\mu$ or $\tau\tau$ at the angles covered by the end cap calorimeters which increases the capabilities for measuring the forward-backward asymmetries. And finally, an improved trigger is more sensitive to complete surprises.

A system of ionization chambers, proportional tubes and wire chambers placed around the Mark II detector and in the SLC tunnels comprise the background monitor system. As the patterns of accelerator-induced background particles in the detector become understood they will be taught to the new trigger system and used for event vetoing.

---

*Work supported by the Department of Energy, contract DE–AC03–76SF00515.

†Presented by T. Glanzman, Stanford Linear Accelerator Center.

[a]Stanford Linear Accelerator Center, Stanford University, Stanford, California 94309 USA.

[b]Lawrence Berkeley Laboratory, University of California, Berkeley, California 94720 USA.

2. HARDWARE

The Mark II detector and data acquisition system have been described elsewhere.[4,5] A schematic diagram of the new software trigger electronics is shown in figure 1. The calorimeter portion of the trigger consists of 656 energy measurements from the liquid argon and end cap calorimeters.

2.1 Liquid Argon

The Mark II liquid argon electromagnetic calorimeter consists of eight identical modules arranged around the solenoid magnet covering 2π in ϕ and from about 45° to 135° with respect to the beam axis in θ. Each module is a cryogenically sealed compartment containing 18 physical lead/liquid argon layers ganged together into six readout layers (three in ϕ, two in θ and one at a relative orientation of 45°). The total thickness of lead and argon is 14 radiation lengths per module.

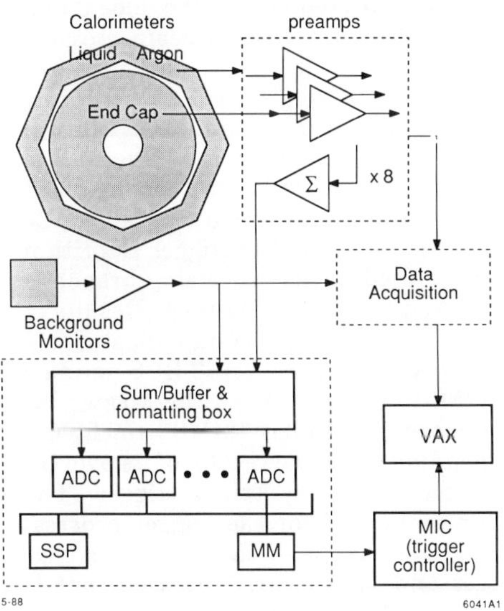

Figure 1. Trigger hardware components.

Analog signals from sets of eight adjacent channels are summed to produce 42 sums per module or a total of 336 sums for the entire liquid argon system. Figure 2(a) shows a cosmic muon event in the plane perpendicular to the beam axis. The calorimeter octants show a grid indicating the segmentation of the trigger information in this projection. Figure 2(b) shows the complete segmentation for the bottom right module in the previous figure. These sums are buffered and digitized by four LeCroy 1885N FASTBUS ADC modules.

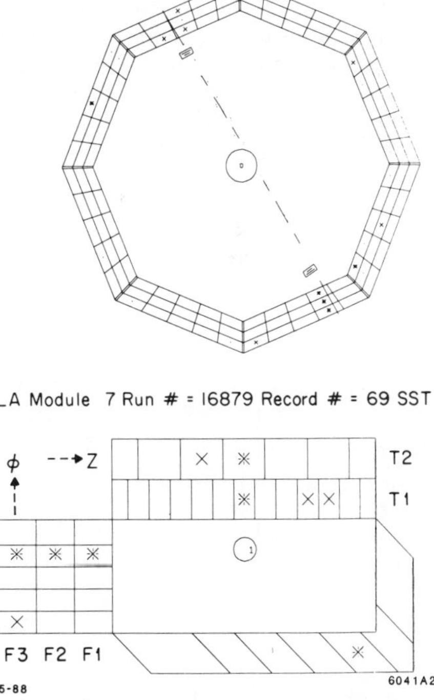

Figure 2. (a) Liquid argon calorimeter ϕ projection; (b) single module segmentation.

2.2 End Cap

Each end of the Mark II contains a retractable door supporting a lead/proportional tube electromagnetic calorimeter covering an angular range from about 15° to 45°. An end cap calorimeter consists of 36 physical layers ganged together into ten readout layers representing four orientations: vertical (three layers), horizontal (three layers) and both 45° diagonals (two layers each).

Total material thickness sums to 18 radiation lengths per end cap. Signals are summed in groups of eight analogously to the liquid argon system. Figure 3 shows a beam background muon event with two muons passing through the detector. The segmentation of the trigger information is indicated by the displays of the end caps four orientations. The 320 sum channels are buffered and digitized by four FASTBUS ADC modules as for the liquid argon system.

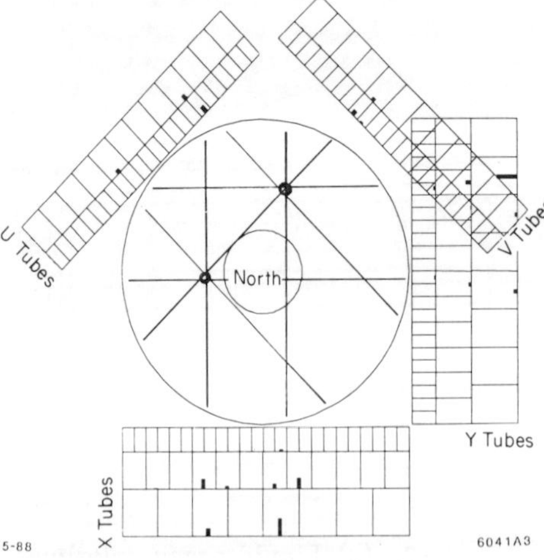

Figure 3. Endcap calorimeter segmentation.

3. READOUT

Data from the ADC modules are read by a single SLAC Scanner Processor (SSP).[6,7] The SSP is a general-purpose FASTBUS device consisting of CPU and I/O sections. The CPU is a 32-bit emulator of the IBM/370 integer instruction set and can be programmed in assembly language or FORTRAN. It is equipped with memory for 4K instructions and 0.5 Mbytes of data memory. As a FASTBUS device, the SSP acts as a master or slave on both the crate and cable segments. The basic clock period of 120 ns is shortened to 80 ns during FASTBUS block transfers resulting in data transfer rates in the range of 20 to 25 Mbytes/s. During event acquisition the contents of SSP memory is read directly into a VAX 8600 for logging to tape.

4. SOFTWARE

Online software for the SSP module is represented by the block diagram in figure 4. The beam crossing frequency at the SLC is sufficiently low that only a single level of triggering is required to maintain an event logging rate of a few Hertz. The SSP is started for every beam crossing (8.33 ms at 120 Hz) at which time all ADCs are read out and the trigger algorithm performed. This processing branch is labeled "TRIG" in figure 4. Results of the algorithm are communicated via a special-purpose FASTBUS module ("MM" in figure 1) to the Master Interrupt Controller (MIC) which, depending upon the results of its own internal logic, may issue a trigger. The time required for the trigger processing will vary with the particular trigger and veto algorithms used, but is expected to be in the range of 5 ms, in addition to the 1.5 ms ADC digitization and readout time. For triggered events, the SSP is started a second time to complete its data acquisition tasks of making gain corrections and formatting the data into a tape buffer; this is indicated by the branch labeled "DAQ" in figure 4.

Various trigger algorithms are being developed. The simplest algorithm computes individual energy sums for each of the ten calorimeter modules. Overall energy deposition and crude topologies based upon the 10 calorimeter modules may then be determined. Another approach is to reconstruct localized energy towers within each calorimeter based upon the patterns of energy deposition. This algorithm compares raw ADC counts with a set of three threshold values on a channel-by-channel basis. Execution of such a tower-finding algorithm consumes a significant number of CPU cycles while looping over different layer combinations much like tracking code. However, techniques to restrict this looping to meaningful combinations yields

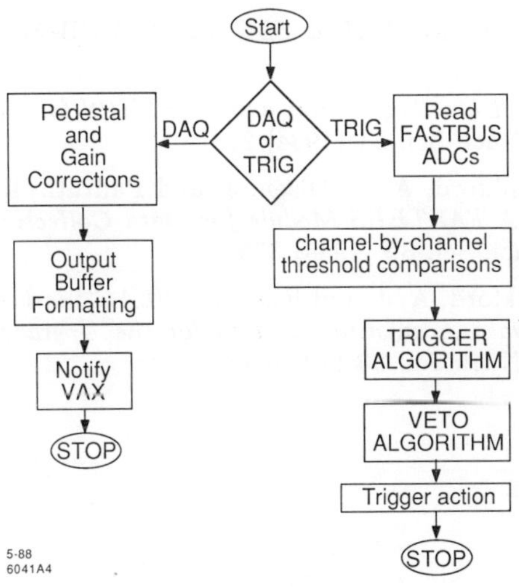

Figure 4. Trigger software components.

an execution time within the desired range. The flexibility of the overall design allows for rapid installation and testing of new algorithms. The algorithms used are expected to evolve reflecting the needs of the experiment and the accelerator environment.

5. RESULTS

Trigger calorimeter data have been recorded for both cosmic ray and beam background events and have been analyzed offline. The first cosmic ray data provided a clean sample of events containing a single μ particle going through the detector. Events were reconstructed using the Mark II central drift chamber and projected into the liquid argon calorimeter. Observed single μ signals were determined to deposit approximately 15 MeV on average per physical layer, representing 2.3–4.0 σ above the equivalent noise charge. Muons in the end cap calorimeter deposit approximately 10 MeV on average per physical layer which is nearly 15 σ above the noise. These results reflect improvements made both to the liquid argon and end cap electronics. They also show the sensitivity to minimum ionizing particles, particularly in the end cap.

Part of the SLC commissioning effort is dedicated to understanding the accelerator-induced backgrounds present at the interaction point. The backgrounds observed in the calorimeters have been due to muons from sources far upstream in the beamline and low energy electrons and photons from local sources. The end cap's trigger data has become an important commissioning tool due to its ability to identify and measure the muon and electromagnetic components. Its high sensitivity also promises to result in a very clean minimum ionizing particle trigger once backgrounds have been reduced.

6. STATUS

All hardware and software components of a new Mark II trigger system have been installed. The triggering scheme has been thoroughly tested with cosmic ray and beam background events. Offline analysis has produced preliminary results on the hardware performance. Work continues to tune the algorithms for physics runs.

ACKNOWLEDGEMENTS

I would like to acknowledge the efforts of the following Mark II physicists and engineers who have made significant contributions to this project and without whom this presentation would not have been possible: D. Burke, D. Fernandes, D. Herrup, B. Jacobsen, S. Klein, A. Lankford, T. Mattison, B. Milliken, D. Wilkinson and L. Wong.

REFERENCES

1. Rees, J., *The Principles and Construction of Linear Colliders,* SLAC-PUB-4073, September 1986.
2. *Proceedings of the SLC Workshop on Experimental Use of the SLAC Linear Collider,* SLAC–Report–247, March 1982.
3. *Proceedings of the Second Mark II Workshop on SLC Physics,* SLAC–Report–306, November 1986.
4. Trilling, G. et al., *Proposal for the Mark II at SLC,* SLAC–PUB–3561 (1983).
5. Lankford, A. and Glanzman, T. *Data Acquisition and FASTBUS for the Mark II Detector,* IEEE Trans. Nucl. Sci. **31** (1984) 225.
6. Brafman, H., Glanzman, T., Lankford, A. J., Olsen, J. and Paffrath, L., *The SLAC Scanner Processor: A FASTBUS Module for Data Collection and Processing,* IEEE Trans. Nucl. Sci. **32** (1985) 336.
7. Barklow, T., Glanzman, T., Lankford, A. J. and Riles, K., *SLAC Scanner Processor Applications in the Data Acquisition System for the Upgraded Mark II Detector,* IEEE Trans. Nucl. Sci. **33** (1986) 775.

From APE to APE-100: Present and Future of the APE project

The APE collaboration

P. Bacilieri[a], S. Cabasino[b], N. Cabibbo[c], A. Castelvetri[a], F. Coppola[d], G. Fiorentini[e], M. P. Lombardo[d], A. Lai[e], P. Marchesini[f], E. Marinari[c], F. Marzano[b], P. Paolucci[b], G. Parisi[c], F. Rapuano[b], E. Remiddi[a-g], R. Rusack[h], G. Salina[c], E. Simeone[d], G. M. Todesco[f], R. Tripiccione[d], W. Tross[b].

a) INFN-CNAF, Bologna, Italy.
b) Dipartimento di Fisica, I Univ. di Roma "La Sapienza" and INFN-sez. di Roma.
c) Dipartimento di Fisica, I Univ. di Roma "La Sapienza" and INFN-sez. di Roma.
d) Dipartimento di Fisica, Univ. di Pisa and INFN-Sez. di Pisa.
e) Dipartimento di Fisica, Univ. di Cagliari and INFN-Sez. di Cagliari.
f) CERN, Geneva, Switzerland.
g) Dipartimento di Fisica, Univ. di Bologna and INFN-Sez. di Bologna.
h) The Rockefeller University, New York, USA.

Abstract

A status report of APE is presented. APE is a parallel processor optimized for numerical simulation of lattice gauge theories. Three APE units, with performances in the range 256 Mflop to 1 Gflop are used for LGT simulations, while a new unit with a top performance of 2 Gflop is planned for the end of 1988. A major upgrade of APE (APE-100) is proposed. APE-100 is a fine grained array of about 1000 processors delivering a top performance of 100 Gflop to be used for LGT simulations on very large lattices.

The APE parallel processor was proposed a few years ago as a dedicated machine for the numerical simulation of lattice gauge theories (LGT). The project was triggered by the belief that performances exceeding those of commercial supercomputers could be achieved by accurately tailoring the architecture of APE to the specific class of applications envisaged.

APE, decribed in details elsewhere[1], is a SIMD (Single Instruction Multiple Data) processor. A number of cells, each containing a memory frame and a floating point unit (FPU) are connected by a circular switchboard. All cells execute the same code in lock-step mode, steered by one control unit, a combination of a CERN/SLAC 3081/E[2] and a dedicated sequencer board.

The multi-cell structure of APE exploits the inherently parallel structure of the computations typical of LGT -numerical data sitting at different lattice points can be manipulated concurrently- while the SIMD mode of operation is adequate to deal with an homogeneous problem -calculations are the same at all lattice sites- avoiding the (sofar unmanageable) problems encountered in an (admittedly more general purpose) MIMD machine.

The simple and efficient APE architecture is supported by a programming language (the APE language) that naturally reflects the parallel structure of the machine. APE can be considered a host processor integrated in a VAX/VMS environment.

Three APE units have been built so far, as detailed in table 1. Two units have 4 cells, delivering a peak performance of 256 Mflop, while the third unit, based on 16 cells, reaches the performance of 1 Gflop. A fourth unit, based on 16 cells, will be ready by the end of 1988. This unit will have two FPU's in each cell, sharing the same data path to the switchboard, for a peak performance of 2 Gflop. Larger memory banks (64 Mbytes) will be used if 1Mbit DRAM chips are available.

More than 10000 APE hours have been logged so far on the three machines, which have been used for LGT simulations since fall, 1986. APE achieves a sustained performance of about 70% of the peak figure on typical LGT codes. As a rule of thumb, one 4-cell APE unit has the same performance as a CRAY XMP-12 supercomputer.

Other processors dedicated to LGT simulations have been built[3]. They deliver peak performances higher than those attained by APE. Our processor is as effective as its competitors in daily usage, however, thanks to its hardware reliability and to the excellent quality of its software.

LGT simulations of QCD aimed at the measurement of the glueball and fermionic spectrum and at the study of the properties of the phase transition of the theory have been carried out. Results have been published elsewhere[4]. They show that APE is indeed a valuable tool in the numerical simulation of lattice gauge theories.

	Completed	Peak speed	Memory Size
APetto#1	1986	256 Mflop	64 Mbyte
APetto#2	1987	256 Mflop	32 Mbyte
APone	1987	1 Gflop	256 Mbyte
APE-21	1988	2 Gflop	256(1000) Mbyte

Table 1

Physical considerations suggest however that a substantial increase in (floating point) performance and memory size would be welcome for greater accuracy and reliability of LGT simulations. Indeed, LGT simulations are affected by statistical errors,vanishing as the square root of the computation time, as well as systematic errors, behaving as 1/(lattice size) with the computational cost growing as $(\text{lattice size})^4$.

For these reasons, we are considering a major upgrade of the APE-processor to be completed early in the 90's. The new machine (APE-100) should deliver about 100 Gflop, with a memory size in the 10 Gbyte range. These figures should allow accurate simulations of lattices of size ~60**4 in a reasonable time. Close matching of the architecture with typical LGT algorithms should be maintained, as well as the parallel and SIMD structure of APE. APE-100 will be however a fine grained processor, since a large fraction of 100 Gflop cannot be squeezed into one FPU. Instead, we keep the performance of one cell to the present level and reduce the complexity of one FPU down to the one chip level, adding significantly to the reliability of a system containing about 1000 nodes.

The architecture of APE-100, sketched in fig. 1, provides a 3-dimensional array of FPU's and memory banks, each node being connected to its nearest neighbours in all 3 dimensions. This structure features expandability and ensures communication capabilities adequate to ensure a level of performance linear with the number of processors for LGT simulations.

The array is partitioned in cells, each cell containing two processors and two memory banks, and fitting into one board. Each cell is logically split into the 8 sites of a tridimensional cube of size 2. Each FPU can be considered as attached to four of the sites in a time multiplexed fashion. At any clock cycle, the two FPU's are

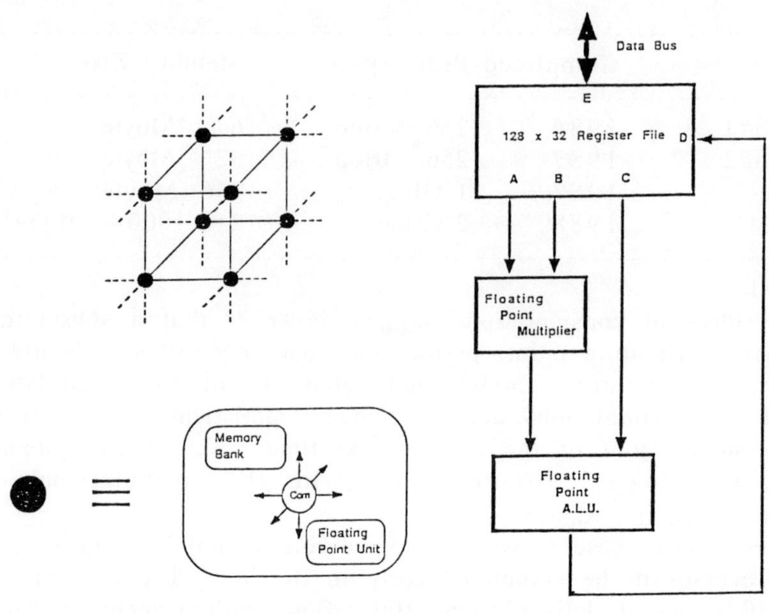

Fig. 1 Fig. 2

connected to sites sitting on one of the long diagonals of the cube. This arrangement reduces the number of connections off-cells, since all logical nodes on the same face of the cube share the same physical connection. We envisage an array of 8*8*8 cells, logically corresponding to 16*16*16 mesh points, and containing 1024 processors, for a peak performance of 64 Gflop. One crate houses 16 boards, corresponding to the x-lines at two given values of y. One rack contains 4 crates making up one x-y plane. Eight such planes (i.e. 8 racks) make up the full processor. Approximately 1500 ribbon cable connections ensure the required electrical connections. The whole array is controlled by one control unit, based on a single board processor very similar in architecture to the well known 3081/E and based on new technology. Each crate contains one controller, all controllers synchronously executing the same code. The controller is responsible for global flow control, integer arithmetic and memory addressing.

As stressed above, each cell is based on a very small number of highly integrated parts, to increase reliability. The FPU itself is a one chip custom-developed device. It integrates a register file and a floating point multiplier and ALU optimized for MAC (multiply and accumulate) operations. The data path of the device is sketched in fig. 2. It is essentially the data-path of the FPU used in APE, restricted to real (floating-point) arithmetic. A new multiply and a new add start at each clock cycle, clock frequency being 33.33 Mhz, in order to achieve a top performance of 64 Mflops. The structure shown in fig. 2 has some advantages with respect to the data path used in APE. Indeed, real and complex arithmetics can be accomplished with the same efficiency, while all registers can be accessed to feed data to all devices. The chip is controlled on a cycle by cycle basis by a very long (87 bits) microcode word, supplied on dedicated pins from a writable control store. This arrangement is the same as in APE. Program compilation down to the microcode level is in fact a key feature to attain very high efficiency in a deeply pipelined machine.

We hope to develop APE-100 on a rather tight time-schedule. We plan to have working samples of M.AD for late spring 1989 and working prototypes of all boards by the end of that year. Full scale production should be undertaken during 1990, and a full scale machine should be ready in 1991.

References

[1] The APE Collaboration, Comp. Phys. Comm 45 (1987) 345.

[2] P. M. Farran et al., Proc. of the Conf. on Computing in High Energy Physics, L. O. Hertzberg and W. Hoogland eds (North Holland, Amsterdam,1986).

[3] J. Beeten, M. Denneau and D. Weingarten, 12th ann. Intl. Symposyum on Computer Architecture (boston, 1985) IEEE Computer Society Press, 1985.

N. H. Christ and A. E. Terrano, IEEE Trans. on Computers 33 (1984) 344.

[4] M. Albanese et al., Phys. Lett. 192B (1987) 163.

M. Albanese et al., Phys. Lett. 197B (1987) 400.

P. Bacilieri et al., preprint ROM2F/88/001, Phys. Lett. B in press.

SPECIAL PURPOSE PROCESSORS FOR HIGH ENERGY PHYSICS

George McPherson

Electronics Division, Rutherford Appleton Laboratory
Chilton, Oxfordshire, England

1. INTRODUCTION

Special purpose processors have been an integral part of the High Energy Physics (HEP) scene for many years. Due to their inherent nature they come in a large variety of flavours, analogue or digital, hard-wired or programmable, arithmetic or logical, fixed point or floating point; from such diverse application areas as sub-microsecond decision making for fast triggering to full event reconstruction for the extraction of physics. They exist because they exhibit some or all of a number of relative advantages such as speed, cost, power consumption, size, reliability or efficiency. That they exist in such rich profusion is a tribute to the inventiveness and ingenuity of those concerned.

The purpose of this paper is to present a brief and (necessarily) selective history of special purpose digital processors, to look at the current state of the art, and to form some conclusions for the future based on the fast-moving events in custom and semi-custom Application Specific Integrated Circuit (ASIC) design.

2. HOW THEY WERE

2.1 Fifteen Years Ago

This was the era of the Bubble Chamber,Spark Chamber and small Multi-Wire Proportional Chamber spectrometer. Data rates and sizes, although low by today's standards, presented a considerable off-line load to the mainframe analysis computers of the time. In an attempt to reduce this load to an acceptable level there was an explosion of ideas for on-line linearisation of data, track finding and pattern recognition. Techniques varied from hardware implementations of specific algorithms [1,2], through to providing the on-line computers which controlled the experiments (PDP-8's, PDP-11's, DDP-516's, HP-2100's) with much needed processing power in the form of add-on programmable arithmetic units, optimised for specific classes of problems.

2.1.1 Programmable machines. For example, vidicon cameras were displacing film for recording events from spark chambers. Experience showed that vidicon scan speed variations resulted in measurement errors of ± 4 mm. in space. Using information from surveyed fiducials scanned by the cameras, coefficients for correcting polynomials could be calculated. It was found that a fifth order correction to the data resulted in a reduction of measurement errors to ±0.1 mm. Unfortunately the on-line host computers of the day rarely had fast multipliers, nor were they optimised for polynomial evaluation. A special purpose processor built for this task [3] gave speed-up factors of several orders of magnitude over the host computer, allowing real-time data correction to take place. It exhibited all the basic ingredients of a modern Digital Signal Processor (DSP) chip. Unlike a DSP chip, however, it occupied a large part of a CAMAC sized crate, dissipated hundreds of watts of power and cost thousands of pounds to build.

Experiments continued to grow in size and complexity, some of the larger proportional chamber spectrometers having tens of thousands of wires. Programmable special purpose processors, usually based upon Harvard (dual bus) architecture with either bit slice Arithmetic and Logic Units (ALU's) [4,5], or more dedicated processing units using multipliers, adders and parallel shifters [6,7], remained popular as an add-on, speed-up extension to the on-line computer for a number of years. They can with hindsight be viewed as the forerunner of the modern coprocessor, having characteristics in many cases clearly recognisable in today's Reduced Instruction Set (RISC) computers; that is, minimised cycle time (typically 100 to 200 nanoseconds), single cycle execution, simple instruction formats, register to register organisation and limited addressing modes.

Several factors eventually led to a decline in their use, although derivatives of some of these early machines are still in operation [8]. One of these factors was the increasing arithmetic performance of the on-line computer. A more decisive reason, however, was because of one of the apparent strong points of the special machines - their programmability. A great deal of system software (assembler, debugger, linker, loader etc) has first to be written in order to support application software on a programmable machine, and even at this level the effort required to write programs for real tasks is huge. Writing high level compilers (eg FORTRAN) would have eased the problem, but for limited production machines this was never seen as being cost effective.

2.1.2 Emulators. One very clever way around the compiler problem was to design a machine which emulated part of the instruction set of a commercial computer for which a FORTRAN compiler already existed, using an architecture dedicated only to the instruction sub-set in order to significantly reduce the cost compared to the original. The first such machine [9] emulated the IBM 168 mainframe. It proved enormously successful for relieving the processing load on scarce off-line computing resources, and the technique has been repeated many times for other commercial computers [10,11,12].

2.2 Ten Years Ago

Ten years ago the size and shape of HEP experiments had started to change. Colliding beam experiments were replacing fixed targets. 4π detectors which used central tracking chambers for particle track reconstruction, and calorimetry techniques for energy measurement, replaced the multi-arm wire chamber spectrometer. Increasing event rates and sizes made it more important to apply careful criteria to selecting which events were worth recording. The emphasis on special purpose processors shifted to trigger processors, which examined sub-sets of the data to decide whether or not to take an event. Due to speed requirements this class of processor generally assumed a highly parallel format, using arrays of logic to effectively sieve trigger data, rejecting several thousand bad events for every good one sent to tape for subsequent analysis off-line.

Although early examples of this type of trigger processor [13)] exhibited a certain degree of flexibility through thoughtful design, rapidly advancing technology was soon to provide a generic solution in the form of large, cheap Random Access Memory (RAM) chips which could hold look-up tables for extremely rapid data transformation. The original UA1 trigger processor [14)] was one of the first to exploit the advantages of cheap RAM look-up tables, and it is worthwhile examining it in a little detail

2.2.1 The UA1 trigger processor. For the purposes of triggering, the 2440 photomultiplier channels of the calorimeter were added to form 288 analogue trigger channels, 136 electromagnetic and 152 hadronic. Each channel was digitised to 8 bits accuracy in 0.5 μsecond using a fast Wilkinson count-down type analogue to digital convertor (ADC), clocked at 500 MHz. Every digital channel then addressed its own look-up table RAM to transform the data from total into transverse energy, at the same time correcting for channel-to-channel gain variations and pedestal differences. Adding adjacent channels and comparing results against pre-loaded thresholds yielded hit patterns in the calorimeters whose multiplicity represented various combinations of electron and jet triggers. By selectively adding all channels, energy deposition profiles within the calorimeter were available. The output of the trigger processor was an 8 bit pattern, each bit of which could be selected as a trigger, a veto or a 'don't care' condition. Total time taken, including digitisation, was 1.3 μseconds, which resulted in zero dead time with the 3.8 μsecond bunch crossing period of the CERN p^+p^- collider.

In practice, two such processors were used for the UA1 first level trigger, one to measure transverse energy and one to measure total energy. Excluding the fast ADC's, the system used approximately 20,000 integrated circuits, had 250,000 interconnections, dissipated several kilowatts and cost £100,000 to build. The UA1 experiment has since been upgraded and the new trigger processor, reported elswhere at this conference, is larger, costlier and more powerful by at least a factor of 5.

3. HOW THEY ARE

3.1 Five Years Ago

Five years ago planning had started for the experiments (ALEPH, DELPHI, L3 and OPAL) which will initially be installed on LEP, the Large Electron Positron collider being built at CERN. An early review of their trigger and data acquisition plans [15] shows that they mostly need very fast (1-2 μsec), highly parallel processors for the first level of trigger rate reduction, followed by a second, somewhat slower (50-100 μsec) processing stage allowing greater refinement and selectivity.

A large fraction of these processors are based upon the 'look-up table in RAM' theme, combined with some hardwired arrays of arithmetic or logical processing using medium scale integration (MSI) chips (adders, registers, comparators and the like). At the second trigger level, where time permits their use, fast programmable machines also appear to be making a comeback using commercially available DSP chips [16]. This is hardly surprising given their ready availability, low cost, high speed and fully supportive development software [17]. Many of the LEP trigger processors are reported on elsewhere in the course of these proceedings, and rather than discuss them here it is interesting instead to note the increasing use of two relatively new trends for this generation of experiments; embedded processors, and the increasing introduction of ASIC's.

3.2 Embedded Processors

First, the embedding of microprocessors into the data acquisition system (DAQ). Although processing farms using multiple microprocessors [18,19,20] have been talked about since the early 1970's, and microprocessors have indeed been used to control data flow in some earlier CAMAC-derived DAQ systems [21], it is only in the last 5 years that a combination of factors has conspired to permit their effective use at a number of levels within the DAQ system.

One of these factors is obviously the advent of fast, powerful 32 bit microprocessors like the 68000 family which, particularly in the newer 68020 and very recently announced 68030 versions, have enough power left over after controlling the data flow to permit some processing and re-formatting of the data as it passes through the system.

Another factor has been the widespread use of FASTBUS as a vehicle for data acquisition. FASTBUS actually needs such embedded microprocessors in order to function efficiently in a system-wide environment.

For these reasons there has been a prolific outpouring of microprocessor controlled FASTBUS master modules for front-end crate readout including such processing tasks as hit list generation, zero supression and data formatting [22,23], and at a higher level for service request and interrupt handling together with sub-event and event building tasks [24,25]. In terms of raw computing power, the LEP experiments probably have more of it embedded within their DAQ's than that available to them for off-line analysis. For example, ALEPH alone has over one hundred 68020's in its DAQ system, all of them programmable in FORTRAN and C. This represents several hundred MIP's (millions of instructions per second) of computing power - comparable with the 48 (IBM 168) units to be provided by CERN in 1990 for all four LEP experiments to share.

3.3 Application Of ASIC's

The second clear trend is in the adoption of ASIC's in some of the processors. This is being approached with a certain amount of caution at present because of lack of experience and confidence, but designs which are committed to silicon are realising significant improvements in speed, size, power consumption and even cost. The following are just a few examples.

3.3.1 DMA controller. The ALEPH Event Builder contains a DMA controller specifically designed for high data rate transfers between FASTBUS and an event buffer memory; including various timeout counters it requires 52 MSI chips to implement. An exercise in custom design using a 2 micron CMOS 'optimised gate array' technique [26] yielded a five chip set (four 8 bit slices plus a controller chip) which consume 100 mAmps at a maximum working frequency of 33 MHz, compared with the 3.5 Amps and 20 MHz of the original; it also releases a significant amount of board space for extra processor memory. The production cost for the 100 prototype chips (20 sets) worked out at approximately £600 a set, roughly an order of magnitude more than the chip cost of the MSI version. This, however, reduces to parity for an order of 1000 chips (200 sets).

3.3.2 32 bit coincidence array. The first level Central Detector trigger processor for OPAL is described elsewhere in these proceedings. It will contain a custom chip in the form of a 32 bit coincidence array which is also described in these proceedings. Although the processor will use only 40 such chips, the savings in board space permit a more effective modularity of processor construction. It also makes the processor far more understandable to the user.

3.3.3 Contiguity array. The DELPHI second level Time Projection Chamber trigger plans to use a highly parallel processor [27] based upon a 'contiguity array' implemented using a gate array chip. Again, more details appear elsewhere in these proceedings.

3.3.4 Histogrammer chip. As part of a program of research and development at the Rutherford Appleton Laboratory, a histogrammer chip [28)] has been designed and fabricated in 2 micron CMOS technology. It has 256 channels (cascadable by adding more chips), each channel holding a 24 bit word. It can be read and written in normal random access fashion, but in addition any channel can be incremented by addressing it and applying a common strobe; there is also a common clear for all channels. A global threshold value can be set which detects any channel exceeding that value; each channel also has local zero detection, allowing thresholds to be set by loading an initial offset into the channel. Data acquisition rate is 8 Mhz and the projected cost (100 up) is £90.

4. HOW THEY WILL BE

Planning is now well advanced for the requirements of experiments on the proposed Superconducting Super Collider (SSC) [29)] in the USA, and the Large Hadron Collider (LHC) [30)] which may be built on top of LEP. Both machines will be operating with bunch crossing rates of 25 to 33 nanoseconds, an order of magnitude faster than the CERN p^+p^- collider, and nearly two orders of magnitude faster than LEP. This, allied with the proposed increase in data (hence trigger) channel numbers, from the hundreds of thousand in current LEP experiments to typically one million or more for SSC/LHC experiments, will create significantly greater trigger and DAQ processing problems.

From the triggering point of view it is no longer practical to scale up existing solutions to overcome these problems; the cost and the physical size of the new UA1 trigger processor, for example, are already daunting. There is a desperate need to reduce size, cost, power consumption and cabling requirements, and to increase reliability. This can only be achieved by a vast increase in the use of custom designed microelectronic chips. In order for these chips to be properly targeted, future systems must be designed in a strictly 'top-down' fashion, identifying the areas which will benefit the most from integrating onto silicon either because of the numbers involved, or because of the general usefulness of the device across experimental boundaries.

Data processing within the environment of the DAQ is not yet a well developed subject. Experience with the LEP experiments will show where its limitations may lie. It must be realised, however, that the 1 to 2 MIP processors currently embedded in the DAQ system could soon be replaced by today's RISC processors yielding a few tens of MIP's, with the promise of a few hundred MIP's from their successors. It is not unreasonable to assume that this trend will continue to thousands of MIP's, especially if processing arrays using Transputer-like devices are developed to their full potential. One thing is certain in this context; no matter how much processing power is made available, it will never be too much!

5 CONCLUSIONS

High Energy Physics instrumentation is rushing headlong into the age of very large scale, very high speed integration. The design tools (low cost workstations, software design and verification suites) are readily available from industry, and they are backed up by manufacturers of large (>100K) gate arrays and mature silicon foundries offering, amongst other things, mixed analogue and digital functions on a chip. Those in the HEP community who are not yet prepared for this new age had better get started now.

REFERENCES

1) Solomon, J. "Pattern Recognition Processor", Track Analysis Seminar for Rapid Cycling Bubble Chambers, Rutherford Lab.Report RHEL/R 271, 133-153 (1973).

2) Verkerk, C. "Special Purpose Processors", Proceedings of the 1974 CERN School of Computing, 223-262.

3) McPherson, G. M. and Wilde, P. "A Polynomial Evaluator", Track Analysis Seminar for Rapid Cycling Bubble Chambers, RHEL/R 271, 175-184 (1973).

4) Lingjaerde, T. "A Fast Microprogrammed Processor", CERN Report DD 74-36.

5) Lecoq, J. D. et al, "A Fast Arithmetical Unit user with Microprogrammable Processors", Nuclear Instruments and Methods, 213, 329-332 (1983).

6) Maclean, C.et al, "Special Purpose Computing Hardware", Proceedings 2nd ISPRA Nuclear Electronics Symposium,Stresa, 307-313 (1975).

7) McPherson, G. et al, "A Specialised Processor for Experiments in High Energy Physics", Euro IFIP 79, London, 685-691 (1979).

8) Lingjaerde, T. "XOP, A Second Generation Fast Processor for On-Line Use in High Energy Physics Experiments", CERN DD-81-11 (1981).

9) Kunz, P. F. "The LASS Hardware Processor", Nuclear Instruments and Methods 135, 435-440 (1976).

10) Halatsis, C. et al, "Architectural Considerations for a Microprogrammable Emmulating Engine using Bit Slices", Proceedings 7th Annual Symposium on Computer Architecture, 278-291 (1980).

11) Brafman, H. and Notz, D. "The 370/E Emulator at DESY", Proceedings of the Symposium on Recent Developments in Computing, Processor, and Software Research for High Energy Physics, Guanajuato, 211-216 (1984).

12) Kunz, P. F. "The 3081/E Processor", Proceedings of the Symposium on Recent Developments in Computing, Processor, and Software Research for High Energy Physics, Guanajuato, 197-209 (1984).

13) Jaroslawski, S. "Data Reduction Systems for HEP Experiments", Nuclear Instruments and Methods 176, 263-269 (1980).

14) Astbury, A. et al, "The UA1 Calorimeter Trigger", Nuclear Instruments and

Methods A238, 288-306 (1985).

15) von Rüden, W. "Trigger and Data Acquisition Plans for the LEP Experiments", Proceedings of the Symposium on Recent Developments in Computing, Processor, and Software Research for High Energy Physics, Guanajuato, 253-279 (1984).

16) Crosetto, D. "FDDP, A Fast Digital Data Processor for the FEMC Trigger", DELPHI-87-70-DAS-58 (1987).

17) Wiegand, J. "EDN's DSP Project", EDN Volume 32 16 111-118,17 183-196,18 137-186 (1987).

28) Buckley, T.F. et al, "Microprocessor Applications in Physics", Europhysics News 7, 7/8, 6-8 (July 1976).

19) Hertzberger, L. O. et al, "The Fast Amsterdam Multiprocessor (FAMP) System Hardware", Topical Conference on the Application of Microprocessors to High Energy Physics Experiments, CERN, 81-07.

20) Nash, T. et al, "Fermilab's Advanced Computer Research and Development Program", Proceedings, Three Day In Depth Review on the Impact of Specialised Processors in Elementary Particle Physics, Padova, 227- (1983).

21) Cittolin, S. and Lofstedt, B. "A REMUS based Crate Controller for the Autonomous Processing of Multichannel Data Streams", Topical Conference on the Application of Microprocessors to High Energy Physics Experiments, CERN 81-07.

22) Amendolia, S. R. et al, "The MC68020-Based FASTBUS Read-Out Processor of the ALEPH Time Projection Chamber", IEEE Transactions on Nuclear Science, NS-34, 128-132.

23) Charpentier, P. et al, "The Fastbus Intersegment Processor (FIP)", IEEE Transactions on Nuclear Science, NS-34, 137-141.

24) Marchioro, A. et al, "The ALEPH Event Builder", IEEE Transactions on Nuclear Science, NS-34, 133-136

25) Muller, H. "The Evolution of the FASTBUS General Purpose Master (GPM)", IEEE Nuclear Science Symposium, San Francisco (1987).

26) Battaiotto, P et al, "The Design of a Custom Chip Set for the ALEPH Event Builder", CERN/EF 87-10 (1987).

27) Darbo, G. and Heck, B. W. "The TPC Trigger for the DELPHI Experiment", IEEE Transactions on Nuclear Science, NS-34, 227-231.

28) Alsford, J. R. and Slorach, F. "A RAM Based CMOS Histogrammer IC", IEEE Nuclear Science Symposium, San Francisco, (1987).

29) "Proceedings of the Workshop on Triggering, Data Acquisition and Computing for High Energy/High Luminosity Hadron-Hadron Colliders", Fermilab, November 11-14 (1985).

30) "Proceedings of the Workshop on Physics at Future Accelerators", La Thuile (Italy) and Geneva (Switzerland), January 7-13 (1987)

THE SECOND LEVEL TRIGGER OF THE L3 EXPERIMENT

Y.Bertsch, H.Bonnefon, F.Chollet, A.Degré, G.Dromby
J.C.Lacotte, J.Lecoq, R.Morand, M.Moynot, G.Perrot

Laboratoire d'Annecy-Le-Vieux de Physique des Particules
BP 909 - 74019 Annecy-Le-Vieux Cedex - France

1. INTRODUCTION

The L3 Data Acquisition System [ref.1] is driven by three levels of trigger. Its working mode and main parameters are shown in figure 1. At nominal LEP luminosity ($10^{31}cm^{-2}s^{-1}$), the expected e^+e^- events rate is less than one Hz. This is compatible with the tape writing rate by VAX 8800 (5 Hz max.). A typical hadronic event produces 100 kbytes of raw data and 8 kbytes of trigger data.

The level-1 trigger decision is made within 22 μs by hardwired processors. A positive decision starts the digitization of the different detectors. This phase introduces a non interruptible dead-time of 500 microseconds. The expected level-1 rate of 100 Hz (500 Hz max.) produces a 5% dead-time (25% max.).

The second level trigger [ref.2] is expected to increase selectivity by a factor 10. It relies on 4 fast XOP [ref.3] processors fully integrated in Fastbus. Their programmability provide flexibility of the selection criteria. At maximum rate, a processing time of 8 ms per event does not increase dead time generated by raw data digitization. A positive decision starts the Time Expansion Chamber data reduction (1 Mbytes to 70 kbytes) and flaggs events to be transferred to the third level of trigger.

The third level trigger builds flagged events by merging the data delivered by the detectors. It makes a decision (accept/reject) by processing each event in a 3081/E IBM emulator and provides a further reduction by a factor 10. To not introduce additional dead-time, at maximum rate, they have to make a decision within 60 ms.

2. LEVEL 2 TRIGGER OPERATION

The level-2 trigger scheme and its associated test set-up are shown in figure 2.

The Multiport Multievent Buffer (MMB F-682B) [ref.4] is an 8 events deep FIFO input memory. After each beam crossing, all data (4K-16bits words) delivered by the front-end trigger digitizers are stored in parallel in the 60 ECL input ports at the nominal speed of 60 ns per word. This is done before the next beam crossing (22.5 μs later).

Data acquisition management is completely hardwired and fully synchronized by signals delivered by the experiment. All the data banks (256 words each) currently connected to the input ports are overwritten at each beam crossing until a level-1 signal validates the event. This increments the Input Event Pointer and the status register. An overflow signal is generated when 7 out of the 8 possible banks are occupied. When activated, it produces dead time in the overall data acquisition chain. To decouple completely level-1 and level-2, the MMB memory provides a real simultaneous random Read/Write access. This is realized using a fast memory driven by an internal clock which interleaves input and output cycles each 25 ns.

Four XOP processors work in round robin mode between two Fastbus segments. When idle, each XOP, implemented as a Fastbus master, arbitrates for the mastership of the input crate segment. The winner reads the MMB status word. If positive (at least one event memorized), it builds the trigger event by reading sequentially the trigger information distributed over the 60 ports, and decrements each MMB status register. Readout of each port is done in pipeline at 125ns/word plus an overhead of 1 microsecond.

When readout is completed, (less than 0.5 ms) it releases mastership and starts computation. After computation it arbitrates for mastership on the output segment, in order to write

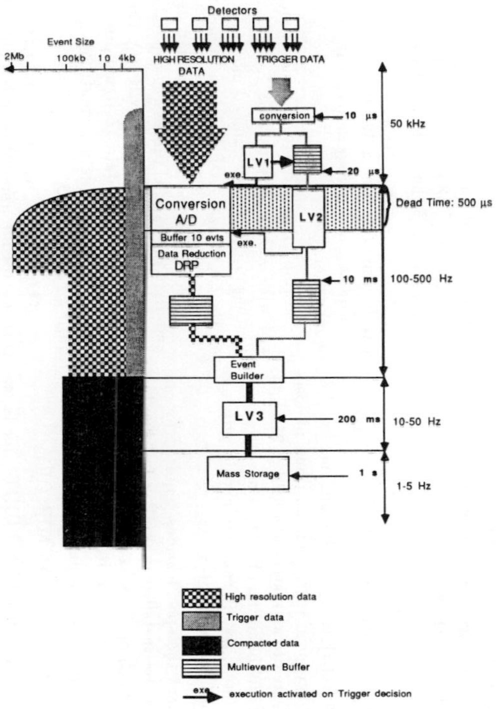

Figure 1. : The L3 Data Acquisition stages and Trigger Levels

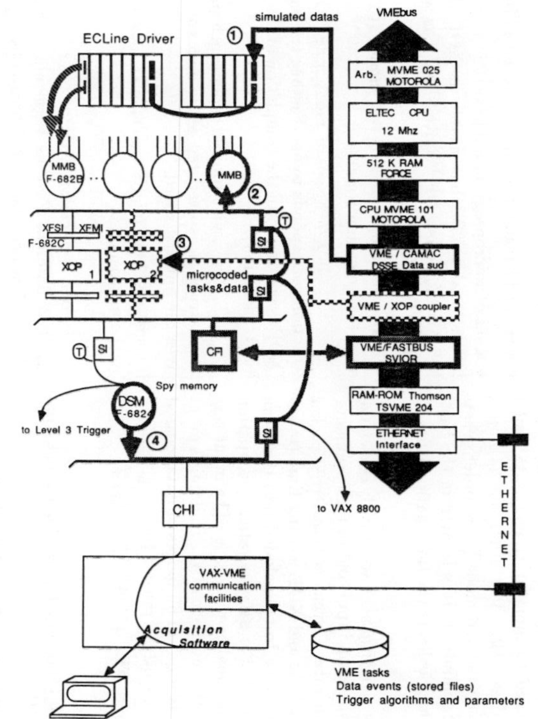

Figure 2. : The Second Level Test Set up

data and results into two Dual Slave Memories [ref.5]. One of these memories is running in spy mode and used to monitor trigger data under control of VAX 750. The other one is implemented in partitionned mode (to re-order events by increasing number). It is located in the Fastbus crate which collects the information from all detectors.

Clearly, performances are stongly correlated to Fastbus speed on input and output segments. At maximum rate, 1 event is transferred each 2 ms. Transfer time and XOP waiting time (to access Fastbus) have to be included into the mean processing time. For safe operation, occupation time has to be smaller than 25%. The input segment is then the most sensitive. Building 1 event requires the readout of 60 blocks of data located on 15 Fastbus cards. Transfer time is given by

$$T = (N \times s) + (60 \times \mathit{Overhead})$$

where,

• T is the total transfer time of 1 event from MMB to XOP,

• N is the total number of words per event,

• s is the transfer speed per word (in Block or Pipeline mode) observed on Fastbus,

• *Overhead* is the software time spent to initiate each of the 60 data transfers.

The transfer speed is optimized by implementation of a fast ECL sequencer. Software overhead optimization is obtained by integrating Fastbus into XOP processor.

3. XOP AS A PERFORMANT FASTBUS PROCESSOR

XOP [ref.3] is a fast (50-100 ns) on line oriented microprocessor. Its multibus philosophy, associated to an instruction field of 256 bits, allows parallel handling of several operations.

Futhermore, its internal modularity provides good extensions capabilities.

To meet the specifications required by our application, we installed in XOP a full Fastbus computer which operates very efficiently on Fastbus according to the powerfull XOP features. As a classical computer, the unit contains:

• a 32 bits INSTRUCTION MEMORY,

• a FASTBUS PROCESSING UNIT, able to execute by hardware all Fastbus transactions (it is based on ECL technology and uses a special sequencer),

• a set of 14 registers dedicated to FASTBUS,

• a set of microprogrammable switches on XOP busses, making Fastbus usable as XOP DATA MEMORY, or as XOP WORKSPACE as well as XOP I/O CHANNEL,

It can be operated in two modes (to optimize the execution time).

– In synchronized mode, one XOP instruction corresponds to one Fastbus transaction (XOP is waiting until the completion of the Fastbus transaction).

– In asynchronized mode, each part of the instruction is executed at its own speed. It gives the optimal speed.

If necessary, both modes can be mixed, controlled by the microcode.

In our application, XOP master has to access two different Fastbus segments. Thus its interface has been built on 3 boards [ref.6]:

• the Master part (XFMI) is an XOP unit,

• the two slave parts (XFSI*s* F-682C) are Fastbus standard units and may be plugged in two different Fastbus crate segments.

Both slave units are coupled to Fastbus via the ECL LAPP coupler [ref.7] and follow the FASTBUS specifications (especially definitions of CSR#0, CSR#8 and CSR#9). A general purpose register (CSR#1) is available. It is used for tests or as a letter box. An extra register (CSR#10) is accessible from the front pannel and can be used either as a 32 bits ECL output register or as a microprogrammable 16 bits ECLine word generator.

Table 1 compares event building time achieved with XOP and with a 68 K CPU accessing Fastbus in DMA/programmed mode. Depending on the event size, XOP is 5 to 20 times faster.

Pushing the comparison further, one has to underline the parallel processing capability of the XOP processor. In parallel with any Fastbus cycle, it may process data. Zero suppression, multiplicity counting, data formatting are possible during Fastbus I/O.

The optimal way to operate the interface is to microcode application programs into XOP. Nevertheless to ensure compatibility with fortran

Trigger event reconstruction = readout of MMB records

- 60 Datas Block transfers (N words from 1 up to 4 kbytes)

- 1 Arbitration cycle
- 30 Master/slave primary address connections
- 60 Secondary address cycles (records internal addressing)
- 60 Random data cycles (fetch word count)

	68000 Master unit	XOP/ XFMI
FASTBUS Access	Programmable+ DMA interface	Processor EXTENSION
Execution time/ Instruction	1-5 μs 32 bits	100 ns 192 bits
Data Transfer	**N x 125 ns** *ECL Technology*	
Secondary Adr. cycle + software overhead	≈10 μs	300 ns maskable
Total Software overhead/event	**few ms**	**60 μs**

Level 2 trigger event reconstruction: Overhead + N x 125 ns	N = 400	N = 4000
with XOP/XFMI	0.110 ms	0.560 ms
with 68 k-DMA	2-5 ms	2.5-5.5 ms

Table 1

Test Conditions and Simulation Performances

ECLine transfer speed: **60 ns/mot**
Data Transfer time (128 words typ.): < **8 μs**
BCR Frequency: > **25 kHz**
Write operation (16 bits Data word) in ECline Driver memory: **5 μs**
Dead time *(Load event in ECLine Driver memory)*: **1.3 ms**
Event rate: **750 evts/s**
Synchronization on STATUS–MMB: **300 ns**
FASTBUS speed (in Pipeline mode): **125 ns/mot**
Readout of an MMB record (128 words typ.): **16 μs**
Event readout *(addressing + data transfers + pointers handling)*: < **60 μs**
Verification: **200 ns /word**
Complete "processing" time per event *(Fastbus readout+data checking)* **200 μs**

Table 2

programs accessing Fastbus, we have microcoded the CERN Fastbus library into XOP. A fortran package allows the user to drive FASTBUS via XOP as well as trough the standard Fastbus interface (CFI ...). The only difference is the execution speed.

4. TEST AND SIMULATION OF EXPERIMENTAL CONDITIONS

Configuration tests are based on built-in devices controlled by dedicated software which reads and writes simulated data events at several points in the hardware (labelled 1 to 4 on figure 2).

ECLine Drivers [ref.8] are CAMAC units providing 256×16 bits datawords memories which can be connected to ECLine bus to deliver data at a rate of 60 ns per word.

The test software is implemented on VME CPU's interfaced to CAMAC, FASTBUS, XOP, Ethernet and operated with CP/M.

The simultaneity of the I/O access to the MMB memory is tested using three different processors. The Motorola MVME101 CPU runs a simulation sequence which reproduces incomming events. Simulated data are written into ECLine Drivers then transferred to MMB units with a Level 1 response signal. At the same time, XOP reads MMB data and process them. The Eltec CPU initiates and controls the test sequence. Running this test, we reached the performances required in L3 experimental environment (cf Table 2).

5. CONCLUSION

The Second level trigger hardware described here meets severe L3 experimental requirements.

The XOP/FASTBUS interface with its software development boosts XOP into a very efficient FASTBUS processor. As fast as the fastest FASTBUS Sequencer, it makes proper use of the major intrinsic XOP performances.

The performances of the system are ensured by test software running on VME CPU's, using built-in hardware facilities.

References

[1] • L3 Technical Proposal CERN/LEPC/83-05, May 1983.

[2] • The L3 Second Level Trigger, note L3-372.

• Thèse de Docteur en Sciences (1987) Université de Savoie (L.A.P.P. Annecy), S.Rosier.

• Thèse de Docteur en Sciences (1988) Université de Savoie (L.A.P.P. Annecy), F.Chollet.

[3] • XOP, A Second Generation Fast Processor for Online Use in High Energy Physics Experiments, T. Lingjaerde, Proceedings of Topical Conference on the Aplication of Microprocessors to High Energy Physics Experiments, (1981) CERN, Geneva, Switzerland, CERN report 81-07.

• XOP Main Documentation, T. Lingjaerde, C. Ljuslin, CERN DD internal note, April 1987.

[4] • F-682B Multiport Multievent Buffer (MMB), J. Lecoq, M. Moynot, G. Perrot, LAPP internal report, 23/01/85 revised 10/04/87.

• Multiport Multievent Buffer Design, CERN/EP note F398-8, 26/03/85, Communication to the 8th CERN FASTBUS meeting by J. Lecoq,

[5] • The DSM, a Fastbus Dual Slave Memory for Data Acquisition, H. Müller, C.M. Story, S. Falciano, CERN-EP Electronics note 85-03.

[6] • The XOP Processor in FASTBUS, J. Lecoq, M. Moynot, G. Perrot, LAPP; P. Baehler, C. Ljuslin, CERN; Proceedings of FASTBUS Software Workshop, (1985) CERN, Geneva, Switzerland, CERN report 85-15.

• F682C XOP FASTBUS Master Interface, J. Lecoq, M. Moynot, G. Perrot, LAPP internal report 1985.

• XFMI/XFSI Spécifications, G.Perrot, LAPP Technical document revised 1987.

[7] • An ECL FASTBUS Slave Coupler, H. Bonnefon, M. Moynot, G. Perrot, J.M. Thénard, LAPP internal note 84-06.

• A FASTBUS Data Transfer Protocol Chip, CERN/EP note F398-6, 10/04/84; An ECL FASTBUS Slave Coupler, CERN/EP note F398-8, 05/02/85, Communications to the 3rd and 7th CERN FASTBUS meetings by G. Perrot.

[8] • ECline Driver (produced by SCAIME - BP 501 Annemasse F-74105), internal report LAPP 1985.

• LAPP/SCAIME, ECLine Driver User's Manual, technical report.

Multiprocessor Event Filtering at the Heidelberg / Darmstadt Crystal Ball

Ch. Ender, R. Männer, W. Ludwig, P. Bauer, A. Wurz
Physikalisches Institut der Universität Heidelberg
ENDER@DHDMPI5V.BITNET

Abstract:

The distributed data acquisition system described here consists of a Fastbus ADC system, multiple read-out channels to a multiprocessor event filter, and an online VAX. Due to system software, it appear as a single system which can be controlled easily by menues. The current filter capacity is 30 MIPS. The complete system will be able to handle up to 10^8 parameters/s .

1. Introduction

The Heidelberg/Darmstadt crystal ball detector [1] consists of 162 NaJ(Tl) scintillation detectors setting up a spherical shell with an inner radius of 25 cm. It is used for recording γ-radiation. Usually, other detectors like parallel plate detectors are placed inside. The architecture of the data acquisition system is shown in fig. 1. The crystal ball detector can be operated with up to 10^6 events/s. Typical event rates are 10^5 events/s, where the event size is in the range of 100 to 1000 parameters. Hardware triggers reduce the typical event rates to 10^3 to 10^5 events/s, which represents a data stream of up to 10^8 parameters/s. Accepted events are digitized by a 12bit ADC system with a conversion time of 10 μs to 300 μs for 10^2 to 10^3 parameters. Zero suppression further reduces the data rate by a factor of 10. The bandwidth required to forward this data stream of 10^6 to 10^7 parameters/s is provided by the Fastbus system. The online computer, a VAX 11/750 accepts, however, only data rates up to $\approx 10^5$ parameter/s. An additional filter step with a data reduction factor of 100 is therefore required to match the speed differences between the Fastbus and the online computer .

Experiments at the crystal ball are scheduled in a typical time scale of two weeks. For each one the experimental setup is different. Similarly, most trigger conditions vary so frequently and are so complex that a hardware realization is not feasible. Each experiment requires specific filter algorithms and data reduction factors, so a programmable, modular and flexible system was designed, which can be adjusted to the actual needs. In addition to the Fastbus ADC system and the online computer the new data acquisition system consists of multiple Fastbus read-out processors and a modular multiprocessor for event filtering, the Heidelberg Polyp.

2. The Multiprocessor System Heidelberg Polyp

Typical filter routines, e.g. a two-dimensional correlation to separate neutrons from γ's, require $\approx 10^3$ instructions to check one event. For an input event rate of 10^5 events/s an average computing power of 100 MIPS is required. To meet this performance figure with microprocessors of the 680xx family, a multiprocessor system with 20 to 100 processors is needed. The Heidelberg Polyp is such a system [2,3], which can be assembled of modules like processors and buses. Its central management concept is to collect identical modules into pools (fig. 1). Each pool is specialized to a certain system function. Pool membership is determined by hardware or by software assignment. To support pool-size transparent programming, the Polyp hardware allows to access available members of a pool anonymously. With this feature it is possible to write system and application software, which is independent of the system size. The pool size, i.e. the number of members within a pool, is, therefore, a system parameter which can be adjusted according to the actual needs. It allows to adjust the computing power by using an appropriate number of processors (general-purpose or special-purpose). For data transfers within the Polyp system, an interconnection network called the Polybus is used. It consists of any number of 32-bit buses operating independently and in parallel. By exploiting the pool feature, Polyp systems become scalable and may be operated fault-tolerantly with graceful degradation in case of errors.

Currently the Polyp consists of three pools, the Polyp input processor pool (PIP) where each module is linked to Fastbus read-out processors, the Polyp filter processor pool (PFP) responsible for event filtering, and the Polyp host processor pool (PHP) which maintains the interactions with the online computer.

3. The distributed data acquisition system

Events are collected by the Fastbus read-out processors. Each one collects a complete event and stores it into its internal buffer memory. It then starts the transfer to the Polyp input processor. To match the speed differences between the Fastbus and Polyp system, multiple I/O channels to the Polyp input processors are used. In addition, this allows to read-out the Fastbus ADC system contiguously.

Events received by the Polyp input processors are distributed to free event buffers in the filter processor pool. There the events are first checked for the conditions currently defined. If an event is accepted it is reformatted for further data reduction and written into an output buffer. If the output buffer is filled with $\approx$ 200 events (depending on the event size) it is flagged to the Polyp host I/O processor. A double buffer technique is used for event output, so filter processing continues with the next buffer, while the previous one is transferred by the Polyp host I/O processor to the online VAX.

There, all output buffers containing the filtered events are stored on tape and analyzed. The Polyp host I/O processor controls also the communication between the Polyp system and the online VAX.

4. System Software

The crystal ball data acquisition system is a distributed processing system constiting of multiple Fastbus read-out processors, multiple Polyp input processors, multiple Polyp filter processors, and the online computer. To handle the data flow through the system, a distributed software management concept is used. The different system components interact with each other in a totally asynchronous way. To become independent of the actual pool sizes in the multiprocessor system and the actual number of Fastbus read-out processors, each processor manages his own in- and output data stream. If one of the processors fails, the sytem continues while the failing processor is inactive. Together with the implemented error handling the data acquisition system becomes fault tolerant.

To manage the distributed computing system, a modular software package was developed consisting of four layers (tab. 1). The uppermost level are the application programs, e.g. for data analysis and experiment control. They issue complex commands to the support level. There several run-time libraries decompose these commands into simpler ones. Examples are configuration and initialization of the front-end electronics and the multiprocessor system, starting the data acquisition, set-up and modification of filter conditions etc.. These commands are forwarded to the communication level. The whole communication between multiple tasks on the VAX host and all members of the three pools (PIPs, PFPs, and the host I/O processors), as well as between different pools, is done via message passing. System software on the VAX and the multiprocessor system routes automatically the commands and data tables through the system into the right module, e.g. the read-out list for the front-end electronics to the Fastbus read-out processors. Routing commands from the VAX to the multiprocessor is controlled by a device driver which is responsible for communication with the host I/O processor of the Polyp. In addition to standard VMS device drivers functions, this driver provides an arbitray number of channels to allow simultaneous links to multiple tasks. In addition, each channel can handle multiple I/Os asynchronously and concurrently. The communication to a similar driver running at the Polyp side is done via a set of mailboxes using a two level handshake protocol controlled by interrupts. This second driver is responsible for routing messages from the VAX to any Polyp module or any pool. In the opposite direction, comparable requests can be issued by any Polyp module. They are forwarded to the host computer synchronously, i.e. the requesting program waits for completion of the operation.

5. User view

The data acquisition system is normally used by small groups without the assistance of a system specialist. They need a simple user interface, which hides all architectural details of the distributed data acquisition system so far as possible. This is provided by an application software package which lets the system look like a simple data acquisition system consisting of an online computer and front-end electronics only. Experiment specific parts, e.g. number and type of digitizers used, however, cannot be hidden from the experimenter, but have to be accessable and programmable at a very high level. A user friendly interface was written which allows to define, to modify, and to control the experiment specific set-ups via a menus. All set-ups are stored in a data base. From this data base, all data definitions required to initialize the distributed data acquisition system are generated automatically. To define and to adjust the event filter, an experimenter uses also menu driven software. Most of the conditions required are implemented as standard routines, e.g. multiplicity, one- and two-dimensional range checks, an anti-compton check for the crystal ball and logical combinations. Only experiment specific conditions, e.g. modelling of the reaction kinematics, needs programming. This can be done in a high level language like Fortran or Pascal; Assembler is also possible. These special routines are loaded into the Polyp filter processors dynamically. This is required to adjust the conditions for the event filter according to the experimental requirements. Simple commands allow to control the experiment and the data flow through the whole system. The event data are monitored and analyzed by the online computer.

6. Current status

Since autumn 1987, experiments have been done with the described hard- and software. One exception is the Fastbus ADC system, which is temporarely replaced by commerially available electronics, as long as the ADC system is still in its prototype test phase. The multiprocessor system consists currently of 30 modules, each one equipped with a 68000 mircoprocessor with 16kB cache memory, 1 MB memory and an interconnection to two data buses. Two of the modules have connections to Fastbus read-out processors (CFI). Due to the CAMAC ADC system currently used, the event rate from the Fastbus system is limited to ≈1000 events/s. Another limitation is given by the Fastbus masters (CFI). Each one of them can collect and transfer ≈800 events/s to the Polyp. To overcome this bottleneck we are just thinking about a new Fastbus master. Due to the limitations mentioned above, the Polyp system runs with a load of 10%. Only three filter modules are required to check and reformat the events. The total filter power of the multiprocessor system attainable with the current hard- and software setup is $\approx 10^4$ events/s. This could be improved by another factor of 5 to 10 just by plugging in more processor boards or by using 68020-based processors.

7. References

[1] Adelberger E., Albrecht A., Fischer R. D., Habs D., Helmer K., v. Helmholt U., Hennerici W., Hennrich H. J., Heyng H. W., Himmele G., Kolb B., Krieger H., Kroth R., Koschorrek O., Kühn W., Lazzarini A., Mühlhans R., Metag V., Pelte D., Repnow R., Schwalm D., Simon R. S., Wahl W.: Eigenschaften des Kristallkugelspektrometers; *Ann. Rep. MPI for Nucl. Phys.*, Heidelberg, Germany (1982) 38 -42

[2] Männer R., Shoemaker R. L., Bartels P. H.: The Heidelberg Polyp System; *IEEE Micro* **7**, 1 (1987) 5 - 13

[3] Männer R., Ender Ch., Ludwig W., Stucky O., Bauer P., Wurz A.: Multiprocessor Data Acquisition and Analysis with the Heidelberg Polyp System; Proc. *Hardware and Software for Real Time Process Control*, Warszawa, Poland, 30 May - 1 June 1988, to be published

Level	Functions	Example
Application	controlling - start, stop data acquisition monitoring - analyze data to check event filter and experiment filter routines	start data taking
Support	run time libraries - decomposition of complex functions and commands into atomics - program interfaces to different system components - interface to communication event manager	- load and initialize Fastbus read-out processors - load read-out lists - load current conditions for event filter - start read-out
Communication	message passing - VAX - Polyp - within Polyp system - Polyp - Fastbus front-end control of communication - multiple channels for multiple task to task communication - asynchronous operation to provide concurrency - fault tolerant operation - error handling	- route commands and data into: Fastbus read-out processors Polyp input processors Polyp filter processors Polyp host I/O processor online computer
System	hardware specific support - control of communication devices - synchronization between different components - interrupt control	move data and commands synchronize requests

Table 1.: Software abstraction levels of the data acquisition system

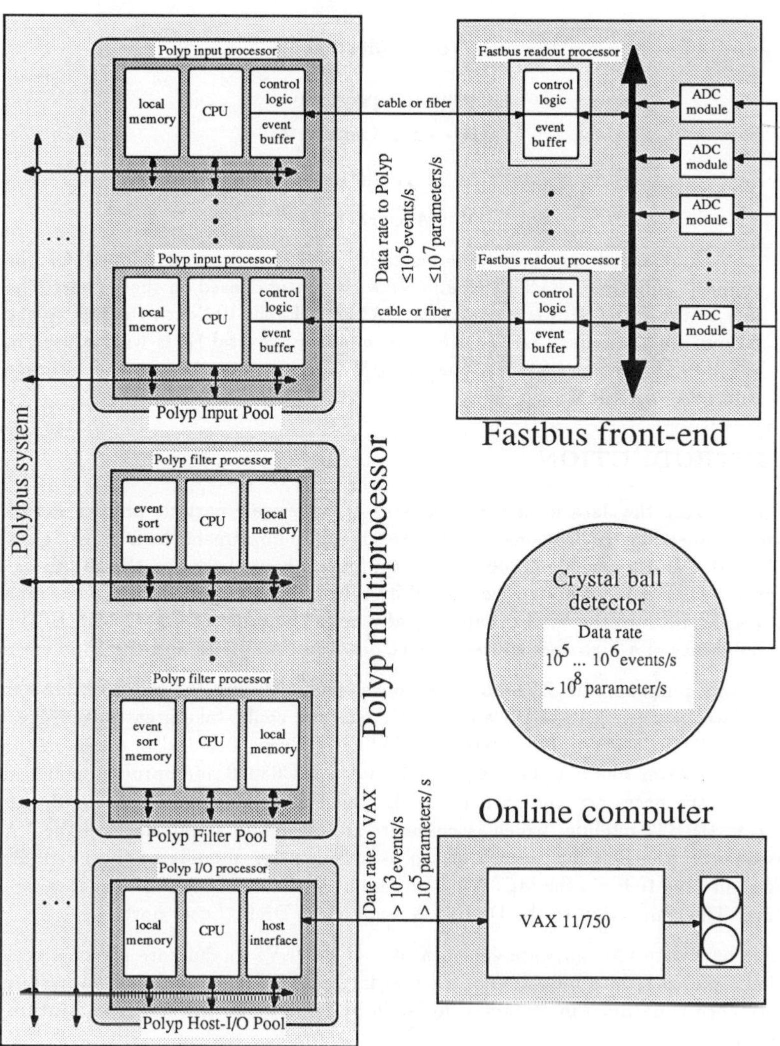

Fig. 1: The distributed multiprocessing system

EVI - a High-Speed Interface Between FASTBUS and VAX-BI

Volker Mertens

CERN - Div. EP
CH - 1211 Geneva 23

Abstract

A high-speed interface module, called EVI, has been developed for the connection between FASTBUS and VAX computers based on the BI bus. The link to the FASTBUS is made via an 'ALEPH Event Builder', the link to the BI bus via a standard DEC DRB32 BI adapter. Several EVIs will be used in the data acquisition system of the ALEPH experiment at the new e^+e^- storage ring LEP at CERN.

1. INTRODUCTION

Considering the data acquisition systems of large-scale particle physics experiments presently under construction one can observe a trend to implement the front end electronics in FASTBUS and to use VAX 8000 family computers being based on the BI bus as online computers. The amount of data generated by the experiments which has to be transferred from FASTBUS into the VAXes will typically be in the order of 0.1 - 1 MB/s. Therefore the availability of a high-speed interconnect between FASTBUS and the BI is essential.

For this purpose, the EVI has been developed. During the design of the module as interface between FASTBUS and the BI bus one could take great advantage of the existence of two other units: firstly the 'ALEPH Event Builder'[1)], a general purpose FASTBUS master module being equipped with a MC68020 microprocessor and already containing the necessary and rather complicated FASTBUS specific part and, secondly, the DEC DRB32 module, which is commercially available as interface to the VAX-BI. Consequently, the EVI has been built up as an extension module to the Event Builder, being connected to it via the MC68020 system bus available at the front panel of the Event Builder. The connection to the DRB32 is made via a DRB32 user port.

The hardware and software specifications of the EVI module are given in section 2. Section 3 sketches the application of the EVI in the ALEPH data acquisition system. The performance is outlined in section 4, followed by remarks on the status and future plans in section 5.

2. SPECIFICATIONS

2.1. Hardware

Fig. 1 shows the basic building blocks of the EVI. The connection to the Event Builder is shown on the right hand side, to the DRB32 on the left hand side. The main data paths are represented by thick arrows, others by medium sized arrows, and control paths by thin arrows. The main data paths are 32 bits wide. The EVI contains a memory

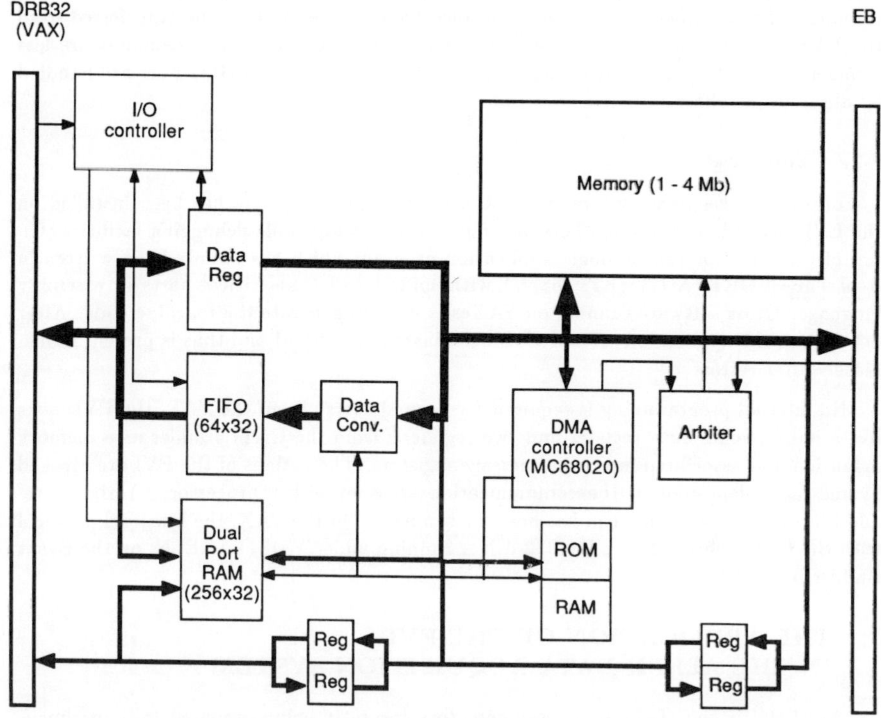

Figure 1: EVI Block Diagram

accessible from the Event Builder with an initial size of 1 MB being expandable to 4 MB. Up to four EVIs may be connected to one Event Builder. As a key feature, a MC68020 microprocessor was employed as a dedicated DMA controller between the memory and the DRB32 port. It performs all internal data transfers by software thus combining high speed with great flexibility. For example, it permits the reformatting/assembling of event data at transfer time and autonomous transfer control without further intervention of the Event Builder. This local processor runs at 20 MHz. Attached to it are a ROM of 256 kB and a RAM of 128 kB. An arbitration unit resolves simultaneous accesses to the memory from the Event Builder and the local processor.

A dual ported RAM of 1 kB serves as a communication area between the VAX and the EVI. The communication between the Event Builder and the EVI takes place via a reserved area of 1 kB inside the memory. Both channels are supported by associated interrupt/message registers in either direction.

The transfer of blocks of data from the memory to the VAX passes via a FIFO buffer of 64 longwords which decouples the speed of the DRB32/BI bus from that of the local processor. A gate array in front of the FIFO allows the byte swapping, word swapping, and single precision float point conversion (IEEE-to-DEC format) on the data. More complex conversion actions may be performed by software. Data to be transferred from the VAX into the memory pass via a simple data register since these demands are less stringent in most applications. The hardware functions on the DRB32 port are handled by a microcontroller.

2.2. Software

In order to keep the system small no proper operating system has been installed on the EVI, instead, an assembly language monitor provides simple debugging facilities. On top of this monitor runs a single application program which has mainly been written in Real-Time FORTRAN77 (RTF/68k[2]) with some small time-critical parts in assembly language. Cross software running on VAXes is used to generate the machine code. After debugging, the application software has been installed in ROM and thus is present immediately after restart.

No internal programming is required for normal users. From the VAX, the EVI looks like a bidirectional data register and two registers; from the Event Builder as a memory extension and associated interrupt/message registers. The actions of the EVI are invoked by messages deposited in the communication areas by either processor. A simple request/acknowledge mechanism has been chosen here. On the VAX, the low level protocol with the EVI is handled by a DRB32 driver running under VMS/VAXELN, on the Event Builder by an OS-9 task.

3. THE APPLICATION OF THE EVI IN THE ALEPH DATA ACQUISITION SYSTEM

ALEPH[3] is one of the four large detectors currently being prepared for experimentation at the new e^+e^- storage ring LEP at CERN. The event rate is estimated to be less than 10 Hz with an average event size of about 100 kB. Fig. 2 gives a partial view of the top level of the ALEPH data acquisition system where several EVIs will be used in different applications. Their positions are marked as shaded rectangles.

ALEPH uses almost entirely FASTBUS as the bus system for the front end electronics. One VAX 8700, a VAX 8250 and two VAX 8200 will serve as online computers. The FASTBUS equipment will remain under ground close to the experiment; the online computers will reside in a building on the surface. The distance between these components will be covered by high-speed (several MB/s) optical links. The link couplers will also conform to the DRB32 port specifications.

At the upper levels of the data acquisition system the data arriving from the front

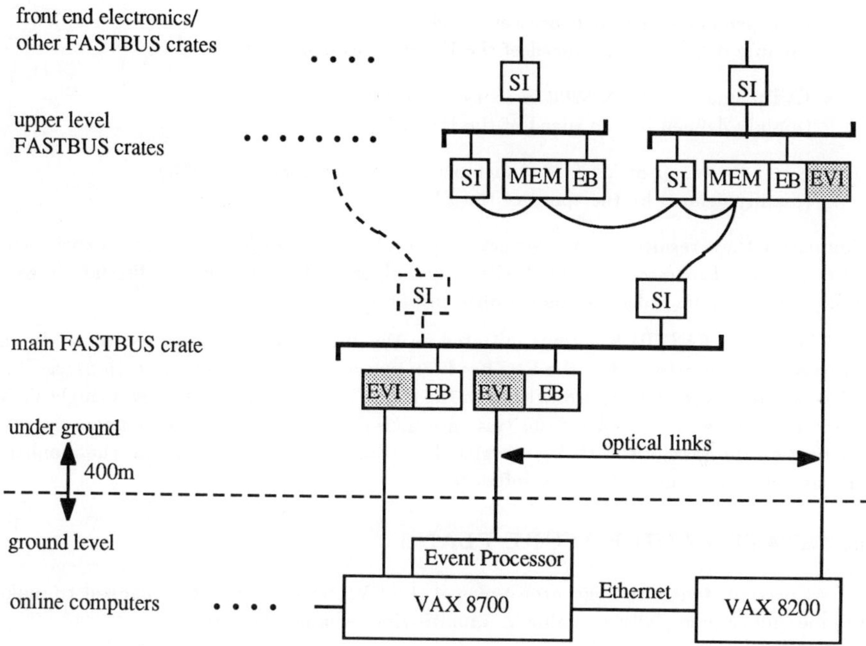

Figure 2: Partial View of the Top Level of the ALEPH Data Acquisition System (EB = Event Builder, MEM = Event Builder Memory (readable from FASTBUS), EVI = Event Builder - VAX-BI - Interface, SI = Segment Interconnect)

end electronics will be collected by Event Builders and stored in the attached memory modules[1]. From here, one Event Builder in the main FASTBUS crate can collect the information of the whole detector. Via an EVI and an optical link the complete events will be transferred to an Event Processor residing inside the VAX 8700. Via own EVIs the other VAXes have access to Event Builders in the upper level FASTBUS crates (e.g. for monitoring parts of the detector). An additional EVI together with another Event Builder in the main FASTBUS crate is reserved for individual FASTBUS actions initiated on the VAX 8700 (general purpose FASTBUS interface). Due to the smaller demands the other VAXes do not have such an extra channel.

4. PERFORMANCE

The following numbers for block transfers of data have been obtained with an EVI being directly connected (without optical link) to a DRB32 inside a VAX 8200, also housing the Event Processor:

- EVI memory → Event Processor = 2.4 MB/s (mainly defined by the speed of the Event Processor),
- EVI memory → VAX 8200 memory = 5.0 MB/s (mainly defined by the speed of the DRB32),
- Event Processor or VAX 8200 memory → EVI memory = 2.2 MB/s (mainly defined by the speed of the EVI).

Comparing these results with the expected data rate to be transferred from the experiment into the Event Processor of about 1 MB/s the link capacity seems to be sufficient. In case of bottlenecks more links may be installed in parallel.

The rate of FASTBUS actions initiated on the VAX 8200 and executed remotely on an Event Builder attached to the EVI has been measured to be about 350 actions/s. This refers to single word FASTBUS actions in the local crate being requested by a single VAX process. In case of the VAX 8700 this rate increases to about 1200 actions/s. In this application the task of the EVI is restricted to simple message passing and the numbers are almost entirely given by VAX software.

5. STATUS AND FUTURE PLANS

At present, two wire-wrap prototypes of the EVI are existing. It is planned to make the module commercially available in a multi-wire version within 1988.

Acknowledgements

I am grateful to Drs. J. Harvey, B. Jost, A. Marchioro and W. von Rüden for many helpful discussions and their support. I would like to thank the organizers of this conference for their efforts.

References

[1] P. Battaiotto, K. Einsweiler, A. Marchioro, W. von Rüden, The ALEPH Event Builder, A Multi-User Fastbus Master, Proc. of the IEEE Nuclear Science Symposium, October 21 - 23, 1987, San Francisco (to be published).

[2] H. von der Schmitt, RTF/68k Real-Time FORTRAN-77 for 68k Processors, Manual of Compiler and Run-time Library, Univ. Heidelberg (in preparation).

[3] ALEPH Collaboration, Technical Report, CERN/LEPC/83-2.

MICROPROCESSORS IN THE LEP CONTROL SYSTEM

P.G. Innocenti
European Organization for Nuclear Research (CERN),
CH - 1211 Geneva 23, Switzerland

ABSTRACT

The design principles of the LEP Control System are recalled. Progress on the various control system's building blocks is reported. Particular attention is given to the less conventional solutions brought about by the size of the machine.

1. INTRODUCTION

LEP, the Large Electron Positron collider presently under construction at CERN, and the LEP control system have been described in a number of publications[1-8]. In this paper features of LEP controls related to technological advances or to specific constraints will be reported.

There are two classes of constraints imposed to the LEP control system. A first class arises from the geometry of LEP expressed by its large circumference of 27 km, by its location completely underground, with eight access shafts suitable for exchange of control information between surface and underground. The limited number of surface links of any type between the eight LEP sites (some sites can be reached only via the tunnel) and the impossibility of using optical fibre cables in the machine tunnel on account of damage due to synchrotron radiation complete the physical environmental constraints.

The second class of constraints comes from the design choice of using the existing SPS accelerator as an injector to LEP and to operate both machines from the same control room, actually the existing control room of the SPS. Although the LEP control strategy has been worked out based on principles similar to these of the SPS, a number of details are different, on account of history, operation practice, technology, emergence of international standards etc. Moreover, the SPS control system is bound to evolve, both in order to satisfy the requirements for LEP injection and on account of the obsolescence of a number of its components. Hence the task of building a control system for LEP is amplified by the evolution of SPS controls in the LEP direction, the only conceivable line for rationalizing the whole set-up in the long range, yet with a number of short-term actions needed.

2. COMMUNICATION INFRASTRUCTURE

The prohibitive cost of control cables over distances of 4 to 10 km, as required to connect the LEP access points with each other and with the LEP/SPS control room, has strongly influenced the choice of multiplexing a large number of signals over a limited number of high bandwidth connections, either optical fibres on the surface or coaxial cables underground.

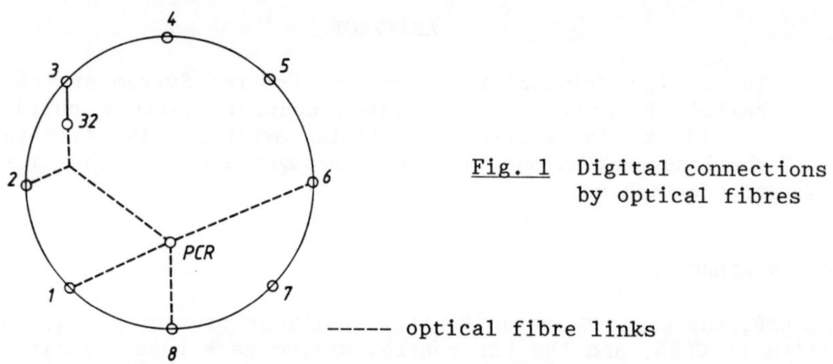

Fig. 1 Digital connections by optical fibres

Only a few tens of hardwired signals, essential for the safety of personnel, are transported over copper without multiplexing. The ultimate bandwidth permitted by the installed cable connections to the sites is between 1 and 5 Gbit/s depending on the importance of the site, as only four of the possible eight collision points house an experiment. The methods of multiplexing are according to CCITT recommendations G700 for Time Division Multiplexing (TDM)[9].

The TDM system carries a number of services such as machine control networks, machine synchronization, connection of experiments to the CERN computer centre and telephone and terminal traffic through a distributed ISPBX.

3. CONTROL SYSTEM ARCHITECTURE

The functional requirements of controlling an accelerator in a distributed environment are satisfied, in general, by a network of computers and by a synchronization system. In LEP the network has two logical levels:

(i) The upper level consists of central or local consoles, central servers and local process computers, running a program or a sequence of programs under operator control.

(ii) The lower level consists of microprocessors imbedded in the equipment, presenting fixed access points to the upper level and performing predetermined tasks either upon request or routinely.

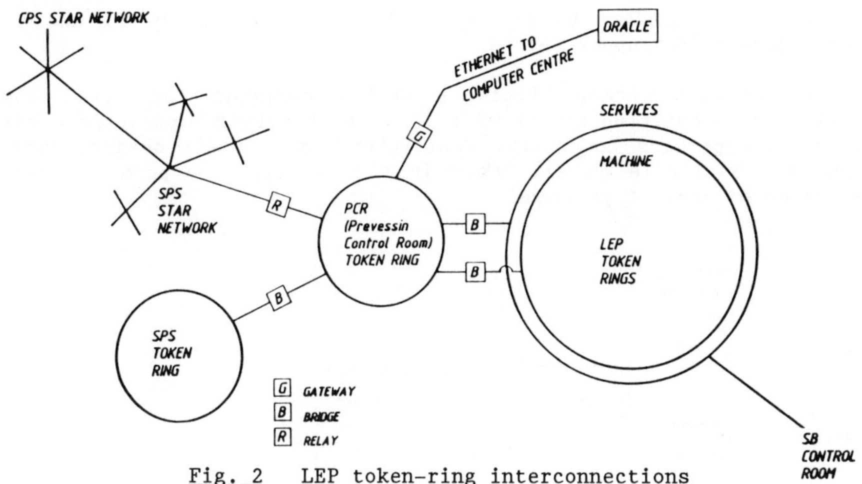

Fig. 2 LEP token-ring interconnections

3.1 Networks

3.1.1 Token passing rings. A number of interconnected token rings provide the physical medium of the upper network, interconnecting consoles, servers and process computers[10,11].

The choice of the token ring has been influenced by considerations such as:

- **Distance.** There is no conceptual limitation on distance and rings have been operated at CERN up to 25 km.
- **Physical medium.** The one-directional transmission permits mixing twisted pairs, optical fibres and TDM channels without complications.
- **Deterministic response** within a maximum time, important as, during the accelerator cycle, there exist periods when many computers want to access the network.
- **Distributed monitoring**, potentially existing in each station.
- **Fault tolerance**, through reconfiguration by "wrap-back" foreseen in the wiring scheme.
- **International standard** ISO 8802.5.

Two independent networks will be installed around LEP, each one consisting not necessarily of a single ring, but of a system of rings interconnected by Medium Access Control (MAC) bridges in some cases. (We have experimented bridge throughput up to 85% of the nominal ring bandwidth.)

- A **services ring** of high availability, as it carries information related to safety (personnel and equipment).All hosts on this network will be located in surface buildings, hence accessible all the time.
- A **machine ring**, for controlling the functions associated with the beams.

Protocols conform to ISO up to level 2 (LLC class 1) and to DARPA (TCP/IP) for upper levels.

A Real-Time Access Protocol (RATP) designed for speed and simplicity is built on top of UDP: it is particularly appropriate for use in the process computers. Facilities for remote procedure calls (RPC) are available on all hosts in the network, to permit remote execution of control programs.

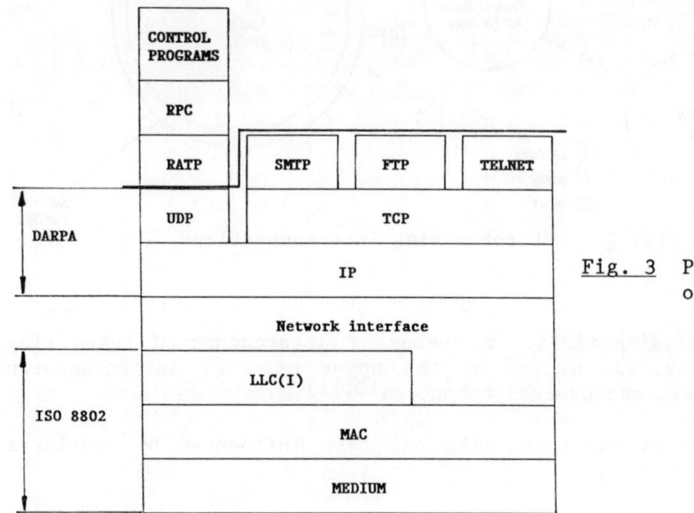

Fig. 3 Protocol organization

Gateways to other networks supporting TCP/IP are possible at the Internet Protocol (IP) level. This is the case of the connection to the four LEP experiments, which have a cluster of VAXes interconnected by Ethernet.

In addition to the Ethernets of the experiments, the LEP system of token rings is connected to the SPS token ring, the CERN computer centre, and to the SPS star networks: the latter connection has required a laborious system of relays as the only peer-to-peer connection possible has to be done at the transport level.

3.1.2 MIL-STD-1553B Multidrop Highway. A multidrop highway provides the connection between the process computers and the microprocessors imbedded in the equipment. The choosen multidrop is the MIL-STD-1553B[11] based on considerations of:

- noise immunity, by transformer coupling,
- speed, 1 Mbit/s over 400 m,
- distance, up to 20 km with repeaters at 125 kbit/s,
- single twisted pair cable,
- deterministic response, through polling by master.

The software protocol has been designed to be simple command response, but has grown more elaborate in order to accomodate requests for more services for the microprocessors in the field[12]. The schedule of the LEP project has not permitted to take advantage of the standardization efforts of the MAP community for a field bus, nor for software protocols related to it. Indeed LEP installations is in progress now.

3.2 Process Computers

The process computers bridging across token-rings and multidrop highways are multimicroprocessor assemblies[13-15], named DLX, based on the VMEbus and the Motorola 68010, to be supplied by CIMSA-SINTRA, a division of Thomson-CSF. They operate under a multiprocessor operating system, Electre, and have a UNIX-like filing scheme enhanced with network features.

Each DLX computer consists of a number of Motorola 68010 performing different functions, such as I/O to the token ring and to the 1553 multidrop or processing. Special care has been devoted to the power supplies, buffered by a battery to ensure reli- able operation on power failure for alarm treatment and transmission, or to permit survival of the beams in case of short power transients.

Programming languages are C and NODAL, the interpreter on which the SPS control system is based[16,17].

3.3 Microprocessors in the Field

All equipment in the field is controlled by an imbedded microprocessor housed in general in a VMEbus or G.64 crate. Preferred microprocessors are the Motorola 68000 or 6809, but there are other types present. The preferred operating systems are RMS68K, OS-9 and AMX. The systems to be controlled vary in complexity from simple arrays of input/output to complex networks with distributed intelligence performing a sequence of operation or autonomous surveillance. Special problems arize for commercial equipment supplied with a turnkey control system: an interface is needed to the 1553 multidrop, usually through a protocol converter.

3.4 Operator Interface

A design goal of the LEP control system is to provide consoles which are upwards compatible from the field to the control room, so that programs run from (simple) consoles in the proximity of the equipment may as well be run from the control room. On the other hand the importance of object manipulation on workstation through windowing is fully recognized as an essential tool for accelerator operation[18-22]. Therefore two types of consoles have been chosen: Affordable PC/ATs, connected either to token ring or 1553 multidrop, running XENIX or Apollo (for the control room only) connected to a token ring, running UNIX.

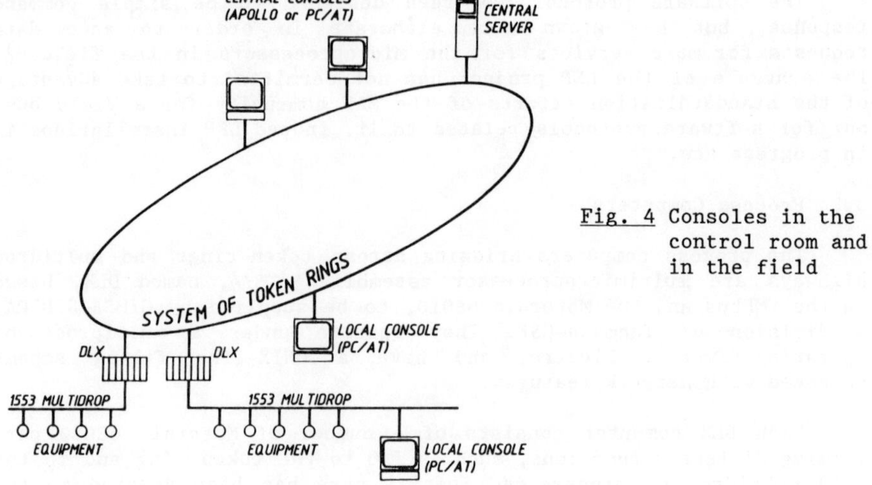

Fig. 4 Consoles in the control room and in the field

Two problems arise in trying to preserve compatibility.

(i) The network protocol on the 1553 multidrop is not based on TCP/IP. Hence for a console on the multidrop to perform the services permitted by RPC, one must bridge the gap between the simple multidrop protocol and the RPC facilities.

(ii) Graphic presentation on the PC/AT under XENIX has been based on GKS. There are indications that packages based on GKS on Apollo, competitive in performance with the proprietary Apollo graphic package (Dialog), will be available in time for LEP operation.

3.5 Machine synchronization

There are two distribution systems for machine synchronization frames. One is providing general machine timing (GMT)[23] consisting of "calendar" pulses at one millisecond intervals (with a precision better than .5 μs) and coded events inbetween. It is distributed to all hosts (consoles, servers and DLX) and to all microprocessor controllers of equipmet which may need it.

Long distance distribution of GMT is carried by a TDM channel, whilst local distribution is done either directly or as a companion to the 1553 multidrop, on a separate pair of the multidrop cable.

The second distribution (Beam Synchronization Timing (BST)) permits tagging each one of the four electron and four positron bunches, within a revolution, lasting 89 μs[24-26]. It is broadcast in a way similar to GMT and has a better precision (.2 μs). It is used primarily by the beam observation stations which are storing data in a cyclic memory in a free running mode and insert time markers to correlate the measurements with the particle bunches. The ability of

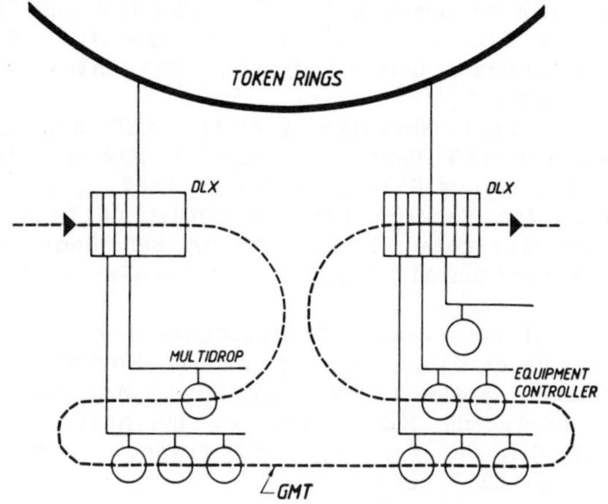

Fig. 5 Machine synchronization (GMT)

recording the position of all the bunches for many subsequent turns at each transit through the 504 position pick-ups is a unique diagnostic tools of LEP. It is made possible by the use of flash ADCs. Signals from the pick-ups are amplified then digitized in less than 400 ms so that resolution in time of bunches of opposite charge is possible starting from 10ms from the interaction regions.

REFERENCES

1) Plass, G., The LEP Project, Status and Plans, IEEE Trans. on Nucl. Sci., Vol. NS-30, No. 4, p. 1978 (1983).
2) LEP Design Report, Vol.1: The LEP Injector Chain, CERN-LEP/-TH83-29, CERN/PS/DL/83-31, CERN/SPS/83-26, LAL/RT/83-09, Geneva June 1983; Vol. 2: The LEP Main Ring, CERN-LEP/ 84-01, Geneva June 1984 (unpublished).
3) Schopper, H., Status of LEP and its Experimental Programme, IEEE Trans. on Nucl. Sci., Vol. NS-32, No. 5, p. 1561 (1985).
4) Wyss, C., Status Report on LEP, Proceedings of the ICFA Seminar on Future Perspectives in High-Energy Physics, Brookhaven National Laboratory, Upton, NY, 5-10 October 1987.
5) Crowley-Milling, M.C., The Control System for LEP, IEEE Trans. on Nucl. Sci., Vol. NS-30, No. 4, p. 2142 (1983).
6) Altaber, J., The Convergent LEP and SPS Control Systems, Proceed- ings of the 13th International Conference on High-Energy Accelerators, Novosibirsk, 7-11 August 1986.
7) Beetham, C.G., et al, Development of the SPS/LEP Control System, Proc. of the 1987 Part. Acc. Conf., Washington D.C., Vol. 1, p. 741 (1987).
8) Innocenti, P.G., The LEP Control System, Proc. of the EPS Conf. on Control Systems for Experimental Physics, Villars-sur-Ollon, 28.9-2.10.1987..
9) Parker, C.R.C.B., The Communications Infrastructure for the LEP collider, Proc. of the EPS Conf. on Control Systems for Experimental Physics, Villars-sur-Ollon, 28.9-2.10.1987.

10) Rausch, R., A Large Token-Ring Network for the CERN LEP and SPS Accelerators and Collider Rings, Proceedings of the 7th IFAC Workshop on Distributed Computer Control Systems, Mayschloss/Bad Neunhar, 30 Sept.-2 Oct. 1986.
11) Rausch, R., Real-time Control Networks for the LEP and SPS Accelerators, Proc. of the EPS Conf. on Control Systems for Experimental Physics, Villars-sur-Ollon, 28.9-2.10.1987.
12) Bland, A., The Communication Package for the Equipment Network of the LEP and SPS Accelerators, Proc. of the EPS Conf. on Control Systems for Experimental Physics, Villars-sur-Ollon, 28.9-2.10.1987.
13) Altaber, J., et al, Multimicroprocessor Architecture for the LEP Storage Ring Controls, Proceedings of 6th Annual Workshop on Distributed Computer Control Systems, Monterey, 20-22 May 1985.
14) Altaber, J., et al, Replacing Mini-computers by Multimicro-processors for the LEP Control System, IEEE Trans. on Nucl. Sci., Vol. NS-30, No. 4, p. 2287 (1983).
15) Van der Stok, P.D.V., DLX, the Multiprocessor Assembly for LEP/SPS Controls, Proc. of the EPS Conf. on Control Systems for Experimental Physics, Villars-sur-Ollon, 28.9-2.10.1987.
16) Crowley-Milling, M.C., and Shering, G., The NODAL System for the SPS, CERN 78-07.
17) Van der Stok, P.V.D., NODAL Reference Manual, LEP Controls Note 63, 30 January 1986 (unpublished).
18) Theil, E., et al, The Impact of New Computer Technology on Accelerator Control, Proc. of the 1987 Part. Acc. Conf., Washington D.C., Vol. 1, p. 529 (1987).
19) Paxon, V., et al, A Scientific Workstation Operator-Interface for Accelerator Control, Proc. of the 1987 Part. Acc. Conf., Washington D.C., Vol. 1, p. 556 (1987).
20) Anderssen, P.S., Personal Computers in Accelerator Controls, Proceedings of the European Summer Schools on Computing Techniques in Physcis, Tabor, 9-11 June 1987.
21) Anderssen, P.S., The New Control Room Infrastructure for the LEP and SPS accelerators, Proc. of the EPS Conf. on Control Systems for Experimental Physics, Villars-sur-Ollon, 28.9-2.10.1987.
22) Bailey, R., et al, The New Applications Software for the CERN Super Proton Synchrotron, Proc. of the EPS Conf. on Control Systems for Experimental Physics, Villars-sur-Ollon, 28.9-2.10.1987.
23) Beetham, C.G., et al, Overview of the SPS/LEP Fast Broadcast Message Timing Systen, Proc. of the 1987 Part. Acc. Conf., Washington D.C., Vol. 1, p. 766 (1987).
24) Borer, H., and Rabany, M., Description of the Beam Synchronous Timing for the LEP-BM System, LEP Note 575 (1987) (unpublished).
25) Bovet, C., Beam Diagnostics for LEP, Proceedings of the 13th Int. Conf. on High-Energy Accelerators, Novosibirsk, 7-11 August 1986.
26) Borer, J., et al, The LEP Beam Orbit Measurement System, Proc. of the 1987 Part. Acc. Conf., Washington D.C., Vol. 2, P. 778 (1987).

DATA ACQUISITION HARDWARE FOR THE D0 MICROVAX FARM†

David Cutts and Jan S. Hoftun

Physics Department, Brown University
Providence, Rhode Island 02906
U.S.A.

Christopher R. Johnson and Raymond T. Zeller

ZRL
8 Rushton Drive, Cranston, Rhode Island 02905
U.S.A.

ABSTRACT

We describe the MicroVAX–based data acquisition system used by the D0 experiment. We give operational characteristics of the data flow in this system: a "farm" of MicroVAX processors each receiving an event over multiple input dual–port memory channels. We discuss the individual hardware components and their performance at the Fermilab NWA test beam and other sites, as well as some further developments.

The D0 data acquisition system is based on an array or "farm" of MicroVAX computers to handle the high data rates and trigger rate reduction required for the D0 experiment at the Tevatron–I collider. The D0 detector, which will become partially operational in 1989, uses excellent calorimetry and lepton identification to study the new physics at 2 TeV. Because of the high potential data rate (100,000 interactions/second each on average supplying 250,000 bytes of digitized information), the data collection system uses a hierachial trigger structure. The "Level–0" trigger provides from scintillator arrays an accurate interaction time as well as a rough measure of the interaction position in the z–direction. Using this position measurement, the "Level–1" trigger can calculate transverse energy; and with this and other quantities obtained from fast ouputs from the calorimeters, muon counters, and other elements of the detector, Level–1 forms a hardware–based selection in the 3.5 microsecond time between crossings. This selection results in a reduction of the trigger rate to 200–400 events/second; these events are then fully digitized subject to a final, software–based filter process (each event in a single MicroVAX) which reduces the events recorded by the host system to 1–2 per second. The hardware used in the acquisition stage of the event flow is the subject of this paper. For aspects of the data acquisition software see Ref. 1.

The D0 data acquisition system comprises an event's path from the 84 VME crates, where the digitization of the detector's signals takes place, to the data management structure (ZEBRA) in the memory of a specific MicroVAX Level–2 node, and out again, if accepted, to a corresponding structure in the host computer. The key feature of this system is that the data, after being digitized in parallel, flows at high speed over parallel lines to a single, selected MicroVAX which then performs the complete Level–2 filter analysis. As shown in Figure 1, the digitization crates are divided by detector type into 7 sections; the crates in each section share the use of a data cable which operates at 40 MBytes/sec with a simple data dump protocol. Each of these 7 cables, plus an eigth which carries data from the Level–1 hardware trigger, is connected to an external port of a dual–ported memory in every MicroVAX Level–2 node. Thus the data flows from the VME crates directly into MicroVAX memory at 320 Mbytes/sec. As indicated in Figure 1, this data flow is controlled by a separate MicroVAX system, the acquisition "Supervisor", which has interface lines both to the VME data sources and to each of the nodes, and which directly enables the dual–port channels in an available node before initiating the event transfer.

The data acquisition system is centered on this simple yet high speed transfer of an event from the digitization crates to a specific MicroVAX node. There are a number of specialized hardware components, principally ZRL dual–port memories, which make this scheme possible. All of the components, except a "Sequencer" module — which allows use of a single data cable by multiple VME crates — have been in active use by various D0 groups for over a year. We describe the characteristics of each of the principal components below.

Data Cable — The data cable is the unidirectional bus connecting the VME crates with the Level–2 MicroVAX nodes. It is specified to run at up to 40 Mbyte/sec., and has 32 data lines, 2 control lines (SYNCH and BUSY), 2 parity lines, and 9 expansion lines. The data cable protocol calls for BUSY to be asserted at the start of transfer and held throughout, and data is latched by the receiving memory on the trailing edge of each SYNCH pulse. Signals are transmitted using Futurebus open collector transceivers.

Dual – port Memory — The dual–port memory provides event buffering into the Level–2 MicroVAX nodes. Each quad–width Q–Bus dual–port board consists of two channels of 256 KBytes memory. The external ports are 32 bits wide, use the Futurebus transceivers, and can be operated as either input or output, with 100 nsec read/write cycle time (for 40

MByte/sec. transfer speed). On the Q–Bus side, the dual–port boards are transparent Q–Bus memory, and with an associated controller, achieve a 3 MByte/sec transfer into the MicroVAX's private memory. This transfer does not degrade the performance of the cpu, so that event analysis processing of one aspect of the event (say the Level–1 trigger data block) proceeds while other data is transfered.

VME Buffer/driver — This module buffers event data in a VME front end electronics crate and outputs it on a data cable. To the crate electronics it appears as normal VME memory. It supports 8/16/32 bit transfers, and can act as a DMA master — incorporating list processing capability and supporting up to 30 MByte/sec transfer rate within VME. The board incorporates dual 256 KByte buffers, toggled on alternate events. The output port is 32 bits wide and compatible with the data cable protocols, with parity generation and a 40 Mbyte/sec transfer rate. A non–DMA version has been operational for 18 months; the DMA buffer/driver will be prototyped this summer.

Sequencer — For each data cable a Sequencer module allows the use of the cable by multiple VME crates (each with a buffer/driver). This module is interfaced with the Level–1 trigger system and the Level–2 Supervisor as well as each of the buffer/drivers in its readout section. It contains a downloaded lookup table, so that given the 32–bit pattern of "specific triggers fired" from Level–1, it obtains a list of the crates to be readout. After handshaking with the Supervisor (to have the dual–ports of a selected Level–2 node enabled), the Sequencer sequentially enables the use of the data cable by each VME crate. Presently the functions of the Sequencer are being handled by the Supervisor MicroVAX; a discrete module will be operational this summer.

Data acquisition systems based on the above hardware have been in operational use since June 1986, as part of 5 different D0 systems. In addition, at Brown a total of 32 MicroVAX–II computers of various flavors and 2 MicroVAX–III systems have been used for development work and as a dedicated Monte Carlo farm. Indeed, the production running of GEANT–based simulations of the D0 detector, so vital for design decisions, has to date entirely been run on a MicroVAX analysis farm at Brown. The computing facilities at Fermilab and at our collaborating institutions, including supercomputer and other non–conventional systems, have been unable to provide an implementation of the D0 GEANT package that is competitive with a MicroVAX farm.

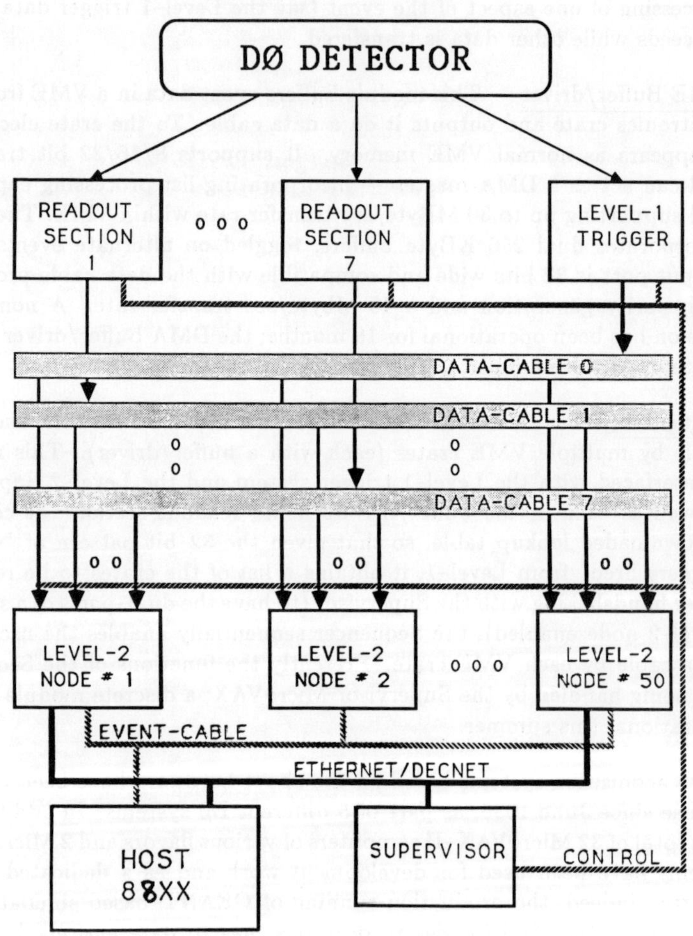

Figure 1: Diagram of the D0 Data Acquisition System.

The most intensive use of the MicroVAX acquisition hardware has been with a system at a large scale detector–component test facility at Fermilab's NWA test beam. This facility has been in continuous use since June 1987, with beam available until February 1988, and expected again at the end of this year. The experience with the hardware has been very positive: the acquisition path of data cables and dual–port memories proved extremely reliable, with no failures except for a data cable which was torn accidentally from its connector. During the initial phase of this work, the Level–1 trigger framework and the VME digitizing systems were being developed; the acquisition system necessarily became more flexible and capable of surviving various error conditions. As the focus of the data flow control, the Supervisor became more complex both in its hardware interfaces and in its software. Even though this paper concerns the data acquisition hardware, it must be mentioned that the real–time framework used in all the nodes, VAXELN, has been very successful. Code that runs on the VMS host, in various high–level languages (including FORTRAN and C), is easily packaged and downloaded to MicroVAX nodes, with concurrent programming, DECNET support, and full symbolic, remote debugging included. Real–time performance is excellent, and creation of device drivers is straight–forward. Features in the lastest release include remote displays, a remote command facility, and a remote performance analyzer. The rapid evolution of the Supervisor program at NWA was possible only because of the ease–of–use and strengths of the VAXELN environment.

To direct the data flow at NWA, the Supervisor has communication lines with the Level–1 trigger, with the VME readout crates, and with the Level–2 nodes. In addition to these control lines, it is interfaced to each of the data cables and thus able to itself send data to a Level–2 node's dual–port memories, for diagnostic testing as well as for error recovery. At NWA these connections, via DRV11–J parallel interfaces, allow the Supervisor to maintain a list of available nodes (by trigger type) and to control the data flow by enabling dual–port memories and VME readouts. Communication with the host VAX is supported for run control and status/diagnostic functions, via a separate job in the Supervisor MicroVAX.

The typical event cycle at NWA is initiated by a readout request from the VME crate (TRGR) associated with the Level–1 trigger. The Supervisor reads the trigger type and the "SYNCH" bits (the 3 low order bits of the event number), accompanying the TRGR readouts request. Using the trigger type, the Supervisor assigns the Level–2 node (enables its dual–ports), initiates the TRGR readout, and proceeds to process other readouts (determined by trigger type) to complete the event. At NWA there are two additional data cables: one for calorimeter data and one for central tracking detector data. For readout requests from each of these sources, the Supervisor reads the accompanying SYNCH bits and compares them with the SYNCH bits associated with the TRGR readout. If the bits agree (the two data blocks are from the same event) the Supervisor enables the appropriate dual–port and VME readout. If the bits indicate that the VME data is for the next event, the Supervisor enables the dual–port but itself sends a short dummy block (with error flag) to complete the present event in the Level–2 node. Or, if the SYNCH bits indicate that the VME has old data, the Supervisor enables the VME readout but not the dual–port — thus dumping that buffer "on the floor" — and waits for a further readout request. Various additional timeouts prevent data flow from being hung in any abnormal condition associated with the Supervisor, and all the data flow actions are logged and visible via a status display.

The typical response time at NWA, from readout request to start of data flow into Level–2 dual–port memories is of order 100 microseconds. This time frame is reasonable for D0, where the event flow itself (250 K Bytes over 8 parallel 40 MByte/sec data cables) will take approximately 1 millisec, and the Level–2 system will handle 200–400 events/second. Additionally, event data is buffered 4 deep, before the Level–2, with two input capacitors on the ADCs and two buffers in the VME data cable buffer/driver. Having the data flow control entirely in the Supervisor software provides flexibility and is very useful, especially when the rest of the digitization and triggering system is developing. Some of the control functions will be migrated into hardware in the future (for example, into the Sequencer), but the basic feature of having the data acquisition under direct real–time Supervisor control will be maintained.

Another important lesson from the NWA beam concerns the data flow rate through the Level–2 system and onto the host. Typical events at NWA are 10 KByte (with 8 K Byte on the calorimeter data cable), so from VME to a Level–2 node takes roughly 250 microseconds. In the Level–2, to put the data into output ZEBRA banks (without filtering), takes approximately 15 milliseconds, so the three nodes used at NWA could process 200 events/second. However, in the VME crates we use an early version of the buffer/driver which is not a DMA master, and indeed its buffer from VME requires 15 microseconds per 32 bit word. Thus moving the 8 Kbyte calorimeter data block can be done at no more than 33 events/second. Even worse, the link over which we send data to the host (a MicroVAX Q5 system) is Ethernet; we find a typical operating rate of 150 K Bytes/second, which limits the overall system throughput to 15 events/second. (Also, since the NWA beam is not a collider, we buffer events in the Level–2 nodes and send data to the host also in non–beam time.) For the test beam operation, users are happy with this throughput; typical runs take 5 minutes. However, for D0, where "1–2" events/second to the host are the goal (surely more on occasion), and events average 250 KBytes, it is clear that Ethernet (though fine for system functions) is not appropriate for data.

The solution to the data throughput problem involves the same hardware used for data acquisition: dual–port memories and the data cable. The present generation of dual–port boards are bi–directional, with rates up to 40 Mbytes/second per channel. Each channel can be independently set to transmit or receive; the dualports have an on–board 10 MHz output clock (for 40 MBytes/sec) and allow loopback diagnostics. At Brown we are currently using dual–ports in a separate MicroVAX system to output onto data cables, simulating VME buffer/driver outputs for system development work. For D0 we will use dual–ports in output mode to feed a back–end bus for moving selected events to the host VMS system, using a dual–port in the host to receive events. Drivers for VAXELN and for VMS are under development which will make their use transparent to the existing routines (thus simulating DECNET/Ethernet). We anticipte an average system transfer rate of 8 MBytes/second.

Another potential impediment to the efficient flow of data through the system results from the use of Q–Bus transfers to bring data from the dual–port memories to the processor's private memory. As mentioned above, with an associated DMA controller, this transfer occurs at 3 MByte/second. Because the PMI supports a higher bandwidth than utilized by the processor, these transfers can occur without loss of CPU cycles. Thus once the short data block associated with the Level–1 trigger is loaded, the Level–2 analysis program will start; the subsequent channels will be moved into private memory as appropriate, and the transfer time will not significantly impact the filtering efficiency. (Most events will be discarded without the largest data channels, those from the central tracking, being used). However, it is clear that the system would be more elegant if the Q–bus latency could be eliminated. ZRL is currently developing a new generation of multiport memories with this capability; these memories will be used in future D0 systems.

Much of the strength of the D0 data acquisition system is based in its integration with the DEC/VMS environment. A common software/hardware system is an enormous benefit, most directly in software transportability; but data compatability and integration of real–time control software, part of data acquisition, is also very important. It is commonly thought that these benefits, while perhaps cost–effective when users' convenience is included, come at a large premium. Actually, pricing of the D0 MicroVAX farm competes favorably with other systems of similar performance.

In summary, we have described the D0 data acquisition hardware and discussed its data–taking performance. We have found that this simple and elegant system, an acquisition farm based on MicroVAX nodes equipped with dual–port memories, with data flowing directly on parallel data cables directly to the memories associated with a single node, to be powerful, reliable and easy to use, yet cost effective.

REFERENCES

† This work supported in part by the U. S. Department of Energy.

1. Cutts, D., Hoftun, J.S., Johnson, C.R., and Zeller, R.T., "The MicroVAX–Based Data Acquisition System for D0", IEEE Transactions on Nuclear Science, Vol. NS – 34 (1987).

The ALEPH Event-Processor*

Beat Jost
CERN/EP
Geneva, Switzerland

Abstract

The task of the ALEPH Event-Processor is to reinforce the selectivity of the electronic trigger system in order to match the primary data rate with the data recording capability. A reduction factor of approximately 10 in rate is envisaged by a software separation of the genuine e^+e^- events from background triggers using the full granularity and improved precision of the triggering detector elements.

The talk will describe the multi-processor hardware and the system related software for the event readout from the FASTBUS data source into the trigger processors and and the transfer of the selected events into the host VAX 8700.

1 INTRODUCTION

ALEPH is one of the four large detectors currently under construction in the tunnel of the LEP e^+e^- collider at CERN. The detector consists of several subdetectors (Time Projection Chamber (TPC), Hadron Calorimeter (HCal), Electromagnetic Calorimeter (ECal) etc.) with altogether approximately 500,000 readout channels. After zero-suppression of analog signals this leads to a typical event size of about 100 kBytes.

At the LEP design luminosity the event rate from genuine e^+e^- collisions under the Z^0 peak is calculated to be about 1 Hz. It is estimated that after the fast electronic event selection there will be a background trigger rate up to 10 Hz. This number is only an estimate and could well be a factor of 2 to 3 wrong.

The experiment has a trigger system consisting of three levels (see also Table 1). Levels 1 and 2 are hardware triggers whereas the third level is a software selection. The beam crossing rate of the LEP accelerator is 45 kHz. The output rate of the Level 1 trigger should be below 500 Hz. This trigger is based mainly on calorimetry: HCal and ECal plus information from the Inner Tracking Chamber (ITC). The Level 2 trigger uses information from the trigger pads of the TPC. Its output rate is estimated to be of the order of 10 Hz. In order to cope with the tape writing speed it is necessary to further reduce the number of events per second to about 1 Hz.

The task of the Event-Processor (EP) is to perform this last reduction in rate by using the full granularity of the completely assembled events. The application software running in the EP partly reconstructs the events to confirm the decision of the first two trigger levels. The subject of this talk is to describe the EP. It will deal mainly with the readout of the events from the data acquisition system and the transfer of the selected events to the data logging computer.

---

*The results presented in this talk were obtained in a joint research project between CERN and Digital Equipment Corporation.

Trigger Level	Input Rate	Output Rate	Number of Segments	Based on
1	45 kHz	$\leq$ 500 Hz	72	ITC, HCal, ECal
2	$\leq$ 500 Hz	$\sim$ 10 Hz	1044	TPC Trigger pads
3	$\sim$ 10 Hz	$\sim$ 1 Hz	-	Full events

Table 1: Summary of the characteristics of the three trigger levels.

2 EVENT-PROCESSOR HARDWARE AND SYSTEM SOFTWARE

2.1 Requirements

Since we do not yet know how high the background trigger rate will be and what the background will consist of it is obvious that the system must be flexible from the software point of view as well as expansible in what concerns the CPU power available. We were therefore looking into the possibility of using a modular commercial product for the EP. The main reasons for using a commercial product were

- good technical support for the hardware
- good software support and maintenance even after changes in the operating system of the host
- good programming and development tools
- identical code and results for the application software whether running on the host or on the EP

2.2 Hardware

In order to meet the requirements for the EP we have chosen to use a hardware configuration consisting of single-board computers (SBC) mounted directly on the VAXBI buses of our main data-acquisition computer, which is a VAX 8700 (see Fig. 1). One SBC has the CPU power of a μVAX II and is equipped with 1 MByte of local RAM on board. This memory can be expanded in increments of 2 MBytes to a maximum of 11 Mbytes. The local RAM is connected to the CPU via a private bus and not via the BI; but it is accessible from other nodes connected to the same BI. The SBC's on the other hand have access to all memory connected to the same BI. They can even map over the main memory of the VAX 8700 which resides on yet another bus. The SBC's have in addition a datamover capability implemented in hardware. This utility can be used to transfer data from one SBC to another SBC or to the host.

The expansibility of the system is guaranteed by the fact that as many as 3 BI buses can be filled with SBC's (see Fig. 1). Depending on the amount of local memory needed per CPU, up to 13 SBC's (with 1 MByte of local RAM) can be connected to one BI bus. This corresponds to a CPU power of $\sim$6 IBM-168 units.

2.3 System Software

The system software is based on VAXELN provided by DEC. This is fairly high speed set of kernel routines for communication, device drivers and multi-tasking management in a real-time environment. The nice feature of ELN is that all system related software can be

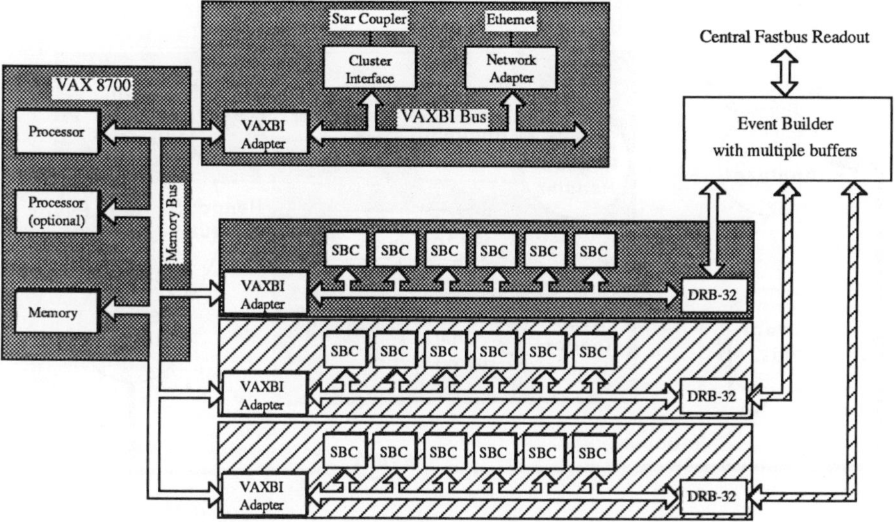

Figure 1: Schematic view of the hardware configuration of the main DAQ VAX with the Event Processor. The gray shaded areas represent the basic components of the system. The diagonal shaded areas show possible extensions. Note that the number of SBC's is limited to 10 processors per BI (for 3 Mbytes of memory per SBC) due to power limitations.

written in high level languages (PASCAL, C, FORTRAN, Ada). Another very valuable feature is the possibility for remote symbolic debugging even of device drivers and routines running in kernel mode (as opposed to VMS where no such possibility exists).

2.3.1 Event readout mechanism

The data will be read out from the Main Event-Builder residing in FASTBUS via an "Event-Builder to VAX Interface" (EVI) [1] and a 32-bit parallel interface (DRB32) from DEC. We have decided to assign one SBC per DRB32 to act as a manager (MSBC) in the sense that this SBC controls the DRB32 and handles the protocol with the EVI (see also Fig. 2).

Tasks in other processors (either in the SBC's or in the host) receive events by sending a request to the manager. This request consists of information about the buffer supposed to receive the event as well as of requirements on the event wanted (event types, trigger masks etc.). Once a readout request for an event with the demanded characteristics is posted by the EVI, the MSBC sets up the DRB32 mapping registers in such a way that the data flow directly into the requesting SBCs memory. No further copying of the event data is therefore necessary.

[1] see the contribution of V. Mertens elsewhere in these proceedings.

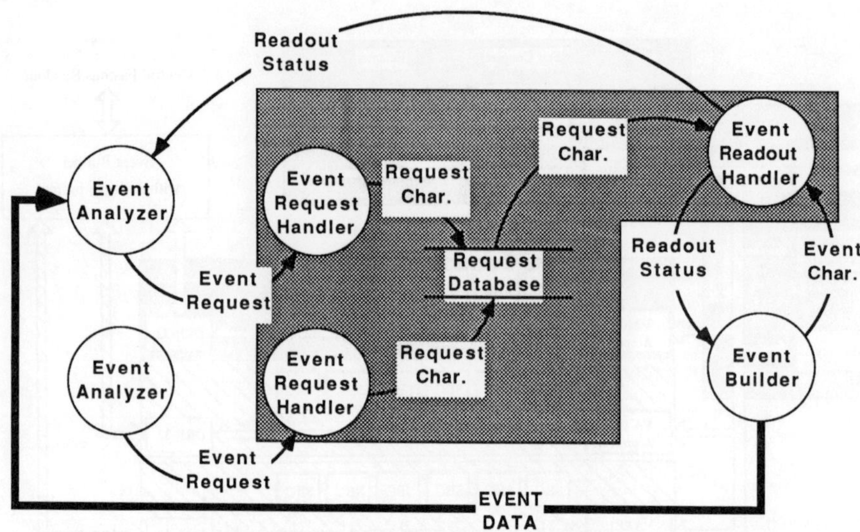

Figure 2: Schematic diagram of the information flow for an event readout. The event analyzer can run in any CPU (SBC's or host) connected to the same BI as the request handler and the DRB32. The functions in the shaded area are implemented in the MSBC. The Event Builder is connected to the BI bus via an EVI and a DRB32.

The performance of the event readout is listed in Table 2. It should be noted that the numbers in Table 2 are preliminary since the software running on the MSBC has not

Event Size	Events Transferred [1/sec]	Transfer rate [kBytes/sec]
4 Bytes	110	0.440
4 kBytes	100	400
100 kBytes	20	2000

Table 2: Performance of the event readout (preliminary). The maximum transfer speed is limited by the SBC hardware to 2.5 MBytes/sec.

yet been optimized for speed. Note also that the total transfer rate can be increased by installing additional EVI's and DRB's. The target SBC will be notified of the arrival of the event by a short message from the MSBC. It then performs the necessary steps to verify the validity of the hardware trigger and upon success transfers the event to the main VAX.

2.3.2 Event transfer to Main DAQ VAX

This transfer uses special multi-processor communication software which implements the communication philosophy of ELN (Ports, Messages, Circuits) under VMS. Therefore for the ELN system software the communication with VMS is completely transparent. Due to this fact , the same software can also be used to transfer the event requests to the MSBC as well as control and status information from the SBC's to the host. Some measurements of the performance of the communication software are listed in Table 3. This performance is adequate for our application since we expect an output rate from the SBC's of the order of 100 kBytes/sec only.

Message size [kBytes]	Host CPU Usage [%]	Transfer rate [kBytes/sec]	Transfer rate [Mess/sec]
0.5	90	70	140
100	80	1600	16

Table 3: Results of performance measurements for message transfers between the host (VAX 8200) and the SBC's. The maximum transfer speed is limited by the SBC hardware to 2.5 MBytes/sec.

Acknowledgements

Many people in the ALEPH On-line group have given assistance in the course of this project. I specially thank G. Lütjens and W. v. Rüden for many valuable discussions and V. Mertens for his collaboration during the design and the tests of the event readout.
I would also like to thank R. Vignoni and his team from DEC Engineering for supplying us with the necessary hardware and software during the project. Very much acknowledged is the support and collaboration of E. Gerelle and J. Arnold from the CERN-DIGITAL Joint Project office at CERN.

A Multiuser Data Acquisition Based on the Macintosh Computer

M.Budinich, and P.Giannetti

(Dipartimento di Fisica, INFN and Microprocessor Laboratory, Trieste - Italy)

Presented by P. Giannetti

1 - INTRODUCTION

The Data AcQuisition system (DAQ in what follows) has been built for the Clue experiment that will study high energy γ rays coming from localized cosmic sources [1]. These cosmic photons produce large electromagnetic showers in the atmosphere. A mirror reflects the Cerenkov light emitted by the shower into a photosensitive gas chamber of 256 pads, located in the mirror focal plane.

The success of the detector depends on the transmission of the Cerenkov radiation in the troposphere which can vary heavily with weather conditions. Moderate altitudes and a tropical climate are definitely preferable. So the detector should be located near the Equator, possibly on a hill with very dry nights, far from big cities.

As a consequence of these requests it is clear that the experiment should be able to take data for long periods of time in isolated and hard to reach locations. Few physicists will be working on the apparatus at the same time and they could be not expert in DAQ problems. The conclusion is that the DAQ system should meet the following requirements:

a) to be small, simple and reliable as possible, yet powerful and flexible;

b) to be easy to boot, initialize and use (also without an expert around);

c) to be easy to repair by substituting any piece (including the DAQ computer);

d) to have a dead time as small as possible to be able to detect bursts of cosmic γ coming in very short times;

e) to allow program development and detector checks or simple data analysis without interference to the real DAQ (i.e. a multitasking real time system).

2 - PHILOSOPHY OF THE DAQ SYSTEM

The basic idea is to build a modular system where the work is subdivided between different cpu's working independently at the same time. The different tasks (that in a standard multitasking/multiuser system are carried out by concurrent processes sharing the resources of a single computer) are assigned to different single-user workstations connected together by network (fig. 1).

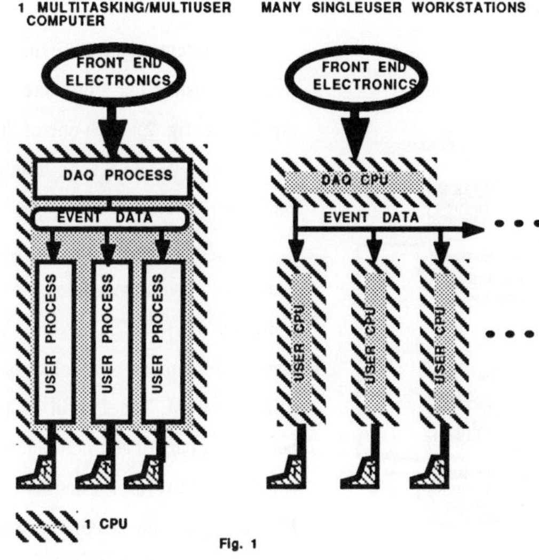

Fig. 1

In such a system the users are able to perform their software development and monitoring jobs without sharing memory and cpu time with the DAQ i.e. without affecting or slowing down the DAQ function.

This is possible since the DAQ task is done by a dedicated cpu that will never be touched by the general user. The data acquired by this cpu are then broadcasted to all the machines along the connecting network with a minimal overhead.

The machines along the network can read on-line data, if they need them, from the network itself.

This architecture reduces drastically the probability of a global system crash or lock and keeps the DAQ clear from user problems. It is also easy to expand the system for a new user by just adding one new cpu in the network.

For the practical implementation of this idea we have used the MacVee system developed at CERN [2], that provides the Apple Macintosh family with compatible direct access to VME bus and CAMAC systems. This system is cheap, fast, reliable and easy to use and profits from a lot of existing software.

For the hardware of a test version of the apparatus we have used the CAMAC standard while VME will be used in the forthcoming experiment in order to build a more sophisticated system with negligible dead time.

For the connecting network we have used the Macintosh standard serial network AppleTalk but a faster communication path is easily implementable (SCSI or other).

In what follows we describe the CAMAC based DAQ we have built for the test that will precede the experiment.

3 - SYSTEM ARCHITECTURE

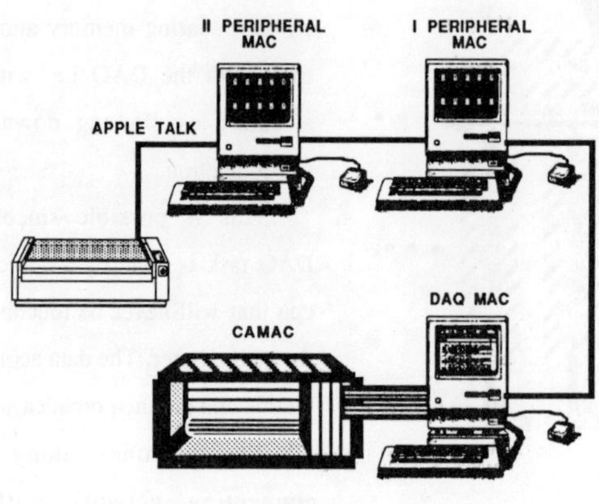

Fig. 2

The system architecture includes 3 Macintosh Plus connected by AppleTalk (fig. 2): each one of them has been dedicated to a specific function.

The first Macintosh, the DAQ Mac, has been modified to be connected to CAMAC. The connection is made by an electronics plinth called MacPlinth which becomes a part of the computer [2] and the CAMAC crate controller, the Mac-CC [2]. The DAQ Mac reads data from the CAMAC processor of the LeCroy 2280 high density ADC System [3] and stores them on its hard disk.

The DAQ Mac can send the events, through AppleTalk, to the peripheral Macintosh computers, that can read them if needed.

The first peripheral Macintosh is dedicated to apparatus monitoring. It provides a continuous sampling of the detector status filling a set of standard histograms with on line data coming from the network.

The second peripheral Macintosh is dedicated to software development and to preliminary offline analysis.

All the Macintosh are equipped with 20 Mbyte hard disks. In the near future the events will be stored in 80 Mbyte hard disks with a capacity of 160,000 events/disk (500 bytes/event). When a disk is full another one takes its place. The acquired events are copied on tape and the disk is ready to be used again.

4 - SOFTWARE

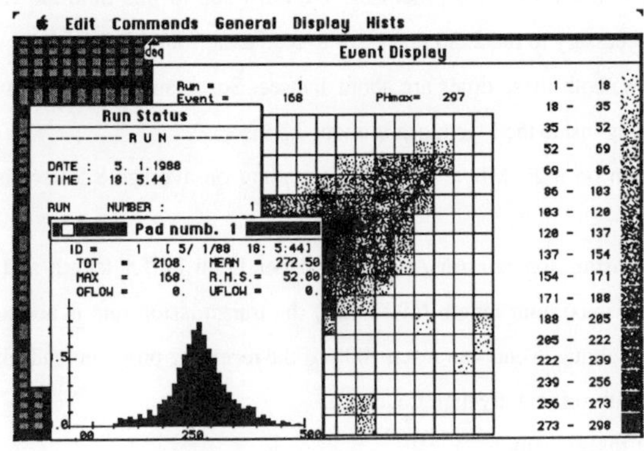

Fig. 3

The integration of Macintosh with laboratory instrumentation systems has allowed the development of some novel approaches for user interaction with experiments. User applications are provided with the interactive screen configuration typical of Macintosh applications. The structures represented may be opened with the mouse; actions or options may be selected by the mouse through standard Macintosh dialogue.

This facilitates the user interaction with the control and monitoring system (fig. 3).

The Clue DAQ system is provided with 4 different programs:

DAQ: takes data from CAMAC, saves them on disk and broadcasts them through the network (runs on DAQ Mac)

CONTROL: Receives data from network and performs monitoring tasks (runs on 1st peripheral Mac)

OFFLINE: Reads data from disk and performs a preliminary offline analysis (runs on 2nd peripheral Mac)

DIAG: diagnostic tool for CAMAC, ADC's and network (runs on all Mac).

These programs were developed starting from a common nucleus that is the DAQ program written at CERN [4]. The language is Fortran 77 and the compiler is the Microsoft Fortran version 2.2. Using this program we have been able to take advantage of the libraries for Microsoft Fortran developed at CERN [4] including an histogram package, a CAMAC package and a window/menu management package.

The software for sending and receiving data through the AppleTalk network has been written in Fortran and Assembler and implements standard Macintosh protocols. In this way the AppleTalk network is also available for standard uses like printer sharing.

5 - PROTOTYPE PERFORMANCES

To have an idea about system performances we measured the timing for the fundamental operations executed by a reduced system that handles only 48 channels (i.e. one single ADC module).

a) **CAMAC operations**: MacVee executes a CAMAC command (CSSA) in 300 μsec and DMA block transfer (CSUBC) in 7.2 μsec/word. Totally the DAQ Mac needs 1.5 msec to check LAM, to read 48 words and to enable again the 2280 processor. We must add to this time the ADC conversion time and the time necessary to the 2280 processor to read data from ADC modules and to organize them in its memory; both these times are about 1 msec. So, to summarize, the time between the trigger and data ready inside the Macintosh is about 3.5 msec.

b) **Writing data**: Data are written on a 20 Mbyte hard disk spending on average 8.5 μsec/byte writing a buffer of 40 Kbytes.

c) **AppleTalk transmission**: transmission rate on AppleTalk depends on buffer length and on network traffic. For a buffer of maximum length (598 bytes) the transmission rate is about 23 Kbytes/sec. The efficiency in receiving events can reach 100% if the receiving program reads data, emptying the dedicated buffer, with sufficiently high frequency.

d) **Histogram filling**: about 60 msec/event.

e) **Event display**: data taking is stopped when the display routine is active. This time, depending on displayed data and user response, can sum up to seconds.

6 - CONCLUSIONS

This working multiuser prototype is cheap, flexible, friendly and fast. It permits to solve in parallel and independently data taking, monitoring functions and preliminary analysis.

We can deduce from the tests that the system with 240 channels will take data spending about 10 msec for CAMAC functions, 4 msec to write data to a 20MB hard disk and, if necessary, 21 msec to broadcast events along AppleTalk.

No estimate is available at the moment for the future VME based data acquisition system.

References:

[1] G. Apollinari et al., *A Wide Dynamic Range Experiment to Measure High Energy gamma-Showers in Air by Detecting Cerenkov Light in the Middle Ultraviolet*, Nuclear Instruments and Methods **A263** (1988), pg. 255.

[2] B.G. Taylor, *Macintosh in the Laboratory*, Microprocessing and Microprogramming **21** (1987), pp. 101-110.

[3] *CAMAC System 2280 High Density ADC Data Acquisition System*, LeCroy Research Systems Corporation, 700 South Main Street, Spring Valley, N.Y. 10977, USA.

[4] A. Contin, *Fortran Data Acquisition Software for Macintosh*, CERN Report, Ep Division, Geneva, July 10 1986.

The LEP OPAL EVENT SELECTION SYSTEM

by Patrick Le Dû

DPhPE- CEN Saclay
91191-Gif sur Yvette- CEDEX
France

Talk given at the International Conference on the Impact of Digital Microelectronics and Microprocessors on Particle Physics - TRIESTE - 28-30 March 1988

1. Abstract

This paper describes the event selection system planed for the OPAL detector which is being constructed for use with the Large Electron Positron collider (LEP) at CERN .

This selection is organized (fig.1) in a three elementary steps. A powerful hardware decision (TRIGGER) using exclusively analog signals within 2 bunch crossing is completed by a microprocessor matrix integrated in the acquisition chain in order to perform a software filtering task (FILTER) and finally an emulator farm is used for the full reconstruction (PROCESS) of selected events prior to transmission to the host computer.

It is proposed to present the final status of the event selection, data handling and on line analysis of this experiment and to describe the read out system based on distributed tree structured architecture of 32 bit MOTOROLA 68020 /68030 microprocessor family .

2. Introduction

OPAL (Omnipurpose Apparatus for LEP) is a large detector [Ref.1] being constructed for use with the LEP collider and is expected to begin operation by middle 89. The detector will be used to study e+e- collisions at energy up to 100 Gev, and is being designed to give 97% of full solid angle coverage with good tracking resolution, particle identification, electromagnetic and hadronic calorimetry.

The principal elements of the detector include (from the inner radius to the outer):

• A vertex detector at a radius 8 cm from the beam, consisting of a radial-stereo drift chamber with a space resolution less than 50 µm.

• A radial (JET) drift chamber with 3750 sense wires and an azimuthal resolution of 100 microns. A good two tracks separation (2mm) and particle identification (dE/dX = 3.5%) is achieved. The third coordinate read out is obtained using charge division (1%) and by an additional ZED drift chamber (100-300 µm)

• A warm aluminum solenoid coil of 1.5 Xo thickness provides a 0.4 Tesla magnetic field.

• A Time Of Flight counter (TOF) with 250 psec time resolution supplement the dE/dX particle identification in the 1-2 Gev/c region . It is used also for triggering purpose.

• A 12500 lead glass blocks calorimeter with presampler chambers provide an energy resolution of 7% to 10% / E and 2 to 5 mm position of the electromagnetic shower.

• An iron calorimeter uses streamer tubes with 1300 pointing towers of pad analog read outs, and 50000 channels of strip digital read out for muon tracking.

• A large external muon chambers system covers both the end cap and barrel area.

More than 150000 analog signals coming from 15 detectors produce 160 kbytes of information per good physical event.

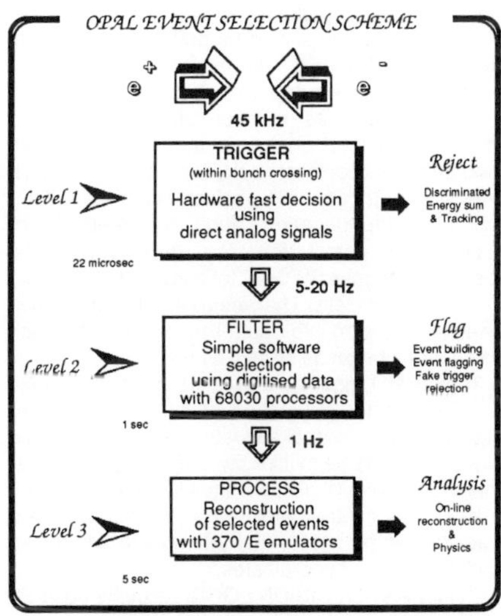

Figure 1

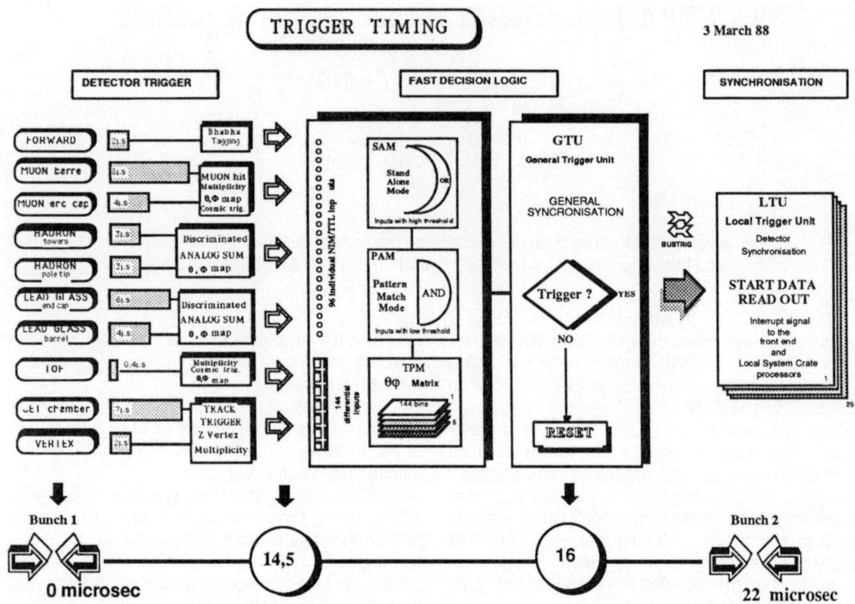

Figure 2

3. The Hardware Trigger

A very important element in the design of this level is the availability of a large number of independent and redundant triggers for most physics reactions. Therefore, triggers should be provided by different detectors elements (such as tracking and electromagnetic subdetectors) or by different trigger elements (such as number of electromagnetic clusters and total energy in the electromagnetic calorimeter , or collinear tracks and track multiplicity). The OPAL trigger will select event based on LEPTON identification and geometrical or topological correlation. To generate these topological trigger, the detector has been divided into 144 "theta-phi" cells. The various subdetectors have been also divided into "theta-Phi" bins which match this segmentation as closely as possible.

The OPAL trigger logic consists of 2 parts (fig.2):

- Subdetector generation of component trigger signals: The individual subdetectors each generate signals such as hit multiplicity and energy sum over threshold, as well their input signals to the "theta-phi" matrix . The central detector has special hardware track finding processor that attempts to find tracks and provide their coordinates.

- The FAST DECISION LOGIC accepts up to 96 signals signals from several subdetectors and generates the final trigger. There are 3 different logics: the Stand Alone Mode (SAM) using only selected individual inputs , the Pattern Match Mode (PAM) which combine inputs and the Theta Phi Matrix (TPM). The hardware implementation of the final decision logic is based on Programmable Array Logic (PAL). Control and synchronization with the read out system is build using the VME norm.

4. Data collection and Filtering

The data collection is performed in series of stages (fig 3).

a) The Front End Digitizer Branches

At the digitizer level, standardization has not been insisted upon. CAMAC and VME are used when the number of analog input is small. ECL type arrangements mostly in Eurocrates and also in some FASTBUS crates are employed to handle large volumes of data. Each such detector branch is driven by a Fast Intelligent Controller (FIC) which assembles the information and performs various hardware-oriented data reduction tasks. Zero suppression, pedestal subtraction and detection of clusters

b) The Local System Crate (LSC)

The Local System Crate is the interface between the various norms used for data collection from the individual detectors, and the homogeneous , all VME system used for assembling the event. There is one system crate for each of the 15 detector branches. They acquire the data previously computed by the Fast Intelligent Controller (FIC), format it, prepare for the software filter, and where possible parametrize the raw data in order to reduce the size of the event. Histograms for control and some statistic can in principle be done at that level (Fig. 4) .

The purpose of the system crate is also to manage the trigger interrupt (Local Trigger Unit) and to ensure synchronization between triggering and information gathering. It is at this point that events may be spied on for sample calculation, test or display on dedicated test computer. Access to the Local Area Network (Ethernet) is also provided.This is where local control of individual detector parameters will be performed and stored in a 'unvolatile' data base. The necessary human interface includes an Apple Macintosh acting as a Flexible Local Intelligent Console (FLIC)

c) The Filter VME Matrix

A microprocessor matrix (fig 5) based on a n-fold repetition of a basic unit : an isolated system capable of receiving in an asynchronous mode ONE EVENT directly and analizing it completely. That level combine then the following tasks:

- data collection (EVENT BUILDER) of one event
- second level trigger using digitized data of one complete event (fake event rejection).
- event filter (flag or reject event if the trigger rate is too high)
- general calibration and monitoring of some special event.
- transfer data to the upper level in asynchronous mode (no dead time).

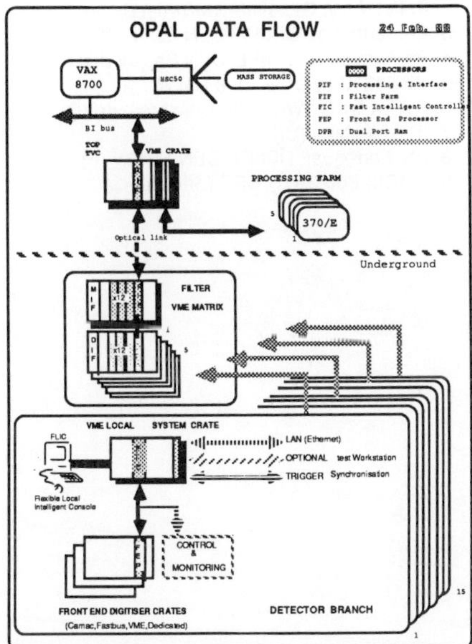

Figure 3

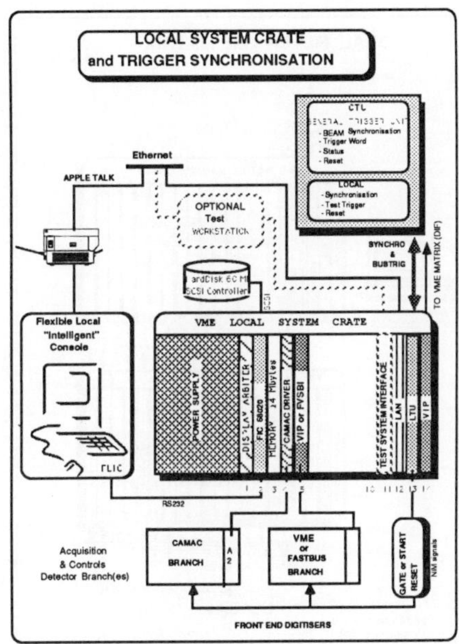

Figure 4

This matrix will initially have 12 parallel channels (columns) . Each of the parallel channel contains in its seven computer memories the data of a complete event. These memories are addressable via the VSB highway by a MC68030 processor with floating point coprocessor (the FIlter Processor FIF) which can therefore assembles the event and work on it (Fake trigger rejection,event selection & Flagging).

Each of the memories is provided with two ports, to allow access both to the VME bus and to the VSB highway. Thus data from a particular event can flow in parallel streams from the detector branches to a single column of the matrix while a number of other events are being processed simultaneously in other columns.

5. The event process

An 370/E EMULATOR FARM (5 units) which performs a first on-line event reconstruction using the ROPE off line analysis program. The connection between this farm and the underground data collection is made via an optical fibre. At the top of the system, a DEC VAX 8700 computer fis used for data storage, general monitoring , networking, run control and data base management.

6. Hardware components

The initial choice was to build this system by using exclusively the VME norm . Only a small number of

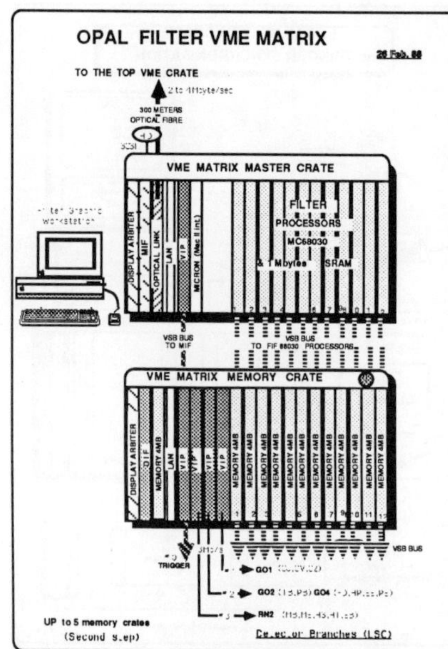

Figure 5

components are used.

1.The Fast Intelligent Controller (FIC) [Ref. 2] is used as the standard 32 bit VME/VSB CPU with a MC68020 microprocessor, 1 Mbyte of SRAM and DMA capability.

2.The Dual Port Memory [Ref. 3] contains 4 to 16 Mbytes of dynamic RAM memory with optional spy and DMA block transfer VSB/VME mode.

3. The VME Interconnect Port (VIP) [Ref. 4] is able to link up to 15 VME chassis, give direct access between VME/VSB master to any slave of any crate, support DMA block transfer (6 Mbytes/s) and includes multimaster protocol.

4.The access to the front end digitizer branch is provided by a VME- CAMAC branch driver [Ref. 5] or VSB-Fastbus interface [Ref. 6].

5. The Local Area Network is connected by using a VME-Ethernet interface [Ref. 7]

6. The Local Trigger Unit (LTU) [Ref. 8] manage all synchronization between hardware trigger signals and the read out tree.

7. Finally, a diagnostic module [Ref. 9] allows a fast detection of most hardware failures and bad connections . It provides also the necessary arbitration for the VME modules in the crate.

7. CONCLUSIONS

The layout adopted for the OPAL data collection system is an attempt to utilize industrial methods and techniques like **VME** in High Energy Physics . The main goals for such a large experiment are to be:

- SIMPLE minimize the number of different hardware modules and fits to only what it is required by avoiding complicated software , too general protocols, cooling and crates supply difficulties etc ..... .

- FLEXIBLE able to handle any type of DATA BUS.

- EVOLUTIVE from the test to the final configuration.

- RELIABLE by using essentially devices commercially supported by the industry.

- UPGRADABLE as the microprocessor technology evolve.

- CHEAP by using widely commercial devices (VME).

- REALISTIC ready in time and compatible with the available manpower.

References

1. The OPAL detector technical proposal (CERN report 1984)
2. FIC8230 by CES Creative Electronics System 70 rue du Pont Butin PO box 107, CH 1213 Petit Lancy, Geneva - Switzerland.
3. Ph.Farthouat, DPhPE-CEN Saclay (F)
4. Ph.Farthouat, DPhPE-CEN Saclay (F)
5. CBD8210 by CES (see Ref.2).
6. FVSBI By M.Weymann,Univ.Freiburg (D)
7. LRT-Filtabyte (UK).
8. Ph. Farthouat, DPhPE-CEN Saclay (F)
9. VMDIS 800 3A by CES (See Ref. 2)

Multiprocessor systems in Fastbus

Alessandro Marchioro

CERN Div. EF
1211 Geneva 23, Switzerland

ABSTRACT

The Fastbus standard has introduced support for multiprocessing in HEP data acquisition systems. Several aspects of multiprocessing in this environment are examined, with emphasis on one practical implementation, the ALEPH read-out system for LEP.

1. INTRODUCTION

With the advent of cheap computing power, in the form of single chip microprocessors, the interest in the applications of distributed processing systems to real time computing problems has been increasing very fast.

The range of multiprocessing possibilities applicable to problems in physics is extremely large, and this paper is concerned merely with reviewing multiprocessing in Fastbus data acquisition systems, typically for High Energy Physics (HEP) environments. We will attempt to identify some typical structures that have been implemented using Fastbus, to highlight some advantages, restrictions and drawbacks that have been observed using this standard.

The requirements when designing a multiprocessor system (MPS) for HEP are quite common between experiments and can be stated as follows:

a) the system must be able to collect digital information in the shortest possible time (to reduce dead-time)

b) to reduce the information to a compacted form (to minimize the amount of significant stored data)

c) to highlight immediately the important pieces of information (peaks, shapes etc.)

d) to find complex correlations between several quantities (track finding, vertex reconstructions etc.)

Multiprocessing can therefore have several meanings, depending on the particular field of interest. It can be understood as a pure replication, at low cost, of computing elements, all running the same algorithm, to satisfy requirements a) through c). It can also be intended as a tightly coupled structure, where massive computing power is necessary to solve, in real time, a difficult problem, as in point d).

2. MODELING A FASTBUS DATA ACQUISITION SYSTEM.

What distinguishes a real multiprocessing system from a a mere ensemble of processors? Three quantities can be used to characterize a MPS in an ideal three-dimensional space:

- the degree of interconnection of the processors and the topology
- the inter-processor control
- the relative size of the processors.

It turns out that the interconnection network is perhaps the most important factor and it imposes severe restrictions on these three variables. We will try to focus how Fastbus fits in this multi-dimensional space.

2.1. Multilevel organization.

Data acquisition systems of modern HEP experiments have many similar characteristics. On one end, one finds a detector with a very high number of electronics channels. On the other, a mass storage device, associated with a given off-line processing capability, determines the maximum amount of data per unit time, and hence the processing capability of the system. Once the Fastbus standard is chosen, two major problems must be solved by the designers of a data acquisition system:

- determine the architecture of the system
- determine how the architecure can be implemented starting with bare modules.

The architecture of a system is determined by the relationships between elements. The dataflow in the HEP case is asymmetrical towards a concentration point and is highly correlated in time. This characteristic has led many designers to opt for a multistage model. Horizontal levels of data collection and reduction have been introduced, implementing a high level of parallelism.

Within an architecture, an allocation scheme must be defined in order to:

1) decompose the problem into a workable set of processors
2) provide a configuration that is modular and expandable
3) minimize the interprocessor communication
4) provide a structure that is replicable in parallel.

Several units with different logical functions can be introduced in a Fastbus system to support a complex read-out protocol. The terminology used here is largely borrowed from the one introduced in the ALEPH experiment, but very similar functions can be recognized in many other systems.

3. THE THREE LEVEL MODEL IN ALEPH

3.1. The lowest acquisition level: Read Out Controllers.

Typical modern HEP experiments and projections for future accelerators and detectors ask designers to cope with a number of input channels from several hundred thousands rapidly approaching the million. Also the volume of detectors is growing fast, giving an enormous number of empty channels per event to be discarded. At the lowest level of multiprocessing, the so called 'Read Out Controller' (ROC) level, data from digitizers are grouped and compacted by a processor sitting directly above them. These processors are given the job of discarding empty channels and of crunching data to a compacted form. They also run background tasks, dedicated to control stability characteristics, detector check-out etc. Some of them are more intelligent incorporating higher level functions, such as signal processing for shape analysis, either in hardware or by specialized coprocessors. Conversely, it is not necessary to implement general purpose programmable processors at this level. Some experiments have opted for hardwired solutions (i.e. scanners etc.), but the logical function still remains the same.

Read out controllers are largely detector dependent. They are optimized for a particular kind of front-end electronics, or for a certain type of data handling (fitting, data adjustment, corrections to the data etc.).

The most critical factor at the controller level is response time. The only way to avoid dead time at this level is to use multiple input buffers in the digitizers in the form of memories. This is not a cost effective solution, because buffers must be multiplied by the number of channels and they are anyway almost always empty.

A large number of variables determines the design of this stage, but detailed analytical theories and simulation tools exist in queuing theory to model it (Ref. 1).

What determines the choice of a certain standard at the digitizer level is the cost of the electronics per channel. The computing power necessary at the ROC level usually contributes a small fraction to the overall cost, especially if some reduction algorithm (like zero suppression and pedestal adjustments) are implemented in hardware.

Fastbus is a very good choice for this level because of its big board size, allowing a large number of channels to be packed per board, with a minimum cost per interface connection. Furthermore, no commercial modules are normally available for the requirement of HEP digitizers, and therefore no direct, industrially mass-produced, competitor exists.

The degree of utilization of the bus at this level can be compared to an I/O bus, because the majority of the modules are just slaves and a very small amount of multiprocessing is required.

3.2. The intermediate level: Subdetector Event Builders.

The next stage of the problem consists in defining an effective way of emptying the controller memories from the data they have collected. At this point the definition of a read out protocol becomes necessary and the transport medium must be effective in implementing the protocol. Broadcast messages at this level are very useful, avoiding the burden of addressing typically dozens of controllers sequentially to distribute the same piece of information. The Fastbus broadcast protocol is ideal at this point. Another useful feature in Fastbus is the T-pin scan read mode. It is a special type of broadcast read operation, allowing to poll an entire segment for data, in one operation.

3.3. The top level: Main Event Builders and dataflow control.

Subdetector data must now be grouped in one single place and resynchonized. This is achieved by the highest collection level, the main Event Builder. Having collected an event for an entire subdetector, some computing power is necessary to work on it. Typically data must be re-formatted and monitored. The protocol between the intermediate and the third level, can be the same as for the levels below. A complete event becomes available for analysis.

The highest level contains also a system that distributes the trigger information to the front-end taking care of the status of all front end modules (busy-free). The 'Trigger Supervisor' performs this activity. The Trigger Supervisor uses both fast, directly connected, hardware signals to distribute time critical functions (start digitization, synchronize timing) and Fastbus, to distribute control information.

Depending on the read-out protocol used, the Service Request Handler (SRH) is an important part. It is used to convert a very low level information as the Fastbus Service Request interrupt line, to a more informative message for a Service Request Server.

In first approximation all these logical blocks can be implemented in different physical modules. Conversely, it is not absolutely necessary to implement all functions on different modules. It is perfectly conceivable that a given computing element, with enough hardware flexibility, can support multiple functions.

3.4. Hierarchy.

A strict hierarchical approach has been used in ALEPH to define the communication between processors. The protocol between levels is based on a master-slave relationship, where the slave can only do what it is told to. There is no connection within a given horizontal level. This scheme also allows easy expansion of the tree, if necessary. New horizontal layers can be added at any stage, using just one module respecting the master-slave correspondance.

The asynchronous pipeline operation has been introduced in the ALEPH dataflow model, to maximize the system's throughput. It smoothes local peaks and provides the shortest latency in treating input data. Its principle is very simple: all along the data readout tree, events don't need to be synchronized horizontally. If enough buffering is provided through the pipeline, the readout scheme can be adjusted dynamically to serve read- out processors in order of arrival, instead of using a fixed read-out sequence. The system readjusts itself automatically to the slowest processors (those with more data) and a full utilization of parallel resources is achieved. Clearly the top level must resynchronize the information, and this is a relatively simple job, if queues are kept strictly sequential through the pipeline. Fig. 1 illustrates its behavior.

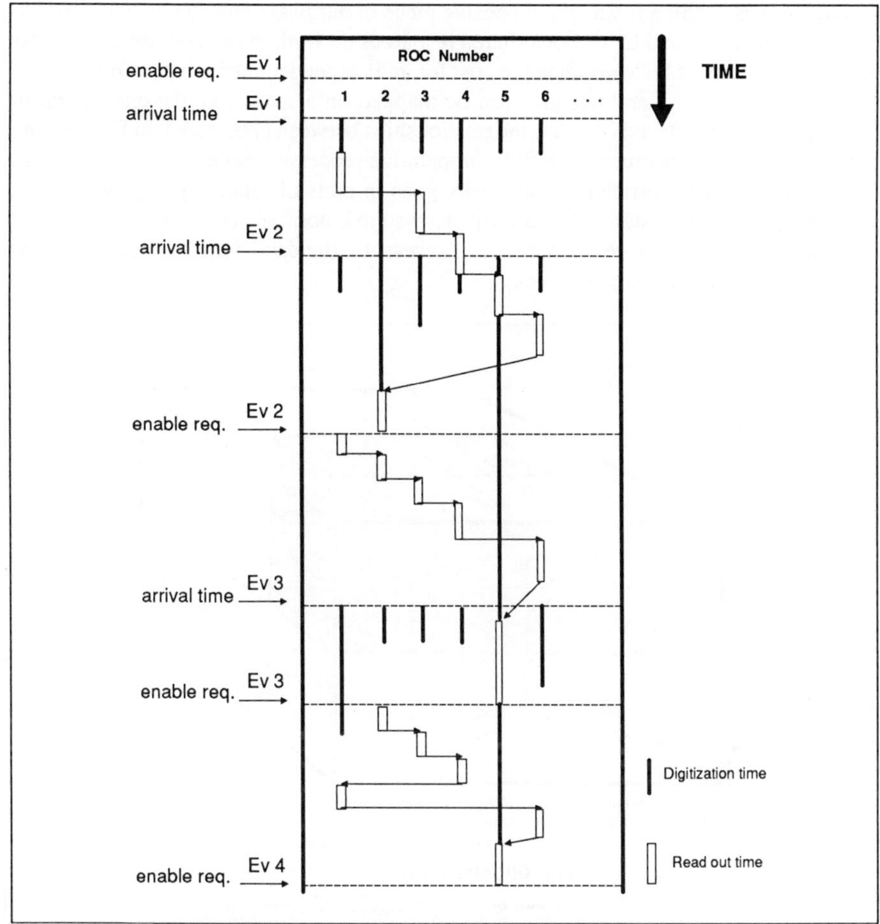

Figure1. The asynchronous pipeline model in ALEPH

4. MULTIPROCESSORS IN FASTBUS

As seen from the previous case, the typical connectivity within an horizontal data acquisition layer is rather loose. In principle, a better utilization of resources could be achieved by allocating dynamically processors within a given layer, or even at the overall system level. This approach has not been implemented in any Fastbus system because it would introduce a much higher level of complexity in the design. At the moment, brute force seems to be the preferred strategy of designers: if computing power is not satisfactory at one point, more processors will be inserted locally. Horizontal communication introduces a communication overhead higher that the benefit of sharing resources and requires a more complex and costly network structure. An horizontal segment connection in Fastbus is still a relatively expensive piece of hardware and can be compared to the cost of an additional CPU. For a shared resources network to be economical, the cost per connection must be lower than the cost for an independent new computing node.

Multiprocessing architectures can be mapped on a space like the one shown in Fig.2 (Ref. 2.). On one axis one has the relationships between processors: in Fastbus one can not achieve architectures with fully cooperative processors because access to shared resources is not very efficient. Granularity goes in units of relatively big blocks, like complete processors systems. The control of a system is normally centralized, with possibility of simultaneous access from many processors; the database itself can be spread over a number of control computers.

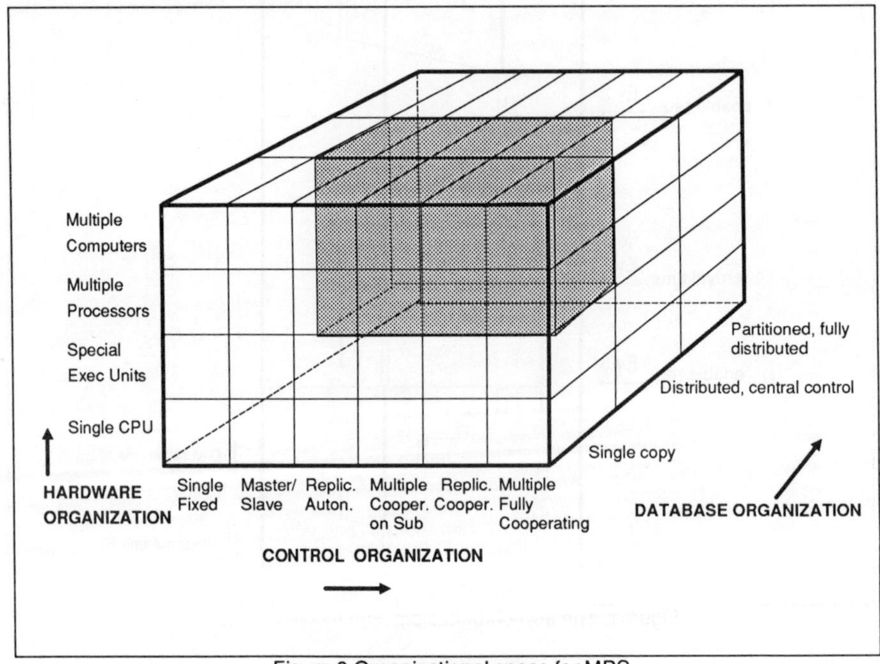

Figure 2 Organizational space for MPS

4.1. Interprocessors communication

Processors in Fastbus share a common backplane (Fastbus segment) and they can communicate using a message passing mechanism. Fastbus interrupt message are very useful at this stage. Nevertheless, the Fastbus protocol to transmit interrupt messages to CSR#100 space provides still a relatively primitive mechanism. A real message passing mechanism would require a higher level of definition in the standard. Finally, mailboxes outside a master are not easy to implement is Fastbus.

4.2. Limitations in Fastbus

In a complex real time system, interrupt messages need to have different priorities while traveling on the bus. A single master in Fastbus is allowed, strictly speaking, only one arbitration level, independently of the function it performs. Therefore a given message can not have an arbitration level attached to the importance of the message, but just to the preassigned priority scheme.

The only way Fastbus slave units can gain attention is by pulling the Service Request (SR) line high. One single interrupt line is a serious limitation. The SR line in Fastbus also has very restricted routing possibilities. In a configuration with two masters and several slaves on the same segment, it is not possible to route a SR from a given slave, to its master. Enabling SR servicing on both masters would lead to a contention situation. Furthermore, in case of real concurrence, i.e. with more than one slave has need for its master at the same time, no priority can be established. An interrupt bus with multiple prioritized lines and programmable sources and destinations would be a better solution. A SR Handler function must be introduced to solve this problem: this is a costly and unnecessary complication.

Multiple logical functions are often loaded in the same module as separate tasks with Fastbus access. In a real time system, different tasks will have different priorities. At the lowest level the bus access priority should be related with the task priority. Fastbus provides only one arbitration level per module, and this is sometimes a limitation.

5. COMPARISON OF FASTBUS AND OTHER BUSSES.

A comparison between Fastbus, VME and DEC's VAXBI™ is given in Table 1. VME has been considered for its popularity in several systems and the excellent industrial support. VAXBI is an example to show how the computer industry would engineer a high performance bus for multiprocessor systems. A few points are worth mentioning:

- Interfacing master and slave devices to Fastbus still requires a major effort from the designer. An easy to use, highly integrated interface, using available state of the art technology, would contribute significantly to the popularity of Fastbus.

- Protocols: a high performance multiprocessor architecture needs a standard message passing system to be defined. The Fastbus user community should urge the standard definition of a high level protocol to effectively support multiprocessing.

What are the major differences that can be recognized between the three standards? The fact of not having a native processor made for the bus, perhaps characterizes Fastbus in the most typical way. Today's micros are able to perform bus cycles in typically 100 ns. The Fastbus access, instead, has been defined with a set of relatively heavy library calls, very difficult to implement in one machine instruction. Typical single word access are of the order of one microsecond, or longer, as seen from the CPU on the Fastbus master. This rules out the possibility of making fast access to shared memories, and a preferred architecture by many designers, has been to put private memories inside their masters, and let Fastbus be only an I/O bus.With respect to the general organizational space, as defined in a previous paragraph, Fastbus occupies a subspace characterized by

Table 1 Comparison of multiprocessor bus systems

	Fastbus	VME	VAXBI™
multimaster	yes	yes	ye
multicrate	yes	no	no
native processor	no	yes	yes
technology	ECL	TTL	TTL
interface implementation	difficult	easy	easy
operation	async	async	sync
mux add/data	yes	no	yes
block transfer	yes	no	limited
distributed arbitration	yes	no	yes
speed (Mb/sec)[1]	40	16	10
parity	yes	no	yes
broadcast	yes	no	yes
standard control registers	yes	no	yes
interface self test	no	no	yes
self configuration	no	no	yes
interlocked access	limited	no	yes
cache support	no	no	yes
interrupt bus	no	yes	no[2]
high level primitives	no	no	yes

[1] these numbers are typical, not best cases
[2] VAXBI uses the address/data bus to transmit interrupt information

loose coupling, coarse granularity and light interprocessor control. With current designs, it is not evident, how a set of Fastbus processors can really partecipate in working onto the same event (on shared data), unless a special private bus is used to bypass it. Shared memory schemes are much more easily implemented in VME. On the other side, again the large board area in Fastbus, allows designers to pack more processors and memories on one board, making the need to share external units less dramatic.

6. PARTITIONING A FASTBUS DATA ACQUISITION SYSTEM.

Big detectors contain usually several subdetectors, each dedicated to a specific aspect of the physics expected in an experiment. During setup and calibration phases, and sometimes during normal data taking, detectors may be operated independently. More than one data taking activity must run at a given time. This leads to the situation where two or more independent 'partitions' of the system are running concurrently over the same network. Here a partition is defined as a complete set of logical blocks necessary (ROC, Event Builders, SRHs etc, as defined above) to collect data from digitizers and store them on a storage medium.
Under such circumstances, the designer will first spend some time in structuring his 're-

source allocation scheme. In a Fastbus partition, resources are determined by the availability of:

- physical modules
- routing schemes
- computing power
- database access
- storage controllers.

Physical modules include all kinds of devices that can be uniquely assigned to a partition, like front end digitizers, ROCs, hardware synchronization boxes, cables for fast signals distribution etc.

The second category, routing schemes, includes instead logical connections between modules of a partition achieved by setting appropriately programmable devices like Fastbus Segment Interconnects or programmable signal distribution boxes (as the Fan In Out module in ALEPH). Two partitions may use two completely independent networks or they may share several branches of the network.

Resource allocation on computing power concerns the assignment of a given amount of CPU time to the different stages of a partition. Sharing of a physical module between different partitions is a very likely event. The way different partitions work is very similar, even if they are dedicated to different parts of the detector: a data driven scheme is advantageous in this situation.

Fastbus supports very nicely the idea of partitioning: broadcast classes are available for fast distribution of messages and the Fastbus network can be reasonably partitioned by properly programming Segment Interconnects.

7. RELIABILITY IN FASTBUS.

Unlike other industrial standards, reliability is of moderate concern for physicists. Failures of electronics are accepted as a fact of life and no intrinsic safety mechanism is built easily in Fastbus systems. It is also true that a serious study of this aspect for large future HEP experiments has never been undertaken, and such a study will be useful for properly engineering future experiments. No explicit support for redundancy is built into Fastbus, and this may turn out to be a severe limitation. Error detection and correction mechanism are also rather poor in Fastbus (one single bit of parity).

8. ACKNOWLEDGMENTS

I would like to thank all the members of the ALEPH Dataflow Group, in which a major part of the design work has been done; they should have the credit for many of the ideas presented. Also I would like to especially thank J. Harvey and W. von Rüden for many useful suggestions.

9. REFERENCES

1. Marin, J., "System Analysis for Data Transmission", Prentice Hall
2. Enslow, P. H. Jr., " What is a Distributed Data Processing System", Computer, Jan. 1978.

USE OF DIGITAL SIGNAL PROCESSORS (DSP) IN HIGH ENERGY PHYSICS EXPERIMENTS

Dario Crosetto
Istituto Nazionale di Fisica Nucleare
Via P. Giuria 1. - 10125 TORINO
ITALY

ABSTRACT

The FDDP - Fast Digital Data Processor - is a modular system for executing parallel digital processing algorithms to perform programmable trigger decisions or programmable on-line data reduction. Typical application involve zero suppression and pulse shape analysis. The characteristics of the system are: modularity, expandibility and flexibility.

INTRODUCTION

The advent, in recent years, of fast analog to digital converters together with ever faster and less expensive Digital Signal Processors is leading to a considerable change in the 'trigger decision box' and in the readout philosophy of fast detector signals.

For both tracking devices and calorimeters, the conflicting trends toward higher luminosity and improved background rejection call for the exploitation of hierarchical, multi-level trigger logic. For the higher trigger levels, depending on the different machine and detector confi gurations, it is sometimes possible to design the 'trigger data box' as a programmable device, with increased flexibility and efficency. It is then conceivable to use large arrays of DSP's working in parallel, obtaining excellent, inexpensive data throughputs, provided that fast and efficient software algorithms can be implemented.

Parallel analysis of readout signals from detectors, has proven to be an efficient way to reduce (zero suppression) the tremendous data flow generated by high speed Flash ADC's in real time. Whereas in the past vast use has been made of highly specialized electronic modules peforming analog signal filtering, shaping and storage followed by an analog to digital converter, we nowdays often find situations where even the details of a pulse shape are treated in digital form from the very beginning of the readout.

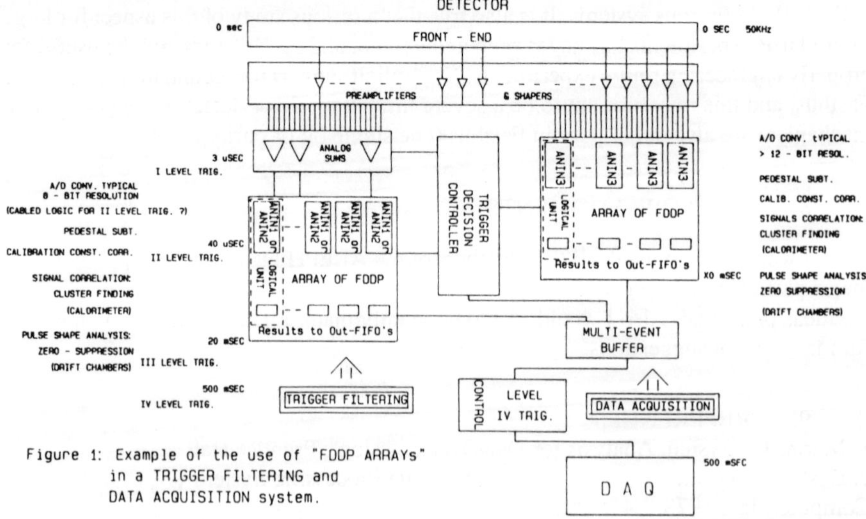

Figure 1: Example of the use of "FDDP ARRAYs" in a TRIGGER FILTERING and DATA ACQUISITION system.

Fig. 1 illustrates a typical "trigger filtering" and "data acquisition" scheme in a modern detector with fast signals that are being filtered and reduced in real time by programmable software algo rithms in arrays of FDDP.

The two FDDP arrays equipped with different interfaces to the front end electronics (ANalog INput boards ANIN1 or ANIN2) due to their low cost, are well suited to be applied as closely as possible to the front-end electronic signals. More powerful programmable devices (such as more advanced DSPs or more powerful computers) instead will better solve the task of "trigger decision controller" or "event builder".

The task of the FDDP trigger filtering array is to correlate signals (cluster finding in calorimeters)[3] or pulse shape analysis for energy calculation in drift chambers. The most important parameter is the speed requirement that is satisfied by:

- reducing the number of front-end signals with analog sum circuits,
- the high parallelism of DSPs,
- some additional cabled logic on the ANIN board.

The task of the FDDP data acquisition array is the "data reduction" among all front end signals that have been converted by analog-to-digital converters with higher resolution than that of the trigger system. The digitizing of pulse shapes has several advantages over simply recording drift time and charge: at the same time it provides a good energy determination and accurate timing which is less sensitive to diffusion than conventional threshold electronics, and furthermore the details of the pulse shapes permit a double track detection.

FDDP ARRAY CHARACTERISTICS

Fig. 2 illustrates an array of FDDP implemented for the II level (energy calculation) and III level (cluster finding) trigger filter of the Forward ElectroMagnetic Calorimeter of DELPHI[1].

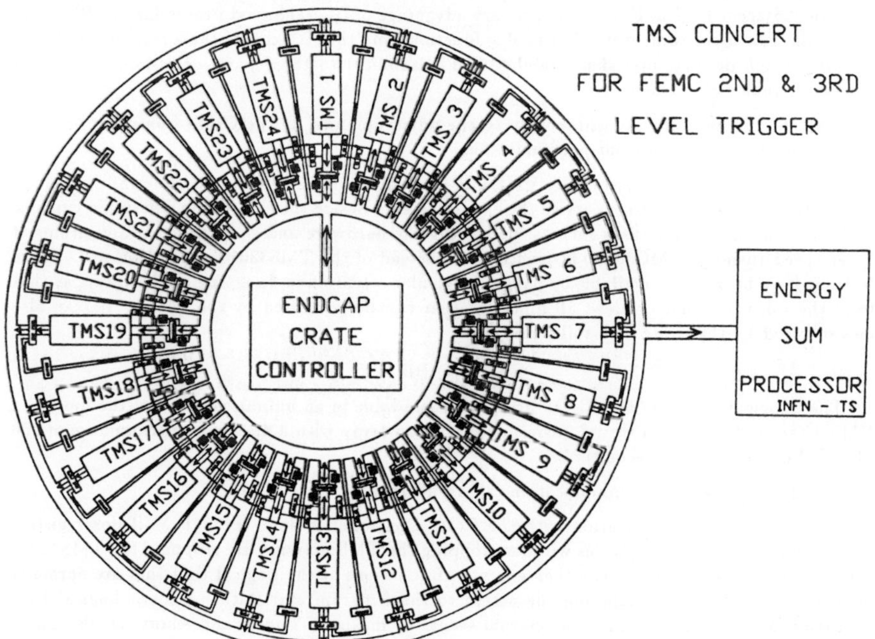

Fig.2

MODULARITY

The FDDP[2] is a modular system based on "Logical Units". The block diagram of a "Logical Unit" is shown in fig. 3.

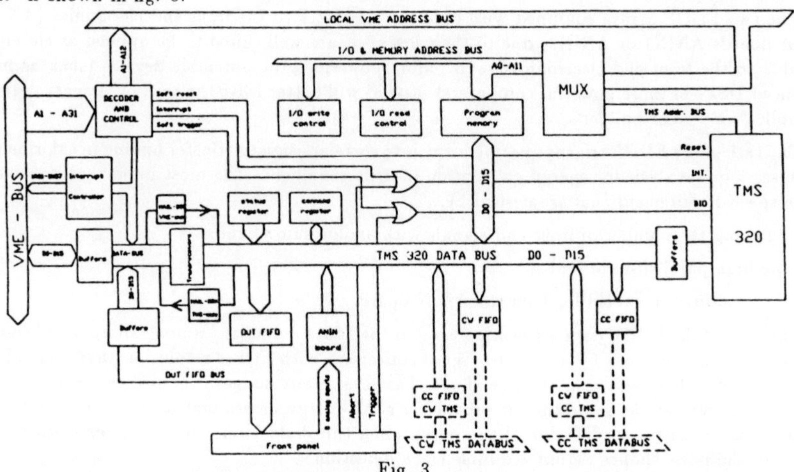

Fig. 3

The main features of a Logical Unit are:

- it can fetch data via its own Input/Output or from an adjacent Logical Unit.

- it has a dedicated TMS32010 DSP that processes the data fetched from the Input/Output or from the adjacent Logical Units. The primary advantage of Digital Signal Processing (DSP) is that it uses discrete algorithms instead of analog functions and thus it extends the advantage of digital circuitry (high density, precision, stability and testability) to those parts of systems previously served by analog components.

- it can communicate asynchronously to the adjacent units, to the Crate Controller or to an external device through a front panel data bus.

A Crate Controller processor (any VME CPU board, see fig. 2) is required to downline load the TMS32010 programs and to supervise the entire system. Beside this, the Crate Controller can also do some other important tasks such as: debugging the hardware and simulating the algorithms at lower speed (using the MC68000 instruction set instead of the TMS32010 instruction set). These features have been made available by paying particular attention in designing the hardware so that from the software point of view, all functions that can be generated by the TMS32010, can also be generated by the Crate Controller processor.

EXPANDIBILITY

The "Logical Units" in the FDDP array are cascadable in an unlimited number (the limit in a VME Crate is 40 "Logical Units" corresponding to an array with a maximum performance of 200 Million Instructions Per Second).

The different types of communication (all asynchronous) are:

- Logical Unit communication with the Crate Controller is done through Mail-Box registers and through Out-FIFO memories without stopping the TMS activity. By stopping the TMS32010 activity, the Crate Controller can then execute all functions at the Logical Unit that are normally done by the TMS32010, including the access to the Program and Data memory. Logical Unit processes can be synchronized with external events by means of Trigger and Abort signals. These signals can be simulated by software from the Crate Controller.

- Logical Unit intercommunication done through FIFO memories.

- Logical Unit communication with External Input Devices done through an optocoupled bus to the ANalog INput board interface.

- Logical Unit communication with External Output Devices done through FIFO memories on a 16-bit three-state bus.

FLEXIBILITY

In order to give more flexibility to the system, the ANIN interface between the Logical Unit and the detector, that is detector dependent, has been built on a separate board, giving thus the possibility to design different personalized analog modules to fulfill better the application needs.

There are essentially two types of application: either a single digitalization of the signal is needed or the entire shape of the signal is required.

In the first case (typically of a Calorimeter, see fig. 4) a single register is sufficient for each input line. The task of the DSP is to process the data (e. g. subtract pedestals, correct with calibration constants, sum data, etc.) and eventually correlate the data with neigbor Units by means of clockwise and counter-clockwise FIFO's.

In the second case (typical of a Drift Chambers, see fig. 5) the ANIN samples the analog input at given rates and store results in data memories. The task of the DSP is to read the memory and make a pulse shape analysis (taking derivatives of the pulse, finding the maximum, minimum, time duration or center of gravity, subtracting pedestals, etc.).

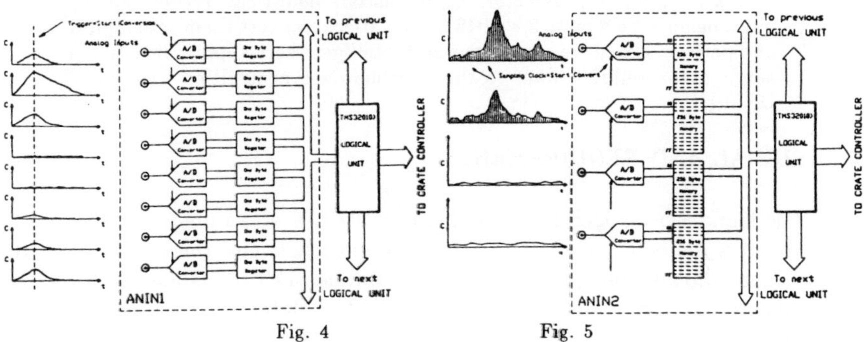

Fig. 4　　　　Fig. 5

Other characteristics of the ANIN board are:

- by designing an ANIN board more or less complex, it is possible to select the number of analog channels that will be associated to each DSP.

- being structurally independent from the digital mother board and with independent power supply lines, all analog circuits can be optically isolated from the digital circuits.

VME BOARD CHARACTERISTICS

The completely assembled FDDP board in VME version contains two Logical Units that consists of a "mother board" hosting two TMS32010 processors, program memories, buffers, registers, ancillary logic, etc. and two piggy back cards called ANIN where are located the analog circuits (operational amplifiers, Flash ADC, optocouplers, etc.) for the data acquisition.

REFERENCES

[1] D. Bulfone et al., "Proposal for the FEMC trigger" DELPHI Internal note 86-100 (1986).

[2] D. Crosetto, "FDDP - Fast Digital Data Processor" CERN Internal Note EP/87-151 (1987).

[3] V. Roveda and A. Werbrouck, "Cluster analysis with a fast microcomputer" Accepted for pubblication in "Microprocessing and Microprogramming (1987).

[4] D. Crosetto, E. Menichetti, G. Rinaudo and A. Werbrouck, "Parallel arrays of DSP as central decision elements for upper level triggers". IEEE Nuclear Trans. (1988).

SIROCCO IV

FRONT END READOUT PROCESSOR FOR DELPHI MICROVERTEX

Nils Bingefors[†]
University of Milano
Via Celoria 16, I–20133 Milano
ITALY

Mike Burns[‡]
CERN, EP
CH–1211 Genève 23
SWITZERLAND

DELPHI Microvertex Group

ABSTRACT

The SIROCCO IV is a Fastbus front end module for the DELPHI Microvertex Silicon-Strip Detector. Each Fastbus board can receive analog pulse heights from up to 2 × 2048 silicon strips, convert them into digital and perform extensive corrections and intelligent zero-suppression before sending the reduced data to the Fastbus data acquisition system of DELPHI.

1. SIGNAL AND REQUIREMENTS

The signals from each individual silicon strip are sampled twice by a VLSI bonded to the detector plaquette. The difference between the samples is time-multiplexed onto an analog bus. Each VLSI handles 128 strips and up to 16 chips are connected in series. In total the Microvertex Detector contains over 55000 channels.

A maximum of 2048 pulse heights are received by one SIROCCO IV input. These must be digitized to 10 bit accuracy and stored in a Front End Buffer at a rate of approximately 4.5 million words/s.

As most of the silicon strips will have no signals from particle tracks it is desirable to suppress these empty channels before sending the data to the data acquisition system. In order to do this, the data must first be corrected for offsets and pedestals, and then a cluster search can be performed. The time allowed for this processing is less than 20 ms for 2048 channels.

---

† On leave from University of Uppsala, Sweden.

‡ Invaluable contributions to the concepts of SIROCCO IV were also given by Roland Horisberger and Jean-Pierre Vanuxem, CERN.

2. ARCHITECTURE

SIROCCO IV is structured as two paths for information, one for silicon strip data and one for event identification, both with two levels of buffering.

A block diagram of one SIROCCO IV channel can be seen in figure 1.

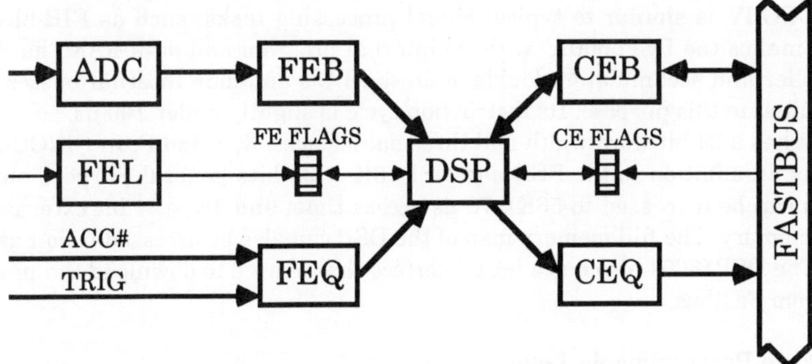

Fig 1. Simplified block diagram of SIROCCO IV.

Analog data is digitized in the ADC and then stored in one of the four Front End Buffers (FEB) of 2048 × 24 bits. The Front End Logic (FEL) controls this operation and sets one buffer flag in a four bit flag register when the buffer is filled. FEL also enters trigger information and the account number (ACC#), an 8-bit number given to each event, into the Front End Queue (FEQ).

The Digital Signal Processor (DSP) can read both of these buffers, process the event and send the reduced event record to one of four Crate Event Buffers (CEB) of 2048 × 32 bits. In the Crate End Queue (CEQ) the processor enters start and end addresses for the event record.

A four bit flag register, indicating filled and empty Crate Event Buffers, completes the handshaking logic between DSP and Fastbus.

3. IMPLEMENTATION

Two SIROCCO IV channels are implemented on a Fastbus board, leaving spacc for two add-on ADC-boards and the Fastbus interface. This requires the use of compact logic design.

3.1 Digital Signal Processor

The digital signal processor, DSP56001 from Motorola[1], was chosen for its very high computing power and for the ease of interfacing to external circuitry that is built into its design.

Much of the processing to be performed by a front end processor like the SIROCCO IV is similar to typical signal processing tasks, such as FIR-filtering, which makes the DSP56001, with its internal program and data RAM, hardware multiplier and accumulator, double address ALU's and four internal busses, very well suited to this purpose. Its instruction cycle is slightly under 100 ns.

It has a 24 bit word width and three data spaces, X, Y and P. In SIROCCO IV there is, in addition to the FEB and CEB, 8K × 24 bits general purpose memory (which can be increased to 56K), 70 ns access time, and 4K × 24 bit external program memory. The full memory map of the DSP can also be accessed from Fastbus.

The DSP56001 also has a host interface that is used to download the program code from Fastbus.

3.2 Fuse-Programmable Logic

Almost all random logic in the SIROCCO IV is implemented as fuse-programmable logic, for reasons of speed, flexibility and, above all, board space. Devices of various complexity from the Signetics line of PLD's[2] are used. Simple gate functions are grouped together in a couple of PAL's (PLHS18P8), memory and I/O decoding and bus request/grant priority encoding are implemented in two PLHS473's, the FE and CE flags occupy two PLS155, the Fastbus CSR's are built with PLS173's.

The front end control logic is programmed into a Programmable Macro Logic unit (PLHS501) saving in the order of 20 chips of board space.

4. SIGNAL PROCESSING

The signal processed by the DSP is a two dimensional, discrete space, discrete time, discrete amplitude signal. The spatial dimension is along the detector circumference (two concentric cylinders make up the Microvertex Detector) where it is sampled in ϕ (DELPHI-coordinate) at a 50 μm pitch. The time samples are taken at beam crossings whenever there is a good T2 trigger (unequally spaced time-samples). The digitization resolution is 10 bits.

An offset for each VLSI-chip is calculated for each event to form a local base line. This offset is found as the maximum in the amplitude distribution or as the median amplitude.

After offset subtraction a pedestal, kept in memory for each strip, is subtracted. This pedestal must then be updated to follow drifts in the detector and VLSI.

The next step is to search for clusters. A particle hit will usually leave charge on several adjacent strips. The probably best way to separate such a signal from noise is to apply an FIR-filter matched to the shape of the expected cluster.

Remaining to do is the formatting of the event record and to extract some statistical quantities from the rejected data. The latter is necessary in order to monitor the detector performance.

It should be noted that processing that can appear trivial in an off-line or Monte Carlo environment can become intricate and critically sensitive when it has to be performed in the front end, in real time. A missadjustment of a threshold cut can either flood the DAQ system with data or the interesting data will be lost irreversibly.

Some demands can thus be made on algorithms for front end processing:

- Possible to code efficiently — both with regard to execution time and program memory, sometimes conflicting demands.
- Predictable and stable — the algorithms should be closely guarded against any instabilities or oscillating tendencies as there is no possibility to reprocess the reduced data.
- Should not distort output data — badly calculated offsets or pedestals could easily distort the important information in the data: position and amplitude of a particle hit.
- Robust — small changes in the running conditions should not change the characteristics of the algorithms, the sensitivity to errors in parameter estimates should be low.
- Self diagnostic/adjusting — if, in spite of the above, errors occur or the algorithms drift into bad regions this should be immediately detected and flagged or adjusted. It should not be possible for distorted data to pose as good data.

Great care has to be taken to fulfill these demands in a compact code.

5. REFERENCES

1) Motorola. DSP56000 Digital Signal Processor User's Manual (Preliminary), 1986, and DSP56001/D Data Sheet, 1987. Contact: M. Pierre Detraz, Motorola Semiconductors, 1211 Genève 20, Switzorland.

2) Philips. Programmable Logic Devices Data Handbook IC13, 1987. Contact: Field Application Engineer Giovanni Violi, Philips S.p.A., Piazza IV Novembre 3, 20 124 Milano, Italy.

3) Bingefors, N. et al, "SIROCCO IV — Hardware and software manual", DELPHI Note under preparation.

A PARALLEL PROCESSOR ARCHITECTURE

FOR IMAGE PROCESSING

S.Ashokkumar, M.Periasamy, M.V.Pitke*

CENTRE FOR DEVELOPMENT OF TELEMATICS
71/1 Miller Road,Bangalore,India

1. INTRODUCTION

Image Processing problem is generally treated as a complex one because of its data quantity and the nature of data or picture elements (pixels) which is so obtuse,i.e the value of any picture element may have little meaning. Meaning is often determined by evaluating a pixel in relation to the value of its neighbours. Furthermore spatial considerations,pattern recognition and the processing operation of neighbour elements complicate the process of determining meaning from a digitized image.

Image formation and enhancement in Radio Astronomy is one the interesting problems in image processing applications. The data in this application is collected from synthesis telescope and the image of the sky is reconstructed. The physical limitation of the telescope and non uniform sampling causes distortion in the picture. Interpolation technique does not really help since the correlation function between adjacent samples is not known. Hence the image formation and enhancement involves gain corrections,griding,mapping and CLEAN algorithm. The main computations of the algorithm involves the 2 dimensional FFT and the search for maximum. The data size is generally assumed as 1K X 1K points on which the computation is iterated. The quantity of data in this kind of application does warrant for parallelism which is anyhow inherent in the problems.

The proposed architecture is based on MIMD principle using 256 Processing Elements(PE). Each processing element is designed to get the processing power of 2.5 MFLOP using off-the-shelf components. The processing elements are interconnected through a self steered network to achieve high communication throughput to match the computation rate of processing elements.

2. ARCHITECTURE

The system is designed to support upto 256 processing elements(PE). Each processing element consists of 32 bit processor with internal cache, memory of 1Mbyte, high speed serial I/O and IEEE standard Floating Point (FP) unit. The PEs are interconnected through a non-blocking self steering switch with its high speed I/O (8Mbps). The Main Controller(MC) controls the switch, the interface and the function of PEs.It has memory of 32Mbytes, interface to mass storage of 1Gbytes and an Ethernet interface to the Host. This is also provided with high speed graphics capability. The parallel processor machine of the above architecture is housed in a single cabinet of 6 feet, having six card frames. Each card frame has 26 slots for 10" x 12" cards with spacing of 30 mm. The system consumes 1.5KWatts power which does not need any special cooling arrangement.A simplified block diagram of the system architecture is shown in (fig 1).

3. PROCESSING ELEMENT

The processing element has been designed independent of architecture. It is mainly based on floating point capability to handle both 32 bit and 64 bit integer and floating arithmetic respectively. High speed inter processing elements' communication with error control is essential to carry on the computation with out waiting for data from other PEs.

---

* Also with Tata Institute of Fundamental Research, Bombay.

Sustained high performance is possible if the processor is used only for computation intensive job and not for I/O bound job. This is achieved in this architecture having four processing elements and a separate Communication Processor (CP) (fig. 2) for all four PEs. Dual port RAM is used between PE and CP to share the computed data. All four serial links of CP thus reduces the communication overhead of PEs. A single high speed link of PE is however used as a cheapernet interface with the main controller for program down loading and for maintainance purpose.

The Transputer T800 of INMOS is found suitable for implementing as processing elements in a single card. It has a 32 bit architecture,4kbytes high speed RAM with 64 bit floating point unit to provide 1.5/2.5 MFlops. It also has four 10/20 Mbps serial link for inter PE communication. It is of CMOS technology and consumes power less than 1 watt.

4. NETWORK DESIGN

The network has been designed with the following assumptions/constrains:

The PEs has a communication channel -a high speed serial link with the speed not less than 8Mbps. This link should provide the the minimum required ISO protocol standards. The interconnection structure of a high performance MIMD system with a large number of nodes must be free of blocking in order to avoid the con siderable performance degradation.

The reconfiguration of interconnection pattern should be fast so that message of any length could be communicated and the throughput rate of the network is maintained. This enables to share the computed data among PEs at faster rate and thus the computation rate and communication rate of PEs are optimised for higher sustained performance. So a self steered switch using Datagram Service may be a solution as against the conventional switching methods.

The network is based on a principle of Time-Space-Time switch (fig. 3), which consists of four modules.

i. Self Steering Logic (SSL)
ii. Time Switch Module (TS)
iii. Space Switch Module (SS)
iv. Time Slot Assigner (TSA)

5. MAIN CONTROLLER

The main controller (fig. 6) is built around Motorola MC68020 with its own memory, interface to winchester, and to the host. Two winchesters of 800MByte capacity are attached to the MC. These are for storing data and programs that are downloaded from the host through a standard an Ethernet interface. While the PEs are performing computation on one job the data and program for the next job can be downloaded from the host and saved in the disks. Similarly the communication between the main controller and the PE is through a cheapernet interface. This is mainly used to down load the program to PEs and occasionally used to report the diagnostics result of PEs.

6. SOFTWARE ORGANIZATION

As of yet not much experience has been gained in the realm of programming the large MIMD machines, as only a small number of experimental systems with rather rudimental software have so far available.

However, the software used in the discussed architecture is split into modules as follow. The operating system of main controller whose task is to handle the global system resource, to take care of the workload distribution and the system initialisation. The OS, however is not involved in the actual programs execution. The actual programs execution is the processing elements which has a mini OS, the floating point arithmetic library software.

A residential software for inter-processing communication protocol handler is provided at the communication processor. The basic application programs are unaltered, a new language layer is added on top of an existing language that discribes multi processing.

•

s.

7. PERFORMANCE COMPARISION

The performance of the system is compared based on the I/O and computing requirement for the application. The I/O path for data transfer is Disk to PE memory through MC memory. The I/O bandwidth of this path is 2.5MB/sec and 4MB/sec respectively. The amount of data to be transferred is estimated as 16MBytes which takes total time of 28 secs.

The time spent on computation is calculated as 4 sec taking 16MBytes data and 200 MFLOPS computation rate. So in effect the total time spent in operation is 32 secs, which is found more suitable for image formation and enhancement in Radio Astronomy application.

In addition the flexibility provided by the self steering switch makes the message passing suitable for inter PE communication to develop application program with a large number of processors. Since the proposed architecture provides the flexibility of inter-connecting the processing elements for various data sharing methods with its fast network reconfiguration capability, the scheme can be used not only for image processing application but also for other applications. The concept of the switch is also very useful for Fast Packet Switching applications.

References

1. Aperture synthesis with a Non-Regular Distribution Inter ferometer baselines J.A.HOGBOM ASTRONOMYS ASTROPHYSICS SUPPL.15 1973.

2.A Production Multi-Tasking Numerical Weather Prediction Control - T.K. Gibson, Computer Physics Communication 37 (1985), North Holland.

3.Supernum - A MIMD/SIMD Supercomputer for Numerical Applications - Wolfgang K. Giloi, GMD-TUB Research Centre for Innovative Computer Systems and Technology, Germany.

4.A Simulation Study of Decoupled Architecture Computers James E. Smith, Shlomo Weiss and Nicholas Y. Pang IEEE Transctions of Computers Vol. C.35, No.8 August, 1986.

5.The Architecture of SM3: A Dynamically Partitionable Multi-Computer System - Chaitanya K. Baru and Stanley Y.W. SU IEEE Transactions on Computers Vol. C.35 No.9 September, 1986.

6.U.K. Metrorological Office's Plans for using Multiprocessor Systems P.W. White and R.L. Wiley, Bracknell, U.K.

6.Introduction Data Communication and Computer Networks - Fred Halsall, Electronics System Engineering Series.

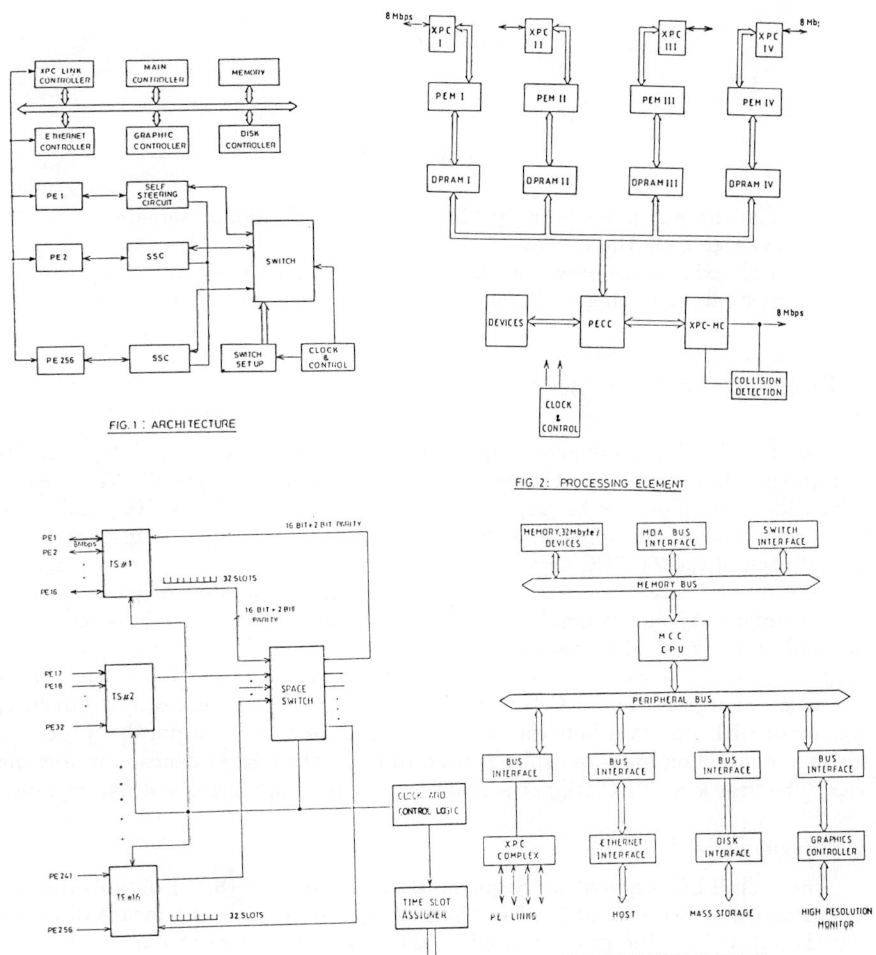

FIG. 1 : ARCHITECTURE

FIG. 2: PROCESSING ELEMENT

FIG. 3 NETWORK ARCHITECTURE

FIG. 4 MAIN CONTROLLER COMPLEX

Charged Particle Trigger for the L3 Detector

M. Bourquin, J.H. Field, G. Forconi
M. Nusbaumer, J. Perrier, N. Produit, H. Stone

University of Geneva

Abstract

The charged particle trigger in L3 will process analog signals generated by the central tracking detector, a drift chamber with axial wires based on the Time Expansion Principle (TEC). Systems are being built for the first- and second-level triggers.

1. First-Level TEC Trigger

The first-level TEC trigger searches for tracks pointing to the interaction region in the plane normal to the beam axis. A segmentation into 96 $R\phi$ slices has been chosen to allow for an adjustable $P_{\perp}$ threshold ($\sim$ 600, $\sim$ 300, and $\sim$ 150 MeV/c for tracks contained in 1, 2, or 3 slices, respectively). It implies a division of each of the 24 TEC segments into four sub-segments. The trigger decision is made on the total number of tracks found, on the number of pairs of tracks with acoplanarity angles smaller than a programmable value, and on more complicated topologies. The various trigger conditions and their combinations are programmable. To meet the time requirements imposed by the drift-time of the ionization charge in the TEC (up to $10\mu s$) and by the eventual 8 by 8 bunch operation of LEP (interval between bunch crossings of $11.1\mu s$), a parallel processing technique based on look-up tables is used to make the trigger decision in less than $1\mu s$. The first-level TEC trigger is realised with modules of three different types.

1.1 "Segment Divider" Modules

For each TEC segment a "Segment Divider" module (S.D.) divides the total drift-times into two bins and solves the left-right ambiguities by means of pick-up signals induced on the grids located at the right and left of the anode plane. A diagram of one S.D. channel is shown in Figure 1. The signal from the anode wire goes through a discriminator, the threshold of which sets the efficiency of the corresponding channel. The pick-up signals from the left and right grid wires are fed to a differential amplifier located in the front-end TEC electronics. At the entrance of the S.D. module, the output signal from the differential amplifier is discriminated for positive (left side) and negative (right side) pulses. Incorrect side determination can arise from large random fluctuations of the differential signal. This source of error is reduced by a coincidence between the anode signal and the left (right) differential pulse. The logic shown in the upper right part of Fig. 1 generates four output bits, *i.e.* 2 time-bins $\times$ 2 sides (left and right). The circuit in the lower right part of Fig. 1 contains a 20 MHz clock and defines the two time-bins. If the differential signal is below threshold, the left-right ambiguity is not solved, and the left and right bits are set to one. This generates one extra hit but no loss in track-finding efficiency.

The S.D. module has been tested in a beam with a TEC prototype. Fractions of left-right ambiguous hits smaller than $\sim$ 5% have been obtained. This can be seen from Fig. 2 which shows the amplitude distribution of the differential signal with and without tracks (the distribution is symmetric *w.r.t.* zero; only half of it is shown). The S.D. module does not add inefficiency to the analog electronics chain.

In each TEC segment, 14 wires out of 62 are used by the first-level trigger. They are distributed approximately as $\sim 1/R$, between $R = 178mm$ and $R = 413mm$. Thus, after processing by the S.D. modules, the first-level TEC trigger information consists of a 96×14 bit matrix.

1.2 "Track Finder" Modules

A flexible "Track Finder" module (T.F.) has been designed which is able to deal with different background conditions and TEC inefficiencies. The 96×14 input signals are used to address Random Access Memories (RAM) which contain the topologies of all relevant tracks. Track candidates or 'masks' are searched for in parallel in all 96 slices.[1] A schematic diagram of the T.F. module is shown in Fig. 3. Three types of tracks are found:

- Tracks with polar angle $42° \lesssim \theta \leq 90°$ and with transverse momentum $P_\perp \gtrsim 600$ MeV/c. This is realised with a 16K$\times$1 RAM addressed by the 14 bits of a single slice (second horizontal branch in Fig. 3). The RAM-output bit is set to one when the 14-bit pattern matches one of the predefined masks. The lower limit of 42° on the polar angle of the track is set by the TEC geometry. Tracks with smaller polar angle leave the TEC through the end-flanges and miss the outer-most wires.
- Tracks with $\theta > 42°$ and $P_\perp \gtrsim 150$ MeV/c. Such tracks cross up to three adjacent slices. The corresponding S.D. modules provide 5×14 bits of information, *i.e.* one times 14 bits from the "reference slice" plus two times 2×14 bits from the two adjacent pairs of slices. To keep the size of RAM memory to a manageable level it is necessary to reduce this information. We have chosen to group the 14 bits from a single slice, radially, two by two. The grouping is a programmable AND or OR, independent for each pair of bits (lower branch in Fig. 3). In this way, if background conditions are good all pairs can be set to OR to maximize trigger efficiency. If background conditions are bad all pairs can be set to AND so as to increase the rejection of spurious tracks. As a compromise, if the background is localised at small radii, the signals in this region can be set to AND, the others, at larger radii, to OR.

  In a given T.F. module, the only masks considered are those which correspond to tracks leaving the TEC in the reference slice. An example of a mask is shown in Fig. 4. Each hit pattern is associated with a mask. There are 29 different types of masks. Each type corresponds to a given $P_\perp$ range and a given sign of curvature. In practice 25 out of the 5×7 bits provided by the S.D. modules are sufficient to address all 29 masks. The amount of RAM per T.F. module is ~ 320 K-bits. Thus, in total, ~ 30 M-bits of RAM are used in the first-level trigger.

- Forward tracks with $\theta \gtrsim 25°$ and $P_\perp \gtrsim 100$ MeV/c. The search for small polar angle tracks is performed with the seven inner-most wires. No AND or OR is done. The system described above is applied seqentially in time, first to find the forward tracks ($25° \lesssim \theta \lesssim 42°$), and secondly to find central tracks ($42° \lesssim \theta \leq 90°$). However, as exactly the same masks are used in both cases, the $P_\perp$ cuts will be lower for the forward tracks. Because of the difference in the maximum drift-time between inner and outer wires, no time is lost in the sequential search. The search for forward tracks can begin after $5\mu s$, $\sim 5\mu s$ before the central track search.

1.3 "Track Adder" Module

The "Track Adder" module (T.A.) counts the total number of tracks found and the number of pairs of tracks with acoplanarity angle smaller than some programmable value. It takes the first-level TEC trigger decision and sends charged track information (24 bits indicating the presence of tracks in 24 $R\phi$ segments) to the other first-level trigger processors.

2. Second-Level TEC Trigger

The second-level of the L3 trigger system is based on fast programmable processors, XOP, which are designed to reduce the trigger rate from about 100Hz at the first-level by a factor of at least ten.[2] The XOP receive, in particular, the first-level TEC trigger data, which consist of the hit pattern in the $R\phi$ projection and of the result of the track search (list of masks), and charge and drift-time measurements from charge division wires. Each TEC segment is equipped with ten resistive wires read-out at both extremities. The hit position along the wire is determined from the charge asymmetry. Charge division resolved in time has the advantage of defining track coordinates in three-dimensions.

The second-level TEC trigger is based on a single type of electronics module ("Rz" module) in which charge and drift-time are digitized and stored (Fig. 5). The charge integrating electronics has a time resolution comparable to the signal widths (typically $100ns$) and multihit capability to ensure high detection efficiency. Such a circuit requires gates directly derived from the signals to be integrated and, in order to overcome their relatively slow reset time of more than $100ns$, the use of three integrators in turn as well as analog storage of accumulated charge. The integrators have been developped at CERN by A. Beer *et al.* .[3]

Tests done with "low" resistivity wires ($300\Omega/m$) lead us to expect a resolution $\sim 2\%$ of the wire length, *i.e.* $\sim 2cm$, with "high" resistivity wires ($\sim 2000\Omega/m$).

3. Readout

Both charged particle trigger systems are implemented in FASTBUS. The first-level TEC trigger will be housed in three FASTBUS crates, each containing 8 S.D. and 8 T.F. modules. Similarly, the second-level will be housed in three crates containing 20 Rz modules each. The trigger data are readout and transferred to the second level input memories by Lecroy 1821 Segment Manager/Interface modules (Fig. 6). A transfer speed of 0.5μs per 32-bit word has beeen obtained.

References

1. The same method was used in the CELLO experiment at PETRA. See H.J. Behrend, Comp. Phys. Comm. 22 365 (1981).
2. J. Lecoq, "The Second-Level Trigger of L3 Experiment", this conference.
3. A. Beer, G. Critin, and G. Schuler, Nucl. Instr. Meth., A234, 294 (1985).

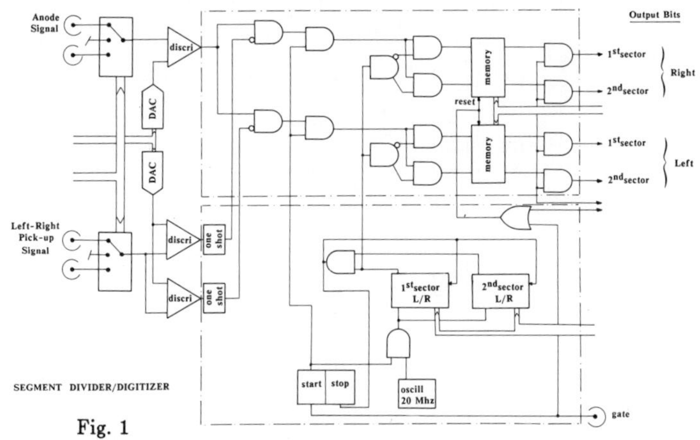

Fig. 1

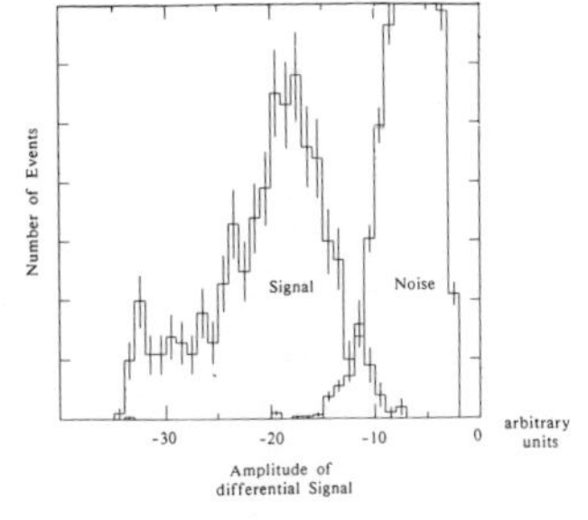

Fig. 2

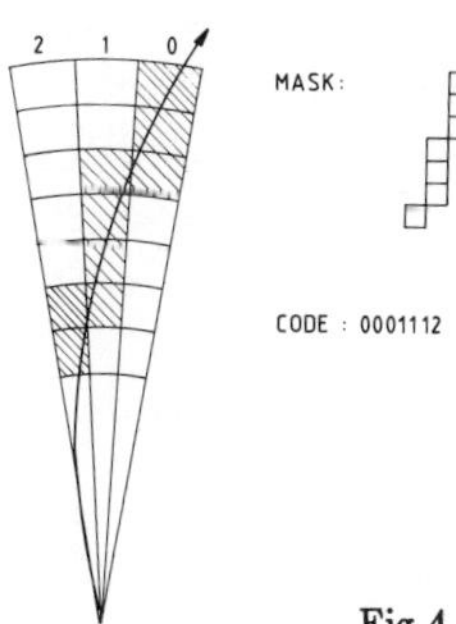

Fig.4

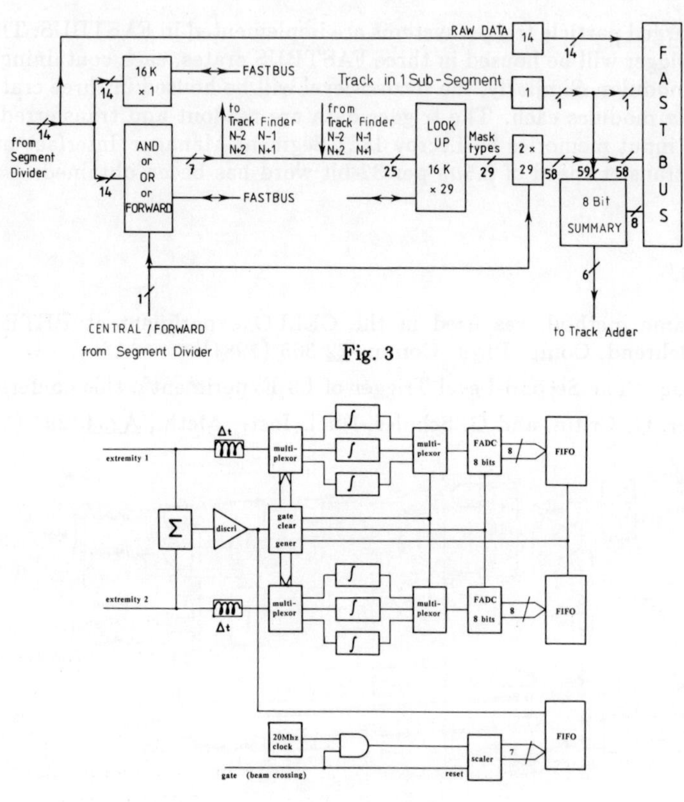

Fig. 3

SECOND LEVEL TEC TRIGGER DIGITIZER

Fig. 5

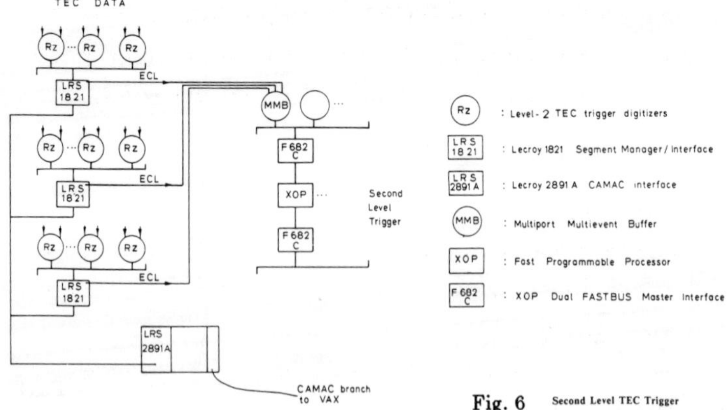

Fig. 6 Second Level TEC Trigger

A FAST TRACK TRIGGER PROCESSOR FOR THE OPAL EXPERIMENT AT LEP, CERN

by M. Bramhall(3), R. Hammarström(1), S. Jaroslawski(3), D. Joos(2), A. Penton(3), C. Weber(2).

presented by S. JAROSLAWSKI, Rutherford Appleton Laboratory, England.

(1) CERN
(2) Freiburg University
(3) Rutherford Appleton Laboratory

A fast programmable trigger processor for the OPAL experiment is described. The processor can handle multihit events. The tracks are found in the R-Z and the R-PHI planes by 24 fast track finder circuits operating in parallel using a novel histogramming technique. A semicustom coincidence array circuit is used to match tracks.

1. INTRODUCTION

Anticipated background rates at LEP from off-momentum electrons have led to a review of the original specifications for a fast hardware processor described in Ref.1. The new requirements (Ref.2) included upgrading the processor to a multihit operation and making the track decisions within 14 microseconds of each beam crossing.

This paper presents the operational features and an engineering overview of criteria used in designing and building of the processor at the Rutherford Appleton Laboratory and Freiburg University. The function of the processor is to examine, between each LEP beam crossing (22 microseconds), the multihit data generated by two central drift chambers and to provide the OPAL first level trigger with information about the topology and the multiplicity of good tracks. The classes of events looked for are in the form of straight tracks in the R-Z plane which originate in the interaction region, Fig.2. The track resolution is required to be 32 bins in theta in each 30 degree R-PHI overlapping segment.

An important feature of the processor is an extensive I/O which allows the operational parameters to be set up and allows a wide range of tests to be carried out from a remote station. A comprehensive set of programs has been written for the purpose. The host computer can set up any required test by communicating with the processor through CAMAC or VME using a simple instruction set.

2. THE ARCHITECTURE

The processor is configured to match the design of the two cylindrical drift chambers; a vertex detector (nearest to the beam pipe) and around it, a large drift chamber denoted "Jet Chamber". Fig.1. Both chambers are divided into R-PHI sectors. There are 36 in the vertex detector and 24 in Jet Chamber. Twelve sense wires in each vertex detector sector and 36 in each Jet Chamber sector are used in the trigger: a total of 1296 wires. The Jet Chamber wires are divided into three groups of 12 wires in each sector. this, in effect, creates three concentric rings which are treated as independent chambers in the processor design.

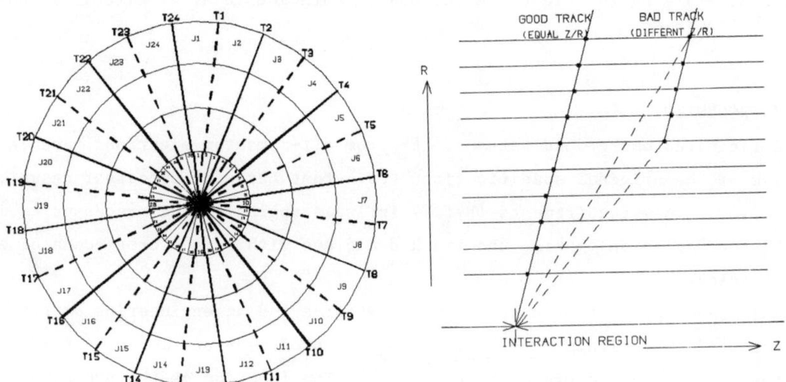

Fig.1. R-PHI projection through two central chambers (not to scale)

Fig.2. Principle of track finding R-Z projection

The adjacent sectors are combined to form 24 overlapping trigger segments. Each such segment consists of two Jet Chamber sectors and three vertex detector sectors and subtends an angle of 30 degrees. Twenty four independent track finder circuits analyse the data generated in the trigger segments. These circuits operate in parallel and are the main building blocks of the processor, Fig.3. The electronics of each track finder is mounted on a separate large wirewrap board and uses over 400 integrated circuit elements (FAST TTL). The track finders are housed in two crates equipped with custom backplanes.

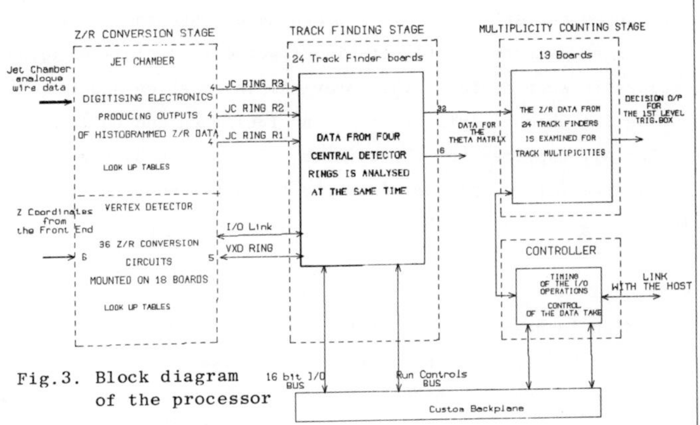

Fig.3. Block diagram of the processor

3. THE OPERATION

The processor carries out the data analysis in three stages linked by ribbon cables as shown in Fig.4. The first stage, the Z/R data to the second stage, the track finding, are controlled by the local circuits. The subsequent data transfers through and out of the 24 track finders are synchronised by a common 10MHz clock generated in the processor's controller. (This module also serves as an intelligent interface in the I/O operations).

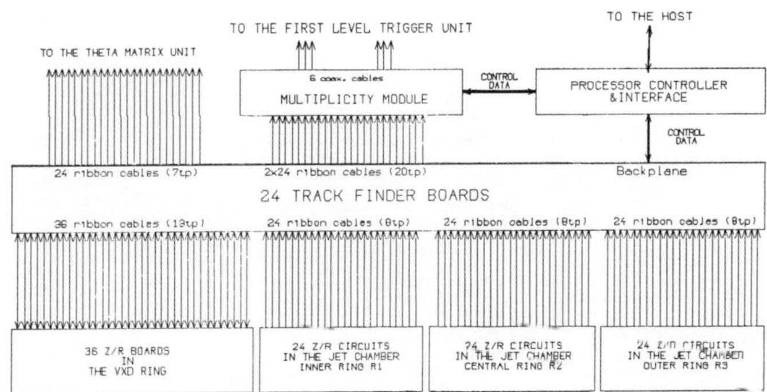

Fig.4. Data Links

3.1 The Z/R Conversion

Look Up tables convert the 6 bit Z coordinates into the 5 bit Z/R (theta) values. The vertex detector theta data is arranged in a block of up seventy two 5 bit Z/R values in each sector. This provides information about up to 6 hits on a wire. The Jet Chamber trigger front end (designed and built by Frieburg University, Ref.4), supplies the theta information as the histogrammed outputs. The data are in the form of blocks of thirty two 4 bit numbers for each sector in each ring.

3.2 Track Finders

Track finding is based on a principle that the tracks originating in the interaction region have equal Z/R values, Fig.2. The good events can, therefore, be detected by examining histogrms of the Z/R values for sharp peaks, Ref.1. This is done independently for the vertex detector and for the Jet Chamber, Fig.5.

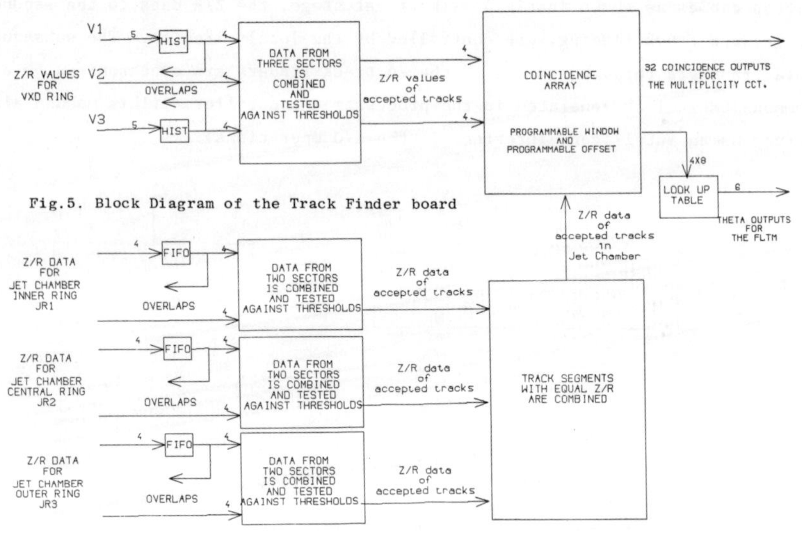

Fig.5. Block Diagram of the Track Finder board

The Z/R values of tracks satisfying the required conditions are entered in the coincidence circuit. The coincidence circuit, designed at RAL, is a semicustom chip based on a 2 micron technology, Ref.3.

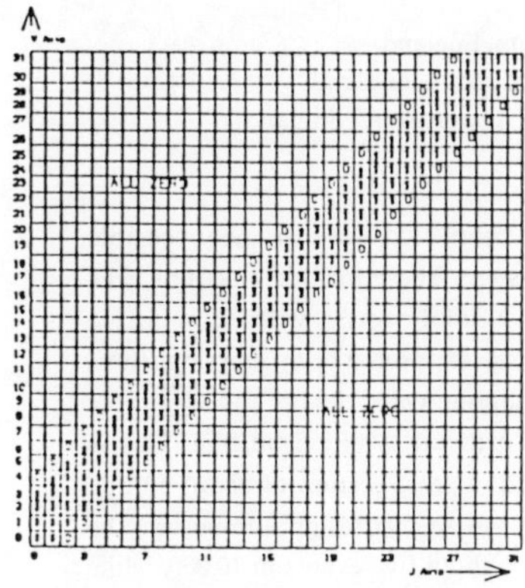

Fig.6. 32 by 32 Coincidence Array

Track matching is done by a 32x32 array of programmable elements and the track decisions are generated as 32 separate outputs. Each output is an OR of the contents of a separate column of 32 coincidence elements. A typical coincidence pattern is shown in Fig.6. Important features of the pattern are the programmable position of the diagonal, which corresponds to the position of the interaction region, and the programmable band of coincidences, which determines the precision in track matching.

3.3 The Multiplicity Counter

The function of this stage is to find track multiplicities of at least 1, of at least 2, and of at least 3 per event. It is intended to investigate patterns of clusters in a 24x32 array formed from 32 coincidence outputs in 24 track finders by examining three projections of the array, including the diagonal, by Look Up tables.

REFERENCES

1. A Fast Track Trigger Processor for the Opal Detector at LEP by D.R. Ward et al NIM A250 (1986) 503-513.

2. Opal Trigger Working Groups internal notes (P. Le Du)

3. 32 Bit Coincidence Array for OPAL. by M. French, F. Slorach: paper presented at this conference.

4. R. Hammarström (CERN), O. Schaile, J. Schwarz, C. Weber, M. Weymann, (Freiburg). Opal Internal Note 2/86.

THE NEW UA1 FIRST-LEVEL TRIGGER PROCESSOR

N.Bains[1], S.A.Baird[3], D.Campbell[3], M.Cawthraw[3], D.Charlton[1], J.Coughlan[3], E.Eisenhandler[2], N.Ellis[1], I.F.Fensome[2], P.Flynn[3], S.Galagedera[3], J.Garvey[1], G.Grayer[3], J.Gregory[1], R.Halsall[3], M.P.Jimack[1], P.Jovanovic[1], I.R.Kenyon[1], M.Landon[2], T.P.Shah[3], R.Stephens[3]

[1] University of Birmingham, Birmingham, England
[2] Queen Mary College, University of London, London, England
[3] Rutherford Appleton Laboratory, Chilton, Didcot, Oxfordshire, England

Introduction

The new first-level trigger processor for UA1 is now nearing completion. An extremely versatile range of triggers will be provided by this fast hard-wired system. These include electrons, jets, total transverse energy, missing transverse energy, and a total-energy veto of multiple interactions. The processor works on calorimeter data from proton–antiproton beam crossings (time spacing 3.8 μs) in the CERN Collider. Allowing for signal shaping, cable delays and electronic fast-clear times, it must reach a decision within 1.5 μs in order to avoid deadtime. This is necessary because the UA1 detector sees an effective interaction cross-section of about 50 mb, which produces an interaction rate of the order of 100 kHz at full ACOL design luminosity. The previous processor[1)], used until 1986, had an acceptance rate of about five events in 10,000, while the new one may have to select as low as one in 10,000 depending on the performance of the second and third-level triggers.

The new UA1 uranium-TMP calorimeter is divided into cells of roughly 10 cm x 10 cm. There are six samplings in depth — four electromagnetic and two hadronic. This is too much information for the first-level trigger to handle, so we use 20 cm x 20 cm electromagnetic cells summed over the four e.m. depth samplings and 20 cm x 40 cm hadronic cells summed over the last two depth samplings. The old hadron calorimeter is also used. This makes a total of over 1700 channels of input data, digitised by 8-bit flash ADCs. Thus, the processor must handle over 400 megabytes of data per second and perform well over 100 000 million 8-bit integer operations every second. To carry out this task we use 18 large crates (9U high x 40 cm deep) of digital logic containing nearly 100,000 high speed TTL and LSI integrated circuits on over 300 Multiwire circuit boards.

We have chosen to use digital electronics throughout for increased versatility, reliability and test-ability. Data is cycled through it up to 8 times per event, enabling different calculations to be performed on each cycle using the same circuitry. To achieve this in the time available a pipelined architecture is employed so that several operations can be performed concurrently. The speed of operation is maximised by using custom-made hard-wired logic controlled by a microcode sequencer. This is user-programmable to allow great flexibility.

Functional description

A conceptual outline of the functions of the new processor in shown in figure 1. At the first stage, high-speed RAMs are used as lookup tables. These enable the raw trigger-ADC information to be converted to energy, pedestals subtracted, and any arbitrary function of the energy from each of the calorimeter cells to be calculated in one operation taking less than 50 ns. Lookup-table pages are cycled up to eight times to give E_t, E, $E_t \sin \phi$ and $E_t \cos \phi$. Following the RAMs there are three main routes through the system. These provide electromagnetic, jet, and energy-sum triggers.

Electromagnetic trigger cells are summed in overlapping horizontal and vertical pairs and compared with transverse-energy thresholds before searching for clusters. The pair sums are latched during the E_t cycle so that eight separate electromagnetic thresholds can be applied sequentially, independent of the RAM cycling. Fast two-dimensional pattern recognition is used to find and count e.m. clusters in the calorimeter. Isolation of the e.m. showers from other significant energy deposition can be imposed, as an option, both laterally and in depth. This is done very quickly and economically for each set of thresholds by the use of programmable array logic (PALs). This novel feature is described in more detail in our poster paper[2]. We therefore have the possibility of up to eight separate e.m. trigger conditions, where each condition consists of a transverse-energy threshold, a shower multiplicity (i.e. ≥1, ≥2, or ≥3 showers in the detector), and the separate options on lateral shower isolation and on veto by energy in the hadron segments.

Jet triggers are similar but simpler. A jet window is the sum of four hadronic cells and eight e.m. cells, and has dimensions 40 cm x 80 cm. Transverse energy is summed over all possible regions of 2 x 2 windows, latched and compared with eight thresholds. Clusters are counted using PALs to do simple pattern recognition. There are again up to eight separate trigger conditions, each consisting of a transverse-energy threshold and a multiplicity.

Transverse energy may be summed over the whole of the calorimeter and compared with various thresholds to provide trigger conditions on unbiased 'high temperature' events. Up to eight 'local' energy sums are also available. Total energy can be used to veto multiple interactions, which are about 15% of the interaction rate at full luminosity even without allowing for substantial trigger bias. Finally, a full transverse energy vector sum is performed over the whole of the calorimeter to provide triggers on missing transverse energy at two threshold levels. This sum avoids the explicit use of signed arithmetic by keeping each quadrant of the detector separate. The addition of the two components of transverse energy in quadrature uses RAM lookup tables.

In the final stage, the eight e.m., eight jet, eight 'global' energy sums, eight local energy sums and two missing energy conditions are combined with up to 32 external conditions such as the fast-scintillator pretrigger and the first-level muon trigger. The logic that does this allows up to 32 combinations of these input conditions, each of which forms an OR'd independent trigger. For each combination the input conditions are treated as one 66-bit word, where each bit can be required to fire, not fire, or not matter. Thus, in addition to very simple triggers like one electron over a threshold of 10 GeV transverse energy, we can demand combinations such as: one electron over a threshold of 5 GeV with lateral isolation of 200 MeV *AND* two jets each over 10 GeV transverse energy *AND* missing transverse energy of over 12 GeV *AND* a muon *AND* a pretrigger *AND* visible total energy of less than 650 GeV. The module that does this uses RAM lookup tables and some simple logic.

Implementation

A block diagram of the processor is shown in figure 2. There are 16 first-level crates, eight for the barrel and eight for the end caps. Each crate contains 10 first-level modules that contain the RAM lookup tables, sum energy over the 13 input channels, and do the transverse energy pair addition and comparisons for the e.m. triggers. There are two e.m. cluster-finding modules and one gated adder module, which serves as the first stage of the adding tree for energy sums. The jet crate contains 14 jet-finding modules that sum the 2 x 2 windows, compare with thresholds and find clusters. In addition, a cluster-counting module adds up the number of jets in the entire system. The third-level crate has a cluster-counting module for e.m. clusters, four gated adders to continue the energy-sum adding tree, an arithmetic module to finish the energy sums and missing transverse energy and compare them with thresholds, and the final decision module that performs the final combinatorial logic. All crates also contain inter-crate connecting modules and a crate controller. The crate controller includes the fast RAM-based timing sequencer that controls the pipelined data flow through the system using 32 clock lines in each crate.

The modules are manufactured using Multiwire, a technique which is topologically similar to wire-wrap but better-suited to automated manufacture and testing, and results in a more robust, thinner module. Prototypes were first made in wire-wrap; the conversion to Multiwire is quite straightforward. With the exception of ECL inter-crate signals, the logic is done using TTL. The crates, which hold up to 21 modules, use three 96-way rear connectors per module. One is linked to a commercial printed-circuit data bus, and the others to a custom-made wire-wrapped backplane.

Control, readout and monitoring

The trigger processor is interfaced to Macintosh II computers and the UA1 data acquisition system via VMEbus. There are nine VME interface cards, each controlling and reading out two of the processor crates. Extensive test, control and monitoring software has been written. In particular, users can select trigger conditions without detailed knowledge of the processor. Incorporated into the processor are diagnostic facilities which enable events to be simulated using Monte-Carlo data. This allows the performance of the processor to be monitored and faults to be diagnosed. When the experiment is running we automatically read out 256 words of data per event from each processor crate into the UA1 data acquisition system. This is double-buffered in the VME interfaces before being passed to the event builder system by a VME-based CPU via a series of dual-ported RAMs. The data are also passed to a separate buffer for online monitoring by a second Macintosh II computer.

Acknowledgements

The final decision module was designed and built by John Oliver and Sham Sumorok at Harvard — we are grateful for their assistance. Peter Sharp has done an excellent job of keeping the project on schedule. The project was supported financially by the U.K. Science and Engineering Research Council.

References

1) A. Astbury et al., 'The UA1 Calorimeter Trigger', Nuclear Instruments and Methods A238 (1985) 288.

2) N.Bains et al., 'Fast Two-Dimensional Electromagnetic Cluster-Finding in the New UA1 First-Level Trigger Processor', poster paper at this conference.

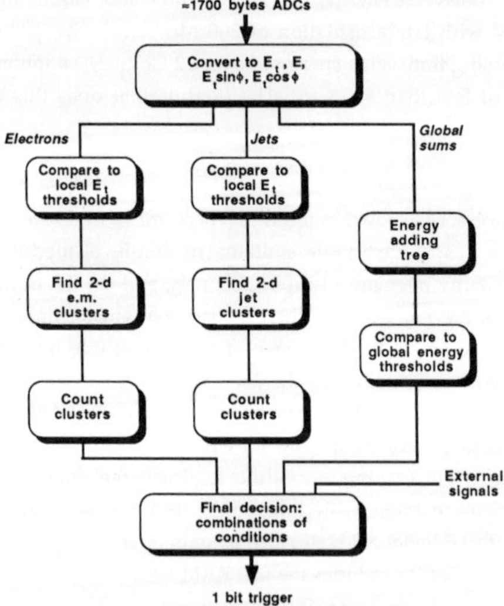

Figure 1. Functions of the trigger processor.

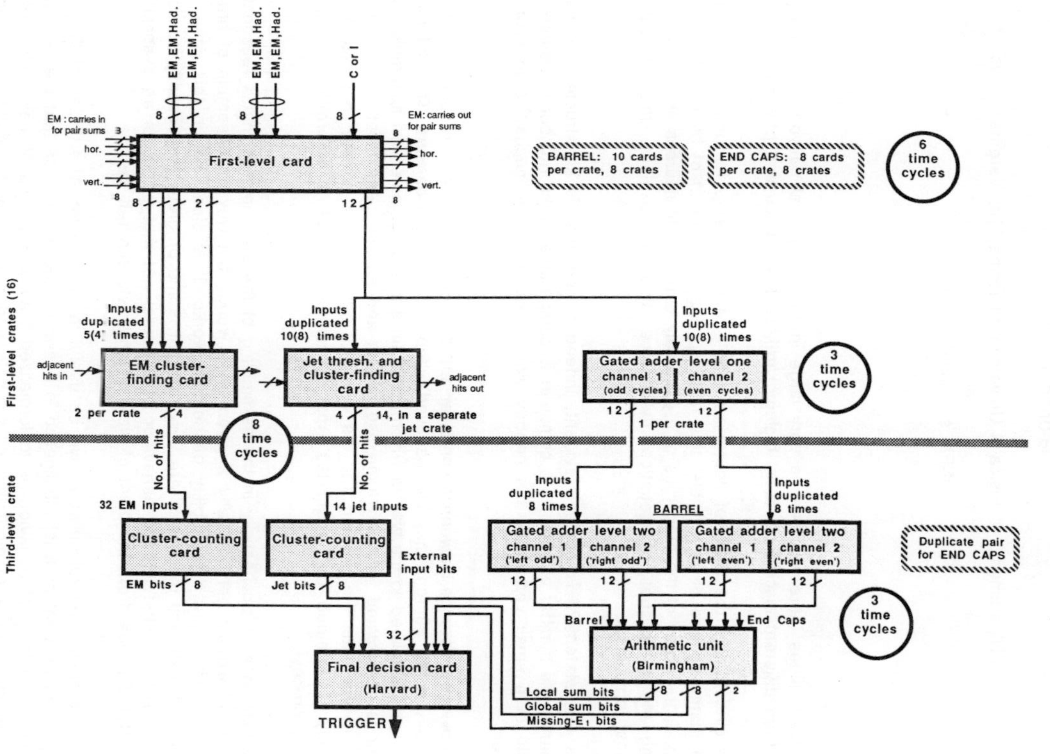

Figure 2. Overall block diagram of the trigger processor.

Why neural network ?

Giorgio Parisi

Dipartimento di Fisica II Universita' di Roma "Tor Vergata"
and
INFN, sezione di Roma

In recent times there has been a very large interest in neural network, from people coming from different disciplines, e. g. biology, artificial intelligence and physics.

It is evident that the study of neural network is a crucial step for understanding the brain of mammals and other animals: this fact explains the interest of biologists. On the other hand, the study of the properties of a extremely large set of neurons should be done by using the techniques of statistical mechanics: this fact explains the interest of theoretical physicists.

This last statement is not self evident: indeed a computer is constructed by many components: the number of gates in a computer is comparable to the number of neurons in the brain of a small vertebrate, but nobody would use statistical mechanics to study a computer.

An obvious difference between a computer and the brain is the larger redundancy of the of the brain; a few neurons may be destroyed without any serious damage. This difference in robustness is related to the way in which the brain and the computer are assembled: the computer is constructed following a precise project and all the gates must be in a given position; in the brain the position and the wiring of each neuron to other neurons is not genetically determined but originates from a combination of a deterministic and a stochastic process.

The existence of stochasticity in the assembling of the brain implies that a reasonable theory should explain not only why a given brain works, but why an ensemble of brains constructed with a given probability distribution works. It is clear that in this case the techniques of statistical mechanics are very relevant: some progress has recently be done on the problem of studying the behaviour of a neural net that works as an associative memory (for a review from the point of view of physicists see ref. 1).

It is less clear if neural networks will be useful in the problem of artificial intelligence. At the present moment artificial intelligence is based on the assumption that the best way to describe an intelligent behaviour is to formalized it as a sequence of logical step: each step corresponds to a manipulation of symbols and can easily (at least in principle) implemented on a computer. In order to solve a problem, we must study the problem and find an appropriate algorithm.

This point of view had many successes in the past, especially in problems that are easy to formalized because they involve some kind of reasoning; unfortunately problems that we solve without thinking, like pattern recognition or learning from examples, can be studied with strong difficulties in the present approach to artificial intelligence (this point is well discussed in ref. 2).

Many people (the connectionist movement) hope that by changing the kind of representation of the data (it is compulsory if we must use a neural network) we are lead to the conception machines which may be able to capture all those aspects of mammal thinking which escape to the reduction to pure logic.

The main progress would be to have machines (or programs) which are able to learn to solve problems by looking to examples: of course we cannot think to have a general purpose machine which solves all kind of problems. The class of problems the machine (or the program) is able to cope efficiently does depend on the architecture of the machine; however, inside a given class of problems, the machine should be able to learn from examples and from previous experience without modifying the software when a different problem of the same class is studied.

This new approach should strongly simplify the software and it would be relative easy to implement on a parallel hardware. It is difficult to say now if this approach will be successful; it seems however that it is worthwhile to investigate it in details.

References

1) M. Mezard, G. Parisi and M. Virasoro, "Spin Glass Theory and beyond", World Scientific, Singapore 1988.

2) A. Hurst and T. Poggio "Making Machines (and AI) see", Dedalus, New York Academy of Sciences (1988). In press.

DATA ACQUISITION AND FILTERING WITH THE ACP MULTIPROCESSOR SYSTEM

Sergio Conetti
Institute for Particle Physics and Physics Department
McGill University, Montreal, Quebec, Canada H3A 2T8

ABSTRACT

The Fermilab Advanced Computer Program, although originally designed for off-line applications, has been adopted in several instances to perform on-line event formatting and filtering. After examining some general issues connected with the use of the ACP in the on-line environment, the existing and planned installations are briefly described.

1. INTRODUCTION

It is well established that the degree of complexity characteristic of Particle Physics experiments – as measured for instance by some combination of the number of channels and the event rate – has grown in parallel with the advances in the field of micro-electronics. More and more experimental setups – and certainly all of the detectors being readied for the new generation of colliders – contain as a standard feature a sophisticated multi-level triggering scheme, where the events are analysed by a succession of on-line filters of increasing selectivity. Typically, the last stage of such a process consists of a bank of micro-processors, each one receiving in turn an individual event, programmed to perform as much as possible of an on-line event reconstruction. It is clear that, for such a scheme to be practical, it is necessary to have access to a moderately priced micro-processor system, easily programmable in a high level language and managed by a user-friendly operating system. Without attempting to perform a detailed analysis of the possible choices, let us just say that the Fermilab ACP multi-microprocessor system, originally developed for cost effective off-line analysis of particle physics experiments, presents itself as a natural candidate for the on-line environment too.

2. USE OF THE ACP FOR ON-LINE APPLICATIONS

In the perspective of the Fermilab group who is responsible for it, the Advanced Computer Program[(1)] is to be viewed as a global ensemble of hardware and software, designed to provide a powerful event analysis machine, while freeing the user from the need of a detailed knowledge of the individual components of the system. In this respect, the program has been very successful, and several ACP installations, organised according to the block diagram of fig. 1, are routinely providing a substantial fraction of the Fermilab's Computing Department number crunching capacity.[(2)] Several experimental groups, both at Fermilab and at some remote installations, are relying on the ACP to reduce the time required to analyse large sets of data, and it has been shown repeatedly that the task of tayloring an existing analysis package to run on the ACP is a very manageable one.

The availability of the hardware and software developed for the off-line purposes makes the ACP a very interesting option for the highest level of real time event selection too, since many of the required features are shared among the off-line and on-line environments. At the same time, the very specifics of real time applications, coupled with the wide variety of event selection and Data Acquisition strategies encountered at different laboratories, leads the interested users to analyse separately the individual ACP components, with the goal of adopting only the immediately useful ones. More specifically, and in a rough generalisation, when designing an on-line processor farm three major questions need to be addressed: how is program development going to be performed, how is run-time control of the farm going to be provided, and how the processing units are going to be supplied with data. One can already guess that the standard ACP hardware and software are likely to satisfy the first requirement, and could be of partial help for the second, but they would definitely need some upgrading and development to deal with the question of bringing the data into the processors. As a consequence, one the most interesting aspects of the existing or planned ACP real time systems is their wide variation in architecture: as described in more detail in the next section, the choice of ACP elements used for each implementation covers the whole range, from the mere adoption of the bare ACP Single Board Processor, up to a system that utilises all of the features of a complete off-line installation.

Trying to identify the major ACP components which are available for an on-line installation, one recognises, hardware-wise, the Single Board Computer processing cell – the "node" – and the Branch Bus with its ancillary modules; from the software point of view one has the low level operating system and general utilities plus the high level event distribution and collection packages. It is useful to describe these elements in more detail, with a particular view to their relevance to the on-line environment.

The Single Board Computer processing node represents the work-horse of the system: based on the popular 68020 processor, the board also includes a floating point unit and up to 6 Mbytes of memory, a most important property for complex experiments featuring very large analysis programs. There is no doubt nevertheless that the most appealing aspect of the ACP node is its cost: priced at less than $2000, and providing the computing power approaching the VAX/780 performance, the ACP node has remained since its appearance the most economical way to obtain lots of number crunching power.

Next, the Branch Bus,(3) designed to connect several crates – VME in the current ACP implementation, but of any type in principle – at data rates of up to 20 Mbytes/sec. Essentially a single master device throughout the duration of each block transfer, the Branch Bus can nevertheless be arbitrated, when idle, among different potential masters. These properties make the Branch Bus extremely useful for on-line applications: the high speed of which it is capable makes it the natural choice for feeding the processors with data from the experiment; the Bus sharing feature will also allow a user-friendly host to access the processors. All of the above considerations notwithstanding, one should not forget that the final Bus performance, due to its passive nature, is fundamentally determined by the intelligence and the intrinsic properties of the Bus Controller(s). In the standard ACP scheme – obviously designed to fulfill the off-line requirements – the only Branch Bus Controller available is the QBBC, i.e. a Bus Controller driven by a Q Bus (or Unibus) device. While appropriate – with some reservations, more later – for control operations,

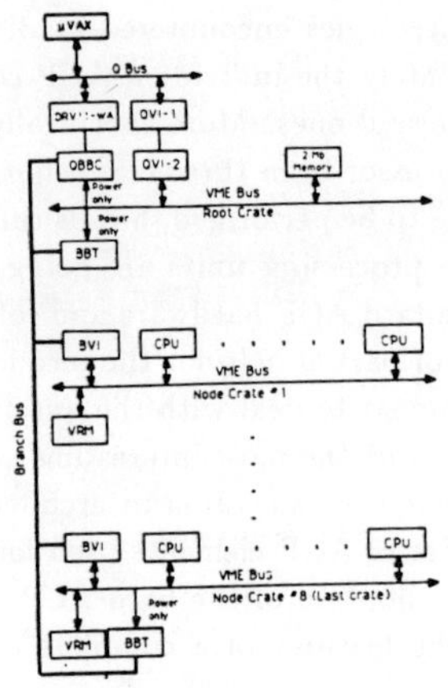

Fig. 1: Standard off-line ACP installation

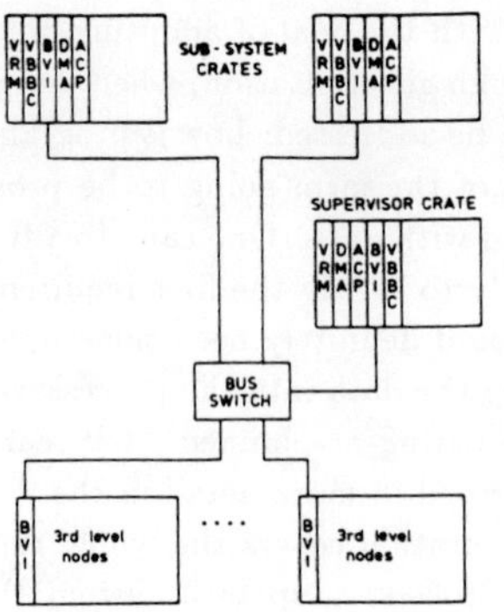

Fig. 2: The "Switch" as an Event Builder

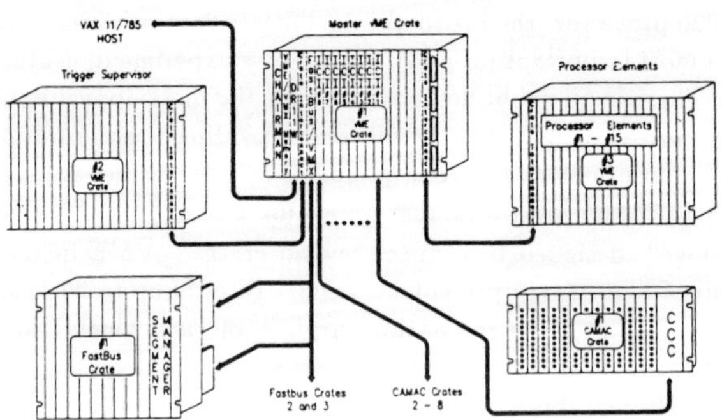

Fig. 3: Data Acquisition system for BNL-802

the QBBC is obviously unsuitable for transmitting data at high rate from the experiment's readout electronics – typically CAMAC or FASTBUS based. Consequently, the Fermilab CDF group, one of the first to choose the ACP for their Third Level Trigger, undertook to develop as a first step, in collaboration with the ACP experts, the FBBC,[4] a FASTBUS driven Branch Bus Controller fulfilling all the requirements of high rate data transmission.

While the FBBC, or some equivalent device for non FASTBUS systems, solves the question of data transmission, there still remains the problem of system control. The most natural picture of a well behaved micro-processor farm calls for the active and efficient intervention of a single "farm manager", whose task is to distribute the load among the farm workers and to monitor their performance. It is also quite uncontroversial that the farm manager should be an easily programmable and user-friendly unit. While these requirements are in principle fulfilled by e.g. a microVAX running the system via a QBBC, it is also clear that there is a considerable overhead involved in such a configuration, as recognised even by the groups that adopted such a scheme. A much more desirable solution would be to be able to run the system "from within", minimising the overhead and allowing to exploit fully the high rate capability of the Branch and VME Buses. Until now such a solution was only possible for small, single crate systems, but, in the near future, the capability of controlling a multi-crate system from within VME will be provided by the VBBC, a full rate, VME-based Branch Bus Controller, soon available from the ACP group.[3]

Another module under design on the part of the ACP group, and also potentially very useful for on-line applications, is the Branch Bus Switch, a 16 branch crossbar that allows up to 16 multi-master Branch Buses to be interconnected. Given that, quite often, one of the first tasks required of on-line processor farms is to perform event building, one can think of a very natural way of accomplishing such a task by means of the "Switch".[5] As sketched in fig. 2, each detector sub-component would be connected, via the Switch, to the micro-processor farm, organised into one or more indipendent Branch Buses according to the overall rate requirements. Under control of the farm manager, responsible for setting up the desired switch connections, the various pieces forming one complete event would be sequentially loaded into the memory of the appropriate node.

Finally, some words about the software. As mentioned earlier, the standard ACP software can be thought of as consisting of a basic operating system – LUNI, intended to emulate on the host VAX the UNIX system under which the ACP nodes run – a set of basic utilities and a data distribution and collection package – employed to transmit events to the nodes and to gather the results of the analysis. It is quite obvious that LUNI and the general utilities will be welcome by most on-line users, since they provide the tools to communicate with the nodes. It is also to be expected, and the existing on-line configurations confirm the expectation, that the event distribution package will be of little use to the on-line applications, and that the system developers will typically need to write their own software, optimised to perform the tasks of data collection and system control as required by the real time environment.

3. EXISTING AND PLANNED INSTALLATIONS

Before giving a more detailed description of the various instances where ACP components were used in the on-line environment, it is worth making some general remarks, drawn from the overall experience of the various users. To begin with, the general consensus is that the individual modules are very reliable, and they perform the way they are advertised. In spite of this, people have encountered some problems with system integration, especially when using architectures which are not the ACP standard. This anyway might be a general VME feature since, even if VME is supposed to be a well defined standard, getting different modules, from different sources, to work together is often a problem, even if all the individual modules allegedly respect the VME specifications. Moreover, the mode of Bus operation encountered in the ACP off-line applications is not very demanding, given that only the BVI (Branch Bus to VME Interface) requires Bus mastership during normal operation. In several on-line systems on the contrary, many different units within a crate are expected to become Bus masters, so that cycle by cycle arbitrations actually take place, under control of the VME protocol, and such a high level of Bus activity has been known to present some problems.

On the software side, a substantial improvement has occurred on what originally was a weaker spot of the ACP system: the well recognised shortcomings of the compiler have been eliminated as and more bugs were removed. Similarly the documentation has been improving with time, and the same goes for debugging aids.

The above reservations notwithstanding, it is undeniable that on-line ACP is in rather good shape, as demonstrated by Table 1, showing all the different projects that have used or are planning to use the ACP for the purpose of event building and filtering. It is significant to point out that successful systems were installed both at Fermilab and away from it, and that in most of the cases the systems were brought into operation with only a limited amount of assistance from the ACP group.

TABLE 1

LAMPF MEGA	Search for $\mu \rightarrow e\gamma$ at Los Alamos
BNL 802	Small but complete spectrometer for heavy ions
BNL 777	Rare decays $K \rightarrow \pi\mu e$
FNAL 705	Large spectrometer, charmonium and direct photon production
FNAL 769	Large spectrometer, charm hadroproduction
FNAL 687	Large spectrometer, broad band photon beam
FNAL CDF	Major detector for 2 TeV $p\bar{p}$ collider
HERA ZEUS	Major detector for e-p collisions at HERA

The list of table 1 spans over a total of eight experiments, differing widely in their parameters: the first three entries represent relatively smaller detectors at lower energy accelerators (BNL and LAMPF), while the next three correspond to rather large spectrometers developed for the Fermilab fixed target program; two very large detectors at the highest energy colliders – CDF at the Fermilab Tevatron and ZEUS at HERA – complete the list. In table 2 the basic performance figures for the various installations are reported: one can observe that the typical input rates most experiments feel comfortable with are up to a few Mbytes/sec, with the ZEUS system expecting to reach up to 10 Mbytes. This suggests that several other factors – maximum rate of logging data, computing power available on-line, etc. – intervene to prevent taking full advantage of the 20 Mbytes/sec capability of the Branch Bus and VME. Two of the experiments, LAMPF MEGA and BNL 777, should be explicitely mentioned for their success in achieving a substantial rate of event selection; it can be expected that the others will follow, although it is understandable that, due to the more complicated events and the larger number of channels characteristic of the higher energy detectors, the process of performing on-line event selection is approached more cautiuosly.

TABLE 2

	INPUT RATE Bytes/sec	Evts/sec	REJECTION FACTOR	TIME/EVENT	# OF NODES
LAMPF MEGA	3×10^6	**2400**	**>100**	**<15 msec**	**32**
BNL 802	$.6 \times 10^6$	**120**	**event formatting**	/	**15**
BNL 777	$.2 \times 10^6$	**250**	**10–20**	**10+300 msec**	**16**
FNAL 705	10^6	**200**	~ 3	**~ 100 msec**	**15**
FNAL 769	1.5×10^6	**450**	**format + compaction**	/	**15**
FNAL 687	3×10^6 1000	~4(?)	?	**16 +...**	
FNAL CDF	2×10^6	20	?	?	> 55
HERA ZEUS	$5 - 10 \times 10^6$	100	10–30	1 sec	~100

NOTE: the bold characters represent an actually achieved performance, regular characters indicate projections.

The entries of table 3 provide more detailed information on the wide variations in the choice of which ACP elements were used, the first two entries representing the two extremes. The LAMPF MEGA installation, described in more detail in a contribution to this Conference(6) resembles in its architecture a standard ACP off-line system, except for the fact that the data is fed into the processing cells via a separate Branch Bus, driven by the FBBC. Even the standard ACP event distribution package is utilised to collect the accepted events: the host microVAX sends dummy events to the ACP nodes and waits for an accepted event in reply, fooled into thinking that it corresponds to the dummy one previously downloaded.

TABLE 3

	LAMPF MEGA	BNL 802	BNL 777	FNAL 705	FNAL 769	FNAL 687	FNAL CDF	HERA ZEUS
Data Source	FASTBUS	FASTBUS CAMAC	CAMAC	CAMAC	CAMAC	FASTBUS	FASTBUS	VME
ACP Node	Y	Y	Y	Y	Y	Y	Y	Y
BRANCH BUS:								
control	Y	N	Y	Y	Y	Y	Y	Y
data	FBBC	N	N	N	N	FBBC	FBBC	VBBC
ACP SOFTWARE:								
low level	Y	N	Y	Y	Y	Y	Y	Y
high level	Y	N	N	N	N	(Y)	Y	?
event distr.	Y	N	N	N	N	N	N	N

At the other extreme, BNL 802 already had advanced plans for a sophisticated Data Acquisition scheme[(7)] when the ACP units came around. The scheme, shown in fig. 3, includes commercial and home built interfaces feeding CAMAC and FASTBUS data into a VME crate. A privileged Single Board Computer, the "chairman", dispatches the data to the processing units, chosen to be the ACP nodes because of their attractive price and performance. According to the designers, a good part of their success in bringing into operation a fairly complex system is due to the use of a commercial, real time multi-tasking operating system, pSOS,[(8)] employed to organise the chairman's software.

The next three entries in the table, BNL–777, FNAL–705 and FNAL–769, represent slight variations of essentially the same scheme, conceived to enhance the on-line performance of traditional, CAMAC based, Data Acquisition. The original idea,[(9)] introduced for FNAL–705, consisted of fast parallel read-out from CAMAC, by means of specially developed Smart Crate Controllers, into a VME based mini-farm, where event building and filtering could be performed. The concept was embraced by FNAL–769, who proceeded in marrying it to the ACP.[(10)] After re-engineering, on the part of Fermilab, of the basic system components,[(11)] one could count on a powerful system, suitable for being treated as an almost "off-the-shelf" item, as was proved by its adoption on the part of BNL–777.[(12)] The basic system architecture is shown in fig. 4, representing the E–769 installation. Data from CAMAC is transmitted in parallel into a bank of Readout Buffers (Rbuf), a double buffered VME memory module. The Rbuf crate is connected, via a passive VME link, to

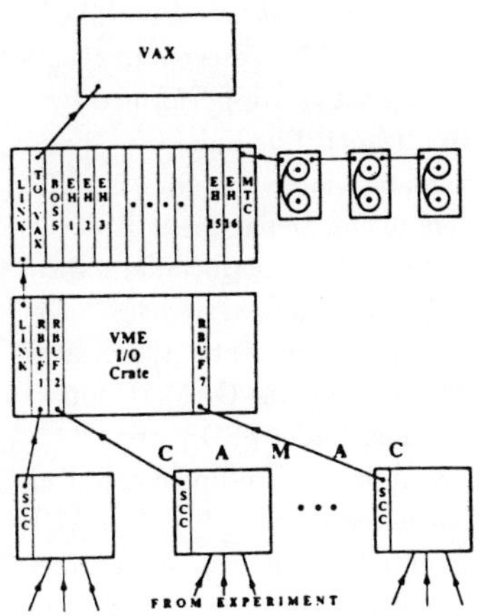

Fig. 4: Fast CAMAC/ACP Data Acquisition at Fermilab

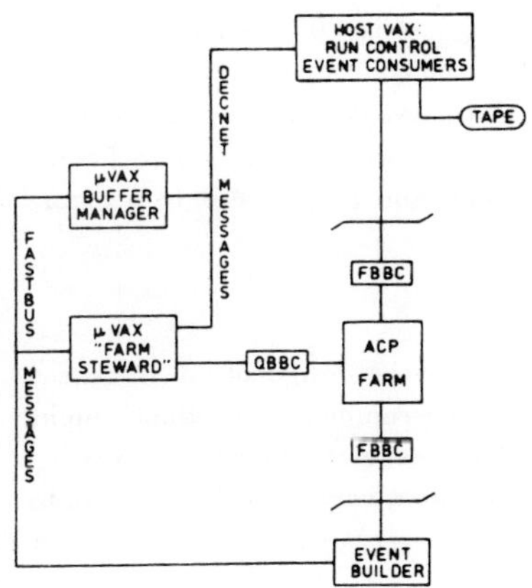

Fig. 5: CDF Third Level Trigger

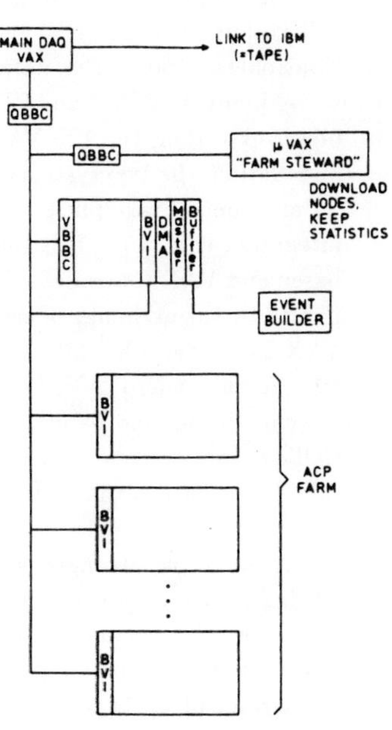

Fig. 6: ZEUS Third Level Trigger

the crate containing the ACP processors: a privileged unit, the "Boss" is responsible for scheduling the activity of the "Event Handlers": at any given time, only one processor will be in the "grabbing" mode, in charge of reading the data from the Rbufs, while all the others will be either "munching" previously acquired events or idling. Accepted events are flagged for logging onto tape, while traditional on-line diagnostics are performed by a VAX, receiving a subset of the events via a standard QBBC Branch Bus link.

Although very effective, the above scheme cannot accomodate more than one crateful of processors. For this and other reasons, the Fermilab Computing Department is developing a more powerful general purpose standard, that will allow high performance Data Acquisition from any combination of FASTBUS and non-FASTBUS sub-systems.[13] Expanding from a system originally designed to meet the E–687 requirements, the system will exploit the experience gained from several directions, and will include CAMAC Smart Controllers, Branch Bus data input via the FBBC, ACP running under pSOS, etc.

Finally, the CDF and ZEUS Third Level Triggers will each provide on-line computing power exceeding the 100 VAX equivalent level. While both based on a multi-crate ACP installation, the two systems otherwise differ rather fundamentally in the way data transfer and control take place. For the CDF case,[14] the ACP multiprocessor farm had to be integrated with a pre-existing, rather sophisticated, Data Acquisition system, FASTBUS based and VAX controlled. To fit in the existing scheme, every event transaction is accompanied by the exchange of several messages between the ACP, the Buffer Manager and the VAX Host (fig. 5), while additional control and monitoring of the farm is accomplished by a further microVAX, the Farm Steward. Its complex structure notwithstanding, the system will become a fully integrated component of the CDF Data Acquisition in the '88 Collider run.

In the case of the ZEUS detector, the design of the Third Level Trigger[15] proceeded in parallel with the global design of the Data Acquisition system and did not have to fit within some pre-arranged scheme. In the proposed design – fig. 6 – heavily based on VME for both data transfer and control, a single Master, the equivalent of the previously mentioned Chairman or Boss, initiates and supervises all the event traffic by means of the VBBC, the VME based Branch Bus Controller. The system will be required to handle input rates of up to 100 events ($\sim$ 10 Mbytes) per second, and to reduce it to no more than a few events/sec.

4. CONCLUSIONS

The ACP has proved itself to be a valuable option in the design of on-line event selection systems. While it is possible that in the near future commercial products might become more and more competitive, the announced set of new generation modules (VBBC, Switch, new node,...) will guarantee that the ACP will remain a very important option to consider.

REFERENCES

1. I. Gaines et al., Proceedings of the Conference on Computing in High Energy Physics, Asilomar, CA, Feb. 1987, and Fermilab-Conf-87/21
2. P. Lebrun, these proceedings
3. R.Hance et al., IEEE Trans. Nucl. Sc., NS-34, 878, 1987

4. M. Larwill, CDF note 453, March 1986
5. R.S. Orr, ZEUS note 88–03, April 1988
6. T. Kozlowski et al., these Proceeding
7. M.J. LeVine et al., IEEE Trans. Nucl. Sc., NS-34, 830, 1987
8. Software Components Group, 4655 Old Ironsides, Santa Clara, CA 95054
9. S. Conetti, M. Haire and K. Kuchela, IEEE Trans. Nucl. Sc., NS–32, 1318, 1985; S. Conetti and K. Kuchela, IEEE Trans. Nucl. Sc., NS–32, 1326, 1985
10. C. Gay and S.Bracker, IEEE Trans. Nucl. Sc., NS-34, 870, 1987
11. R. Vignoni et al., IEEE Trans. Nucl. Sc., NS-34, 756, 1987
12. P. Cooper, Fermilab, private communication
13. Fermilab Computing Department internal note BD-5
14. B. Flaugher et al., IEEE Trans. Nucl. Sc., NS-34, 865, 1987
15. The ZEUS Detector, 1987 Status Report, DESY–PRC 87–02

APPLICATIONS OF NEURAL NETWORKS AND CELLULAR AUTOMATA IN EXPERIMENTAL HIGH ENERGY PHYSICS

Bruce Denby
Laboratoire de l'Accélérateur Linéaire
Bâtiment 200, F-91405 Orsay, France

ABSTRACT

The application of connectionist network methods to track reconstruction, cluster finding, combinatorics, data storage, and triggering, has been discussed elsewhere[1]. In the present work we summarize results on track reconstruction with a simulated neural network, and present a proposal for calorimeter clustering using cellular automata. Possible hardware implementations are also considered.

1. TRACK RECONSTRUCTION WITH A NEURAL NETWORK

Hopfield and Tank[2] used a neural network to solve the Traveling Salesman Problem [TSP], using the distances d_{ij} between cities i and j as coupling strengths between the neurons which represent these cities. As the network evolves, it minimises an energy function which is equal to the trip length, $E = \frac{1}{2}\sum_{i,j} d_{ij} V_i V_j$, where V_i, V_j are the state variables of neurons i and j. The TSP algorithm *almost* works for track finding - in fact it does work for isolated tracks - but for complicated configurations additional information is needed.

We choose as our basic subunit of track information, our 'neuron', so to speak, an arrow from one hit to another. A value $V_i = 1$ for such a segment indicates that the two hits it joins belong to the same track. For the moment we allow interactions only between neurons which share a hit. Neurons with similar orientations will reinforce each other, with a coefficient between neurons i and j defined as $T_{ij} = A\cos^n \theta_{ij}/l_i l_j$, where θ_{ij} is the angle between the neurons and l_i, l_j are their lengths. Because each neuron has a head and a tail, we can define inhibitive, or syntactic coefficients between two neurons which both enter or both leave the same hit, i.e., for this case, $T_{ij}^{syn.} = -B/l_i l_j$. If we have correctly defined the coefficients, the network should, from an arbitrary initial state, evolve to a state in which the shortest neurons are linked together into smooth, continuous, non-bifurcating chains. The energy function which is thereby minimised is like the integrated 'bending energy' of the tracks, $f_k(x)$, i.e., $E \sim \sum_k \int (d^2 f_k(x)/dx^2)dx$.

We iterate the coupled differential network equations[2] , using the coefficients as defined above, and with a time step of $.5\tau$ per iteration, where τ is the characteristic time constant of the network. Convergence normally occurs in less than 10 iterations. A typical result is shown in figure 1, which is a simulated event for the Delphi experiment. The reconstruction good, but the network has made a few mistakes. If we relax the penalty for multiple neurons on a hit, e.g., by setting $T_{ij}^{syn.} = -B$, we find that the new neurons brought in tend to lie along the tracks, and that in fact the track recognition in complicated regions seems to be improved, as in figure 2. Although we have not yet found a set of coefficients that works perfectly for all events, we are confident that this will eventually be possible, and several methods of improvement are now under study. A major problem is that the simulations are quite laborious on a serial machine. A typical simulated Delphi event with 25 tracks can have 2000 neurons and 200,000 interconnections. The mean CPU time is about 45 seconds per Delphi event on a VAX 8600, with the large majority of the time being taken in setting up the neurons and calculating the coefficients.

2. CLUSTER FINDING WITH CELLULAR AUTOMATA

Calorimeter cells with energy above some threshold are considered as automata. The transition rule is as follows. Each of the N cells compares its energy with those of its m nearest neighbors. If the energy of the largest neighbor is bigger than its own energy, it *takes* that larger value as its own. After several iterations, all cells within the same cluster will have the value of the largest cell in that cluster (the method as described obviously only works for non-overlapping clusters). This largest value can be thought of as an *identifier* for the cluster.

The total number of calculations can be shown to be[1] $\sim \frac{1}{2} N m \sqrt{N_{max}}$, where N_{max} is the population of the largest cluster. Clustering can of course also be done with a traditional algorithmic approach. A vector method [1], based on stored neighbor lists, requires $\sim 8 N N_{max}$. Though these estimates are rather crude, it appears that the automaton method is advantageous, and it would be interesting to make a more detailed comparison.

3. POSSIBLE HARDWARE IMPLEMENTATIONS

It is apparent that a simulated neural network does not make a fast track finding algorithm for a serial computer. The speed on a parallel machine will be much better, because 1) the algorithm is intrinsically parallel, and 2) the calculation of the coefficients can also be done in parallel. It would be interesting to experiment with the algorithms on a parallel machine to see if one actually begins to gain over other methods. Even better performance will be had when neurocomputers become available. It may also be possible to build hardware neural track finding networks. Since τ for existing networks is of order $1\mu s$, these should be competitive with trigger level tracking schemes.

Figure 3 shows a schematic track finding network in which each square represents a neuron whose tail-hit is given by its column number and head-hit by its row number. Connections are shown for the neuron $3 \rightarrow 2$, with solid lines representing positive coefficients, and broken lines representing syntactic coefficients.

There are two caveats associated with this network. First, in order to arrive at a list of hit coordinates, we must do compression of the readout data, which may make this type of network unfeasible for triggering. Secondly, the neurons and connections are *different* for each event. If it were necessary to calculate the coefficients for each event on a computer, the network would lose its speed advantage. Rather, the hits must exchange information among themselves to set up the necessary neurons and coefficients for each event, as a first step before the actual network evolution starts. Assuming this is possible, what is the total number of connections required? There are N^2 neurons, each with $2N$ positive connections and $2N$ syntactic connections, for a total of $4N^3$ connections. For $N \sim 300$, this give roughly 10^8 connections.

Figure 4 shows an implementation which does not have the data compression or dynamic connection allocation problems. We have 16 wire planes, with 1080 nodes in ϕ in each plane. At each node, 202 possible neurons interact (triangular zones). The total number of connections is roughly $16 \cdot 1080 \cdot 2 \cdot 10^4 \approx 3.5 \cdot 10^8$. This configuration has worse resolution than the previous one, but it is probably much faster.

A 256 by 256 totally interconnected network (i.e., $6.6 \cdot 10^4$ connections) has been built on a single chip[3)]. In this technology, the connections are fabricated as simple resistors, which are much smaller than transistors, so that it should be possible[3)] to have connection densities of $5 \cdot 10^8$ per cm^2. Evidently, the numbers of connections we need for our neural track finders are not out of line with what is possible with standard VLSI.

Researchers at Caltech[4)] have built a model vertebrate retina in silicon, using a neural network architecture. This device, 5mm by 5mm, extracts the velocity field of a moving pattern projected on its surface. Though smaller than our networks, its success gives us confidence that other VLSI neural network applications will actually work.

5. REFERENCES

1) B. Denby, "Neural Networks and Cellular Automata in Experimental High Energy Physics", LAL 87-56, Université de Paris-sud, Orsay, France.

2) J.J Hopfield and D.W. Tank, Science 233 (1986) 625.

3) L.D. Jackel et al., "Electronic Neural Computing", AT&T Bell Labs, Holmdel, N.J. 07733, presented at Les Houches, April, 1986.

4) C. Mead, "Neural Hardware for Vision", Engineering and Science, (Caltech, pub.), June, 1987.

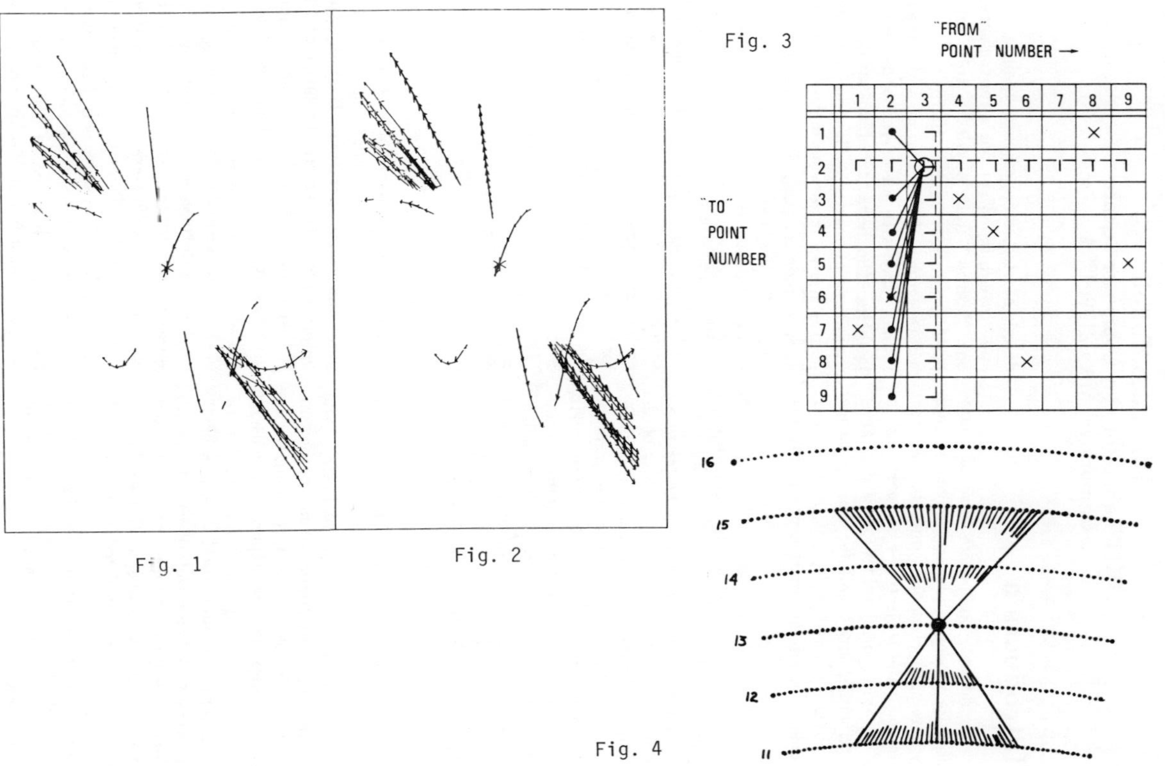

Fig. 1

Fig. 2

Fig. 3

Fig. 4

Cellular Automata Machines as Physics Emulators*

Tommaso Toffoli
MIT Laboratory for Computer Science
Cambridge, MA 02139, USA

1 Introduction

Since their advent, computers have of course been used for modeling physics; indeed, as soon as they were born, electronic computers were set on computing the trajectories of "particles" in "fields"—namely, firing tables for artillery guns.

In principle, any computer can be turned into a physics *simulator*; in fact, a 'computer' is by definition an information-processing machine for which one can program in software whatever operations are not directly supported by the hardware.[1] In practice, though, with any computer the effective accessibility of all but the simplest physical models is severely restricted by performance considerations. In spite of the claim implicit in its name, a "general-purpose" computer is at the bottom a machine optimized for the sequential processing of address fields, and thus particularly good at running subroutines defined in terms of subroutines defined in terms of subroutines ... In sum, a general-purpose computer is designed to be chiefly a *software interpreter*, with all the flexibility—and the *inefficiency*—that that entails.

Can one design a computer optimized to be a physics *emulator* rather than a software interpreter? One whose hardware will directly reflect the "character of physical law[7]," and which will only wait to be given specific initial conditions to merrily go reproducing the details of a complex physical experiment, without having to follow a long script for every operation? A hardware emulator for, say, a 68000 microprocessor is made out of essentially the same kind of stuff as the target chip, but is several times as big, expensive, and power hungry. However, it can be initialized in every single detail, started, stopped, breakpointed, traced, and "core-dumped" at one's convenience. Can one "borrow" from physics a cubic meter of matter, and rearrange its hardware so that it will reproduce—in a blown-up and slowed-down fashion, but with all of the above emulation facilities—the behavior of an arbitrary cubic millimeter of matter? Can one take physical matter and let it do what it does best—that is, at every moment compute its own next state according to the laws of physics—but trade *scale* for experimental *flexibility*?

We have been playing with this concept for only a few years, and within the limits of current technology. The results obtained so far are very encouraging, and appear to be relevant not only for the future of computational physics, but for the future of "general-purpose" computation as well.

2 Fine-grained models of physics

Conventional scientific computers are fast, general-purpose sequential machines equipped with substantial hardware resources for floating-point and vector operations. Basically, they are optimized for the numerical solution of differential equations and for Monte-Carlo simulations involving real-valued variables. To date, tasks of this kind represent an overwhelming fraction of the computational physics business.

The performance of these machines is severely degraded (cf. Section 4) when one turns one's attention to fine-grained models of physics. Here, the relevant phenomenology emerges from the interplay of an *extremely large number* of individually *very simple* microscopic primitives. Faced with such a task, a conventional supercomputer must feel like Gulliver on his awakening in Lilliput—tied down by the million threads of a spider's web.

*This research was supported in part by the Defense Advanced Research Projects Agency, in part by International Business Machines, and in part by the National Science Foundation.

[1] Computation-universality *per se* is so cheap[19] that it would be masochistic to define a 'computer' otherwise.

Yet, such reductionistic models have an extraordinary conceptual appeal. On the strength of Occam's razor, they can speak with great sincerity and authority. But it is hard to make them speak at all! An explicit computer simulation of a system containing as many bits as there are atoms in a gram of matter ($\approx 10^{24}$) is out of the question today as well as for another couple of centuries.[2]

Where, then, do we stand today—or should we relegate the whole business to science-fiction?

Experiments of the above nature, entailing local interactions between state-variables consisting of only a few bits of information at each lattice site, and with a resolution of 10^8 sites, are routinely performed today on conventional supercomputers[21]; note that one can think of 10^8 as the Avogadro number "in one dimension." With dedicated architectures, simulations involving 10^{12} sites and covering a proportionate number of time steps will be feasible in a couple of decades. And within their lifetimes the youngest of us can look forward to computations involving perhaps 10^{16} sites—the Avogadro number "in two dimensions!"

Thus, one can realistically conceive of computational tools that *directly span the gap between the microscopic and the macroscopic world.*

Conceiving is fine—but we want to see how it can be made to happen.

3 Cellular automata

Cellular automata (see [22] for a comprehensive introduction) are discrete dynamical systems whose behavior is completely specified in terms of a local relation, much as is the case for a large class of continuous dynamical systems defined by partial differential equations. In this sense, cellular automata are the computer scientist's counterpart to the physicist's concept of 'field'.

A cellular automaton can be thought of as a stylized universe. Space is represented by a uniform grid, with each site or "cell" containing a few bits of data; time advances in discrete steps; and the laws of the universe are expressed by a single recipe—say, a small lookup table—through which at each step a cell computes its new state from that of its close neighbors. Thus the system's laws are *local* and *uniform.*

Cellular automata were introduced in the late forties by John von Neumann, following a suggestion of Stan Ulam[27], to provide a more realistic model for the behavior of complex, spatially-extended systems than that offered by more traditional paradigms of computation (such as the Turing machine). Within a cellular automaton, objects that may be interpreted as passive data and objects that may be interpreted as computing devices are both assembled out of the same kind of structural elements and subject to the same fine-grained laws; *computation* and *construction* are just two possible modes of activity.

Though von Neumann was a leading physicist as well as a mathematician, explicit physical considerations are lacking in his work on cellular automata: his interest was directed more at a reductionistic explanation of certain aspects of biology. Physical motivation played a more explicit role in Zuse's work[31] on cellular automata—little known because of historical circumstances.

The question of whether cellular automata could model not only general phenomenological aspect of our world (cf. Conway's game of "life") but also directly the laws of physics was raised again by Fredkin (cf. [1]) and myself[22]. A primary theme of that research was the formulation of computer-like models of physics that are *information-preserving*, and thus retain one of the most fundamental features of physics—namely *microscopic reversibility*[11, 2].

The introduction of the first cellular automata machines (see below), and concomitant research work by Margolus[13, 14], Vichniac[28], and myself[23, 24] led to the rapid development of new productive approaches to physical simulation via cellular automata. At the same time, two fundamental paradigms of physical activity reducible to cellular-automaton-like primitives were studied in detail—namely *lattice gases*, by Pomeau, Frisch, and others[12, 9, 15] and microcanonical Ising spin models, by Creutz[4]. The exposure of the physics community to cellular automata was vigorously pursued

[2]In the last four decades, circuit density has been steadily increasing by a factor of 10 per decade. Note that, though at any particular moment one had a rough idea of where the next power of 10 would come from, nobody could confidently predict from what scientific and technological breakthroughs the *following* powers of 10 would come.

by Wolfram[29, 20, 30], though his stress was more on mathematical systems theory (chaos, fractality, etc.) than on physics proper. At present, lattice-gas hydrodynamics is the object of much exciting research [5, 21].

4 Cellular automata machines

A *cellular automata machine* (CAM) directly realizes in hardware the overall structure of a cellular automaton; for many applications, such a machine can be regarded as a volume of simulated "programmable matter" in which a large variety of experiments on spatially-extended systems can be performed rapidly and conveniently. The capabilities of these machines and the new modeling methodologies that are most appropriate for them are amply illustrated in [26] and [17, 18], which should be treated as appended to the present paper.

The novelty of these machines is that they bring a performance advantage of *several orders of magnitude* over traditional computers in this modeling area. Cam-6, which is in commercial production[3], plugs into a slot of an IBM-PC or equivalent unit, and runs cellular automata with a performance comparable to that of a CRAY-1; CAM-8, now at an advanced design stage[18], will provide a performance several thousand times that of a conventional supercomputer of comparable cost, and is upwardly scalable by another two orders of magnitude. Many more orders of magnitude are achievable with further specialization of the class of models one wants to simulate.

These gains are of such magnitude that new classes of *conceptual models* have become *computationally accessible.*

5 Physics emulation

We have seen that cellular automata are intrinsically *uniform* and *local.* The first feature reflects the space- and time-translation invariance of physics; the second recognizes the finite speed of propagation of information, and is a prerequisite for Lorentz invariance.

What we want to discuss here are further specializations of the cellular-automaton paradigm, aimed at capturing other fundamental aspects of physics. When translated in terms of CAM architecture, this specialization should bring two advantages. On one hand, by throwing away cellular-automaton features that are not expected to be essential in physical modeling, we are going to streamline the CAM machinery itself, and thus increase its performance. On the other hand, by specifying in our design computational primitives (state-variables, processing, and interconnection) that are closer to those that physics provides "off-the-shelf," we are going to achieve a better match between our abstract specifications of a CAM architecture and its realization as a *physical* (what else?) computer.

One should recognize that our approach to computational physics is, ultimately, the business of inducing physics to emulate physics itself *at our leisure.* The less coercion we manage to use, the more efficiently will physics happen to do "our thing" while still doing "its thing."

5.1 Things ordinary cellular automata don't know

An ordinary cellular automaton is described in terms of *cell state* and *(local) rule.* The "format" of the rule is that of a gate with several inputs (one for each neighbor) and one output, as illustrated in Fig. 1 in the case of a one-dimensional, three-neighbor cellular automaton; the rule specifies the *entire* new state of a cell as a function of the *entire* current state of all of its neighbors.

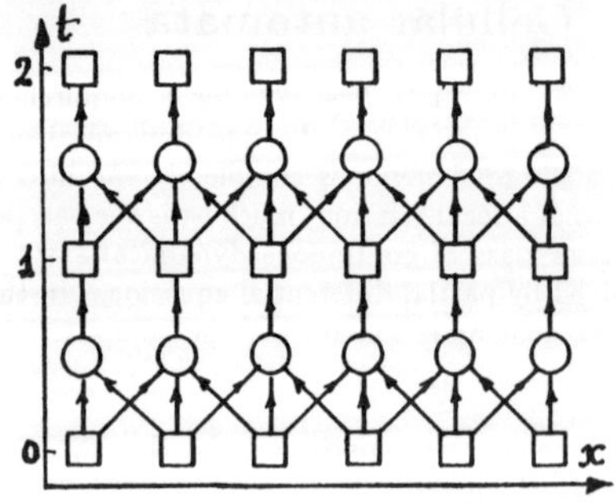

Figure 1: Spacetime structure of an ordinary cellular automaton. Each row of squares represent the array of cell-states at one moment of time; the circles denote applications of the rule.

This arrangement gives more generality than is needed in modeling interactions analogous to those of microphysics, and at the same time creates many headaches:

Conservation principles Suppose, for simplicity, that one of the possible cell states is interpreted as "vacuum" and the others as particles of different "colors," and suppose further that we want to obtain from the rule nontrivial interactions obeying the principle of detailed balance: particles will be reshuffled in different ways depending on the situation, but the number of particles of each color must be conserved.

Note that the neighborhoods of adjacent cells partially overlap (in Fig. 1, each cell shares two of its neighbors with the cell on its right, and similarly for the cell on its left). Thus, at any given place a certain number of copies of the rule ("partners") will go to work as a team on the same input data (the area of overlap); each partner, however, has also access to privileged information (those inputs that are not shared with the other partners). Since each partner will in the end have to produce the next state of just one cell, particle conservation can only be guaranteed by the coordinated activity of the partners. But each copy of the rule is at the same time a partner in many teams: this creates an infinite regress of mutual obligations that in general cannot be fully analyzed and given a provenly safe solution.

Particle motion Suppose, more specifically, that the color denotes the state of motion of a particle, for instance, in the automaton of Fig. 1, R for 'move right', L for 'move left', and S for 'stay in place'. Then, the "propagation" component of the transition function will read, essentially, as follows: "If you see an R particle on your left input, copy it into your cell; if you see an L particle on your right input, copy it to your cell; and if you see an S particle in your middle input (that is, in your own cell), leave your cell unchanged." Since these three conditions are not exclusive, problems arise when more than one particle heads toward the same cell; one could try to add auxiliary cell states to temporarily hold extra guests, but the same kind of problem arises with the auxiliary states, leading again, in general, to infinite regress.

Invertibility Microphysics—whether classical or quantum—is strictly reversible. Given the local rule of a cellular automaton, is there a general recipe for deciding whether the corresponding global dynamics is reversible? This is still an open issue, but the problem is conjectured to be undecidable. Again, the difficulties arise from the fact that each copy of the rule has only partial access to the relevant information.

Relativity The global state of an ordinary cellular automaton is well-defined only if one considers the state of all cells at the same instant of time. Initial conditions specified on a space-like surface "cut" at a different angle than t=const, may leave the subsequent evolution underdetermined or overdetermined (or both). Thus, synchronicity transformations, of which Lorentz transformations are a special case, are in general meaningless.

From the above considerations, one might conclude that cellular automata simply do not have enough expressive power to deal with conservation, motion, reversibility, relativity, etc. On the contrary, ordinary cellular automata have *too much* expressive power for our needs, and make us spend a lot of effort worrying how to extricate them from unwanted situations that they are quite capable of getting into. The above difficulties are eliminated by turning our attention to a *narrower* class of cellular automata.

5.2 Inertial cellular automata

In an *inertial* cellular automaton, the cell state is the product of as many state-components as there are neighbors, and each cell senses only the n-th component of its n-th neighbor, according to the discipline of Fig. 2, which illustrates a one-dimensional, two-neighbor cellular automaton.

Note that this figure can be redrawn much more simply as in Fig. 3, where the graph's arcs by themselves represent the state components (the boxes have been eliminated), and the nodes represent the applications of the rule. The resulting overall picture is one of just *signals* (the arcs) and *events* (the nodes) in spacetime.

Unlike ordinary cellular automata, in an inertial cellular automaton each piece of data is

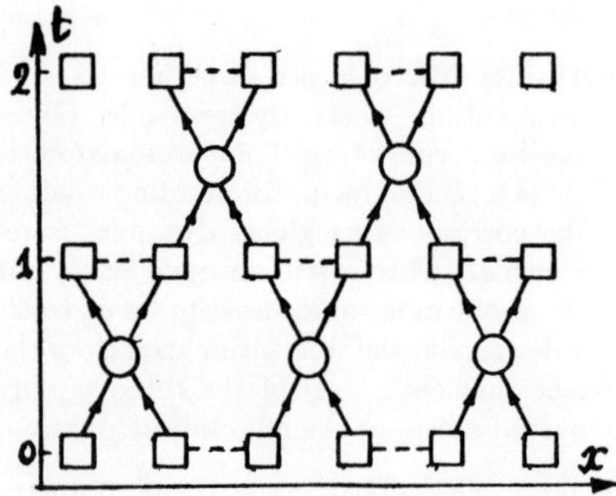

Figure 2: Spacetime structure of an inertial cellular automaton. Here squares represent cell-state *components*; a dashed line shows which components belong to the same cell.

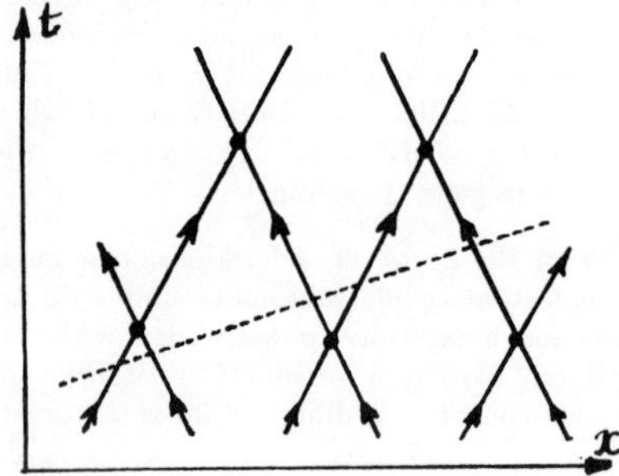

Figure 3: Same as previous figure. Cell-state *components* are represented by the arcs themselves (*signals*), and their interactions by the nodes (*events*). The dotted line represent a space-like surface cut at a slant with respect to $t = \text{const}$.

never seen by more than one neighbor; moreover, each copy of the rule has as many outputs as inputs, and transforms in one step all of the data that it sees without having to worry what the other copies might be doing. The consequences are sweeping:

Reversibility The global dynamics is reversible *if and only if* the local rule is reversible. As simple as that.

Conservation If quantity defined on individual arcs (this is the simplest kind of *locally additive* quantity) is conserved by the local rule, this quantity is globally conserved.

Inertia Let the propagation component of the rule (that is, what the rule would reduce to if interactions were suspended) be, "Pass on to your right-oriented output line (and thence to your right neighbor) whatever comes from your right-oriented input line (and thence from your left neighbor); and similarly for your other input/output lines." Then every particle will move on a straight line at a constant speed, with no conflicts when two particles' paths cross.

According to Aristotle, objects would tend to a state of rest by default, and one had to look for forces to explain their *velocity*; we learned from Galileo to look for forces only when the object was *accelerating*—velocity comes for free, you only pay for acceleration. The situation is analogous in inertial cellular automata. Start with the *identity* rule (what else?) as a default for your dynamics, and you automatically get *inertial* motion. Now that particles are happily zipping by you can start throwing in some interactions. Do you want them reversible? See above! Do you want them conservative? Ditto!

Synchronicity transformations Draw any space-like surface across spacetime (dotted line in Fig. 3), and see where it cuts the network's arcs. Arbitrarily assign a state-component value to each of these arcs, and you'll have a complete and consistent set of initial conditions for the future evolution of the system. If the local rule is invertible, you can consistently travel backwards into the system's past. If your space-like surface happens to be flat and to make a rational angle with the time axis, you can keep updating the system at an even rate at every place, pushing your "present" surface forward and keeping it parallel with the initial line: the resulting evolution is specified by a new cellular automaton, related to the original one by a synchronicity transformation.

The advantages suggested by the above theoretical considerations are confiremd by a wealth of concrete physical models which are most naturally expressed in the inertial-cellular-automaton framework (cf. [26]).

5.3 From theory to practice

It should not come as a surprise that the conceptual advantages of inertial cellular automata are closely matched by implementation advantages. In fact, these and similar considerations have been incorporated in the design of CAM-8 (see [18] for details) yielding an architecture of outstanding cleanliness, flexibility, and performance for the intended range of applications. A CAM-8 installation of cost comparable to that of a conventional supercomputer will handle *billions* of sites (say, an array of 2K×2K×2K 16-bit cells) with an update rate of the order of 0.2 *Tera-event/sec*—or 20 updates/sec for the entire array (the frame-rate of a movie).

6 Conclusions

In the previous section we have explored only some ways of making the architecture of a computer closer to that of physics itself. Other sides of physics that we have not treated here may be as important in view of a better match. Research on quantum-mechanical aspects of parallel computation is still at its beginnings ([8]; [16] and references therein); the emergence of approximate rotational and Lorentz invariance in cellular automata has been observed in a few cases[25], but is far from being under full control; similarly for the emergence of variational principles; etc. Moreover, it is still not clear what one should mean by 'the architecture of physics'; what we are trying to do is play with physics the charade game "If you were a computer, what would you be like?"

Our feeling is that only continual interaction between theoretical physics, theoretical computer science, computer architecture, circuit technology, and materials science will lead to "physics emulators" worthy of the name.

References

[1] BANKS, Edwin, "Information Processing and Transmission in Cellular Automata," *Tech. Rep. MAC TR-81*, MIT Project MAC (1971).

[2] BENNETT, Charles, "Notes on the history of reversible computation," *IBM J. Res. Develop.* **32** (1988), 16–23.

[3] CALIFANO, Andrea, Norman MARGOLUS, and Tommaso TOFFOLI, *CAM-6 User's Guide*, Systems Concepts, San Francisco (1987).

[4] CREUTZ, Michael, "Deterministic Ising Dynamics," *Annals of Physics* **167** (1986), 62–76.

[5] DOOLEN, Gary (ed.), "Workshop on Large Nonlinear Systems," collected in *Complex Systems* **1**:4–5 (August–October 1987).

[6] FARMER, Doyne, Tommaso Toffoli, and Stephen WOLFRAM (eds.), *Cellular Automata*, North-Holland (1984).

[7] FEYNMAN, Richard, *The Character of Physical Law*, MIT Press (1965).

[8] FEYNMAN, Richard, "Simulating Physics with Computers," *Int. J. Theor. Phys.* **21** (1982), 467–488.

[9] FRISCH, Uriel, Brosl HASSLACHER, and Yves POMEAU, "Lattice-Gas Automata for the Navier-Stokes Equation," *Phys. Rev. Lett.* **56** (1986), 1505–1508.

[10] FRISCH, Uriel, et al., "Lattice gas hydrodynamics in two and three dimensions," *Complex Systems* **1**:4 (1987), 633-707.

[11] FREDKIN Edward, and Tommaso TOFFOLI, "Conservative logic," *Int. J. Theor. Phys.* **21** (1982), 219–253.

[12] HARDY, J., O. DE PAZZIS, and Yves POMEAU, "Molecular dynamics of a classical lattice gas: Transport properties and time correlation functions," *Phys. Rev.* **A13** (1976), 1949–1960.

[13] MARGOLUS, Norman "Physics-like models of computation," *Physica* **10D** (1984), 81–95.

[14] MARGOLUS, Norman, "Physics and Computation," Tech. Rep. no. MIT/LCS/TR-415, MIT Laboratory for Computer Science (1988).

[15] MARGOLUS, Norman, Tommaso TOFFOLI, and Gérard VICHNIAC, "Cellular-Automata Supercomputers for Fluid Dynamics Modeling," *Phys. Rev. Lett.* **56** (1986), 1694–1696.

[16] MARGOLUS, Norman, "Quantum Computation," *Annals New York Acad. Sci. bf 480* (1986), 487–497.

[17] MARGOLUS, Norman, Tommaso TOFFOLI, "Cellular Automata Machines," *Complex Systems* **1**, 967–993 (1987).

[18] MARGOLUS, Norman, Tommaso TOFFOLI, "Cellular Automata Machines," revised

version of preceding reference, to appear in *Large Nonlinear Systems*, Gary DOOLEN ed. (1988).

[19] MINSKY, Marvin, *Computation: Finite and Infinite Machines*, Prentice-Hall (1967).

[20] PACKARD, Norman, and Stephen WOLFRAM, "Two-dimensional cellular automata," *J. Stat. Phys.* **38** (1985), 901–946.

[21] SHIMOMURA, T., G. DOOLEN, B. HASSLACHER, and Castor FU, "Calculations Using Lattice Gas Techniques," *Los Alamos Science*, special issue no. 15 (1987), 201.

[22] TOFFOLI, Tommaso, *Cellular Automata Mechanics*, Tech. Rep. no. 208, Comp. Comm. Sci. Dept., The University of Michigan (1977). 195–204.

[23] TOFFOLI, Tommaso, "CAM: A high-performance cellular-automaton machine," *Physica* **10D** (1984), 195–204.

[24] TOFFOLI, Tommaso, "Cellular automata as an alternative to (rather than an approximation of) differential equations in modeling physics," *Physica* **10D** (1984), 117–127.

[25] TOFFOLI, Tommaso, "Lorentz-invariance of computations," in preparation.

[26] TOFFOLI, Tommaso, and Norman MARGOLUS, *Cellular Automata Machines—a new environment for modeling*, MIT Press (1987).

[27] ULAM, Stanislaw, "Random Processes and Transformations," *Proc. Int. Congr. Mathem.* (held in 1950) **2** (1952), 264–275.

[28] VICHNIAC, Gérard, "Simulating physics with cellular automata," *Physica* **10D** (1984), 96–115.

[29] WOLFRAM, Stephen, "Statistical mechanics of cellular automata," *Rev. Mod. Phys.* **55** (1983), 601–644.

[30] WOLFRAM, Stephen (ed.), *Theory and Applications of Cellular Automata*, World Scientific (1986).

[31] ZUSE, Konrad, *Rechnender Raum*, Vieweg, Braunschweig (1969); translated as "Calculating Spaces," *Tech. Transl. AZT-70-164-GEMIT*, MIT Project Mac (1970).

Use of Optical Data Transmission in HEP

Paolo Cennini
CERN EP Div. - 1211 Geneva 23
SWITZERLAND

Zbigniew Kulka
Department of Nuclear Electronics - Institute for Nuclear Studies
Swierk Otwork - Warsaw
POLAND

Emilio Petrolo
Istituto Nazionale di Fisica Nucleare - Sezione di Roma
Piazzale Aldo Moro, 2 - 00185 Roma
ITALY

Presented by: Emilio Petrolo

ABSTRACT

The use of optical data transmission in High Energy Physics is discussed and few basic concepts on optical fibers are given. Comparison with normal electrical transmission is made and some relevant examples of application are presented.

1. BASIC SIMPLE CONCEPTS ON OPTICAL FIBERS

Optical fibers are made by a silica cylinder (core), surrounded by another silica cylinder of different refraction index (cladding). Only the light coming inside the acceptance cone, of angle $2\Theta_m$ (fig. 1), can propagate, since the light coming outside of the cone will be refracted in the cladding.

The ratio $(n_1 - n_2)/n_1$ is of the order of 1%, while the numerical aperture NA = sin Θ_m is $0.15 \div 0.20$ for multi-mode fibers.

The core diameter is 10 μm, or less, for single-mode fibers and 50 μm or 100 μm for multi-mode fibers. The cladding diameter is 125 μm or 140 μm. The most popular fibers are 50/125 μm for multi-mode transmission and 10/125 μm for single - mode transmission.

The speed of propagation of the light in the fiber is $v=(c/n_1)*\cos\Theta$, this gives a propagation delay of $5 \div 6$ ns/m since n_1 is close to 1.5 and $\cos\Theta$ is close to 1.

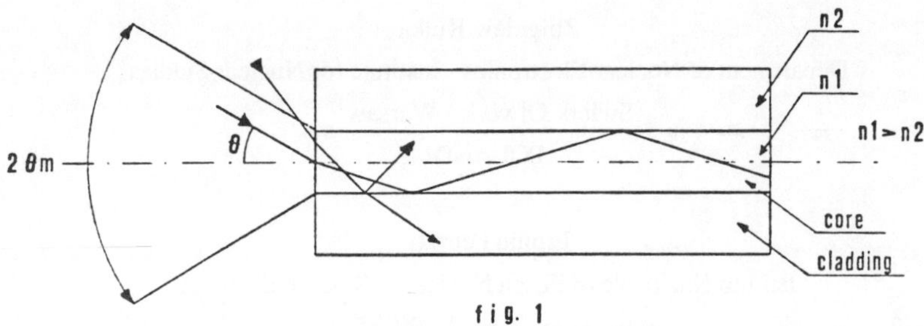

fig. 1

1.1.1. Multi-mode fibers

The light on a fiber can propagate in different "modes", each having a difference of phase between two successive reflections multiple of π rad. The mode that gives a difference of phase of π rad is called "fundamental mode". The possibility of propagation of different modes gives a dispersion in time of the propagation of the light.

1.1.2. Single-mode fibers

Making the core diameter of the fibre very small, less than 10μm, only the fundamental mode can propagate. In this case the dispersion in time of the fibre is very small.

1.2. Attenuation and Frequency Dispersion of Fibres

The attenuation and dispersion of fibers depend from the light wavelenght. Transmission takes place at 3 different wavelenghts called "windows".

1.2.1. 820 nm window (1.st window)

Is intended for transmission on short and medium distances, up to $4 \div 5$ km, using multi-mode fibers. The typical attenuation is $2 \div 5$ db/km and the dispersion 50 ps/nm∗km.

1.2.2 1310 nm and 1550 nm windows (2.nd and 3.st windows)

Intended for long distances and high frequency transmission on single-mode fibers. Fig. 2 shows the attenuation and dispersion on fibers working at 1310 nm and 1550 nm.

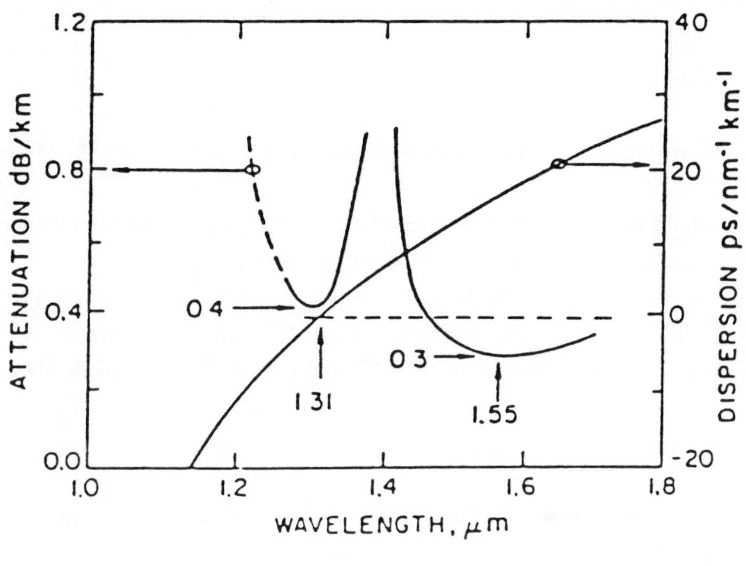

fig. 2

If we consider that the typical laser linewidth is 5nm and the dispersion of the fiber at 1310 nm is 4 ps/nm∗km, over a 50 km link we obtain 5∗4∗50 = 1 ns of dispersion and 0.4∗50 = 20 db of attenuation.

To take advantage of the minor attenuation at 1550 nm, narrow spectral linewidth lasers (10^{-2} nm) can be used.

1.3. Sources and Detectors for Fibers Transmission

1.3.1. Sources

For frequency modulation up to 200 Mb LED's can be used. The LED's commercially available for optical transmission are made by semiconductor materials of the type GaAs, $Ga_xAl_{1-x}As$ or $Ga_xIn_{1-x}As_yP_{1-y}$.

For frequency modulation up to 5 Gb (commercially available) LASER's are used.

1.3.2. Detectors

Only two type of detectors are used: APD's and PIN photodiodes . Avalanche photodiodes have a larger gain∗bandwidth product but they need high and very stable voltage supply. PIN photodiodes are preferred, because they are supplied with low voltage and they have a lower noise .

1.4. Connectors

Various types of connectors are commercially available: SMA, MINI-BNC, BICONICAL, etc.

On the apparatus side, the source or the detector plugs directly into the connector in a very simple self-adjusting way (no centering operation is needed).

On the cable side, specialized man-power and tools are needed for making connectors. Typical prices for commercially available connectors are: 10 $ + 15 $ man-power for multi-mode fibers and 40 $ + 80 $ man-power for single-mode fibers.

1.5. Cables

Fibers are first protected and then arranged in cables. A mono-fiber cable is a cable containing only one fiber. Many fibers, after protection, can be arranged together in multi-fibers cables; this solution is adopted for long and medium distances. For short distances is preferred to put many single-fiber cables together; the higher price of this solution is compensated by the simpler way of splitting the fibers at the ends of the cable.

The price of cables depends from their make-up and number of fibers. Typical prices for mono-fiber cables are 1 ÷ 2 $/m.

2. COMPARING FIBERS TO ELECTRICAL CABLES

The advantages in using fibers with respect to electrical cables are: absence of noise due to electromagnetical pick-up; no ground-loops; no cross-talk; smaller cable surface; lighter weight; possibility of high rates over long distances and lower price.

The disadvantages are: difficulty or impossibility of splitting and broadcasting signals: sensitivity to radiation damage and need of specialized people and tools to make connectors.

3. APPLICATIONS

A large number of applications of fibers to transmission in HEP are in progress. Only a limited number of relevant examples will be presented.

3.1. UA1 U-TMP Calorimeter Optical Link

PATROL (PArallel TRansmission Optical Link), is a link developed for UA1 experiment and used on a large scale (400 pairs). A couple of PATROL RX/TX are used like a pseudo parallel register, loading data into one side and outputting it in the other, having, in between, up to 3 Km of fiber. The transmission speed of PATROL is 2 Mbytes/sec and both TX and RX are housed in standard 28 pin metal box.

The links operate over a distance of 150 m. Each link connects an ADC board, located on the apparatus, to a receiver card (ORCA Optical Read-out CArd), containing four receiver channels, and located in the counting room.

50/125 μm multi-mode fibers, arranged in 18-fiber cables, are used. The price of the link is 680 Sfr/ch; for the electrical solution the price would be 1080 Sfr/ch (60 % more!). The cable surface is 5 mm^2/ch and for the electrical solution it would be 90 mm^2/ch.

VME and CAMAC applications of PATROL are also available.

3.2. DELPHI Forward Electromagnetic Calorimeter Link

In this apparatus the front-end electronics is very sensitive to electromagnetic interferences and ground-loops (the <ENC> of the system is 300 e). For this reason the optical isolation of the front-end electronics from the FASTBUS acquisition system was adopted.

The digital data link developed for UA1, PATROL, was used for communication over 7 m of 50/125 μm multi-mode fiber.

3.3. Analog Data Link at SLD

Also in this case the noise due to electromagnetic interferences and ground-loops is a limiting factor of the front-end electronics performance. Optical isolation of the detector from digitization circuits has been chosen. 170.000 channels of read-out electronics are multiplexed in groups of 256 ch/fiber at the transmission rate of 0.67 MHz.

For the optical solution Hewlett-Packard HFBR series of component were used.

The system has a linearity of ± 1% over a 6250:1 dynamic range and operates on a distance of 70 m using 100/140μm multi-mode fiber. The price is 175 $/ch.

3.4. LEP Fiber Optics Communication

The expected radiation dose inside the tunnel of LEP precludes the use of optical fibers. Optical fiber cables will be layed over surface routes connecting some of the LEP sites. The points that cannot be reached by surface routes, will be connected by coaxial cables in the LEP tunnel.

The fibers will transmit television pictures, time division multiplex data (TDM) and other informations. Special fibers will transmit the signals for the LEP radio-frequency timing and syncronization.

The implemented system is based on TDM technology, following the CCITT 703 series of recommendations. Both single-mode fibers and multi-mode fibers will be used. The maximum unrepeated link lenght is 8 Km.

3.5. Gran Sasso Laboratory Optical Communication System

In this case the underground laboratory will comunicate with the external control room by means of a 100 single-mode fibers cable. The fibers will transmit voice, television pictures, experimental data and will support the ETHERNET LAN (fig. 3).

Also this system is based on time division multiplex (TDM) and it will be implemented on four levels, starting from LEVEL 1. An experimental LEVEL will be implemented in parallel.

3.5.1. Experimental LEVEL

Is a project, STARNET, sponsored by INFN of Italy, for the realization of a Metropolitan Area Network (MAN). The topology of the system is based on a "Star-Bus", with a maximum of 5 stars without repeater and extensible with repeaters on the bus.

A maximum number of 8 hosts will be connected to each star. The system will operate at 140 Mb, over a maximum distance of 50Km.

3.6. ALEPH FASTBUS to FASTBUS Optical Link

The link will connect the TPC read-out to the VAX acquisition system of the ALEPH experiment of LEP. It is a FASTBUS to FASTBUS link implementing the standard FASTBUS protocol by means of two FASTBUS cards each on the segment to be connected.

The link will operate over a distance of 500 m using 50/125 μm or 100/140 μm fibers. The rate on the A/D lines is 2 Mbytes/sec.

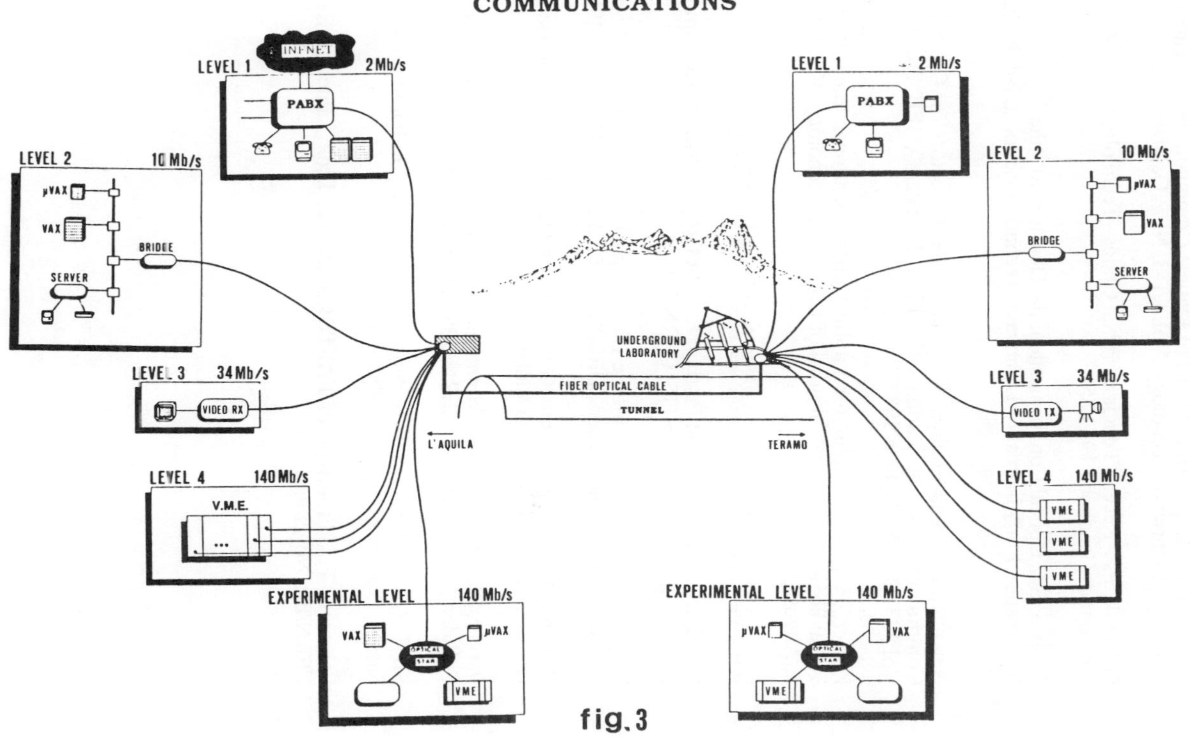
GRAN SASSO LABORATORY
COMMUNICATIONS
LEVEL 1
2 Mb/s
INFNET
PABX
LEVEL 2
10 Mb/s
μVAX
VAX
BRIDGE
SERVER
LEVEL 3
34 Mb/s
VIDEO RX
LEVEL 4
140 Mb/s
V.M.E.
...
EXPERIMENTAL LEVEL
140 Mb/s
VAX
μVAX
OPTICAL STAR
VME
L'AQUILA
FIBER OPTICAL CABLE
TUNNEL
TERAMO
UNDERGROUND LABORATORY
LEVEL 1
2 Mb/s
PABX
LEVEL 2
10 Mb/s
μVAX
VAX
BRIDGE
SERVER
LEVEL 3
34 Mb/s
VIDEO TX
LEVEL 4
140 Mb/s
VME
VME
VME
EXPERIMENTAL LEVEL
140 Mb/s
μVAX
VAX
OPTICAL STAR
VME

fig.3

REFERENCES

Midwinter, J. E. ,"Current Status of Optical Communication Technology", Journal of Lightwave Tecnology, Vol. LT-3, N° 5, October 1985.

Polishuk, P. ,"Recent Trends in Fiber Optics Technology and Markets", Fiberoptics Handbook 1984/85.

Catania, B. ,"Optical Communication: the Doorway to the Information Era", Journal of Lightwave Technology, Vol. LT-4, N° 7, July 1986.

Newman, D.H. and Ritchie, S. ,"Sources and Detectors for Optical Fibre Communications Applications: the First 20 Years", IEE Proceedings, Vol.133, Pt.J, N° 3, June 1987.

Kaiser, P. ,"Fiber Cables for Public Communications: State-of-the-Art Technologies and Future", Journal of Lightwave Tecnology, Vol. LT-4, N° 8, August 1986.

Motorola ,"Optoelectronics Devices Data Book", 1983-84.

Cennini, P. , Kulka, Z. and Petrolo, E. ,"PATROL -PArallel TRansmission Optical Link", UA1 TN/86-115, CERN/EP.

Lippi, I. et al. ,"Proposal for the Data Acquisition System of the Forward Electromagnetic Calorimeter", DELPHI Internal Report, CERN/EP, 1988.

Cisneros, E.L. and Burgueno, G.F. ,"Analog Data Transmission Via Fiber Optics", SLAC-PUB-4105, October 1986.

Parker, R.C.B.C. ,"The Communications Infrastructure for the LEP Collider", Internal Report, CERN/SPS, September 1987.

SIP-DR/AM , "Sistema di Telecomunicazione per il Laboratorio del Gran Sasso", Pescara - Italy, December 1987.

Formenti, F. ,"Prototype of FASTBUS Optical Link", Internal Report, CERN/EP, November 1987.

THE DATA ACQUISITION SYSTEM FOR THE CRYSTAL BALL AT LNS

P. Finocchiaro, C. Maiolino and P. Piattelli

Istituto Nazionale di Fisica Nucleare, Laboratorio Nazionale del Sud
Viale A.Doria (angolo Via S.Sofia), 95100 Catania
ITALY

ABSTRACT

The open architecture of the data acquisition system for the 4π Crystal Ball multidetector is described. Noticeable is that this system is not bound to any particular kind of computer or software, due to its simple architecture that guarantees its eventual implementation with different components. The only requirements are the use of a multiprocessor bus (VME) for the computer part, and a very simple ECL differential bus (FERA) for the fast data transfer.

1. INTRODUCTION

The Crystal Ball multidetector system, now under construction at the Laboratorio Nazionale del Sud, is an array of 420 Barium Fluoride scintillator crystals, arranged around the target in the shape of a ball plus a ring in the forward direction [1]). This system was designed to be used in experiments with the heavy ion beams of the LNS as a γ-ray and light charged particle detector. Due to the number of crystals and the high event rate sustainable by the detector, the realization of a dedicated data acquisition system becomes necessary.

The data acquisition system that we have designed [2]) is based on a "one broadcasting / many receivers" data transfer architecture, to achieve a number of relevant features, such as:

- subdivision of the whole system into independent subsystems
- high modularity at subsystem level
- ease of synchronization
- ease of upgrading, with respect both to the hardware and to the software
- on-line and off-line monitoring of both accepted and rejected data.

The analog to digital conversion system we use is FERA [3]), which is firmly embedded into the preanalysis system described in a separate contribution to this conference. Such a preanalysis system acts also as the data transmitter, using an ECL bus as broadcasting medium. Each receiver subsystem, VMEbus based, is implemented as a multiprocessor system dedicated to special functions.

2. SYSTEM FEATURES

The basic system configuration consists of the preanalysis system and of three receiver stations, of which at least one is enabled as "active listener": in other words it can synchronize the data acquisition rate of the whole system to its own processing rate, forcing a veto condition during the dead time. Conversely a "simple listener" may only sample data at its own maximum rate, without interacting with the data synchronization. The three receivers perform the following functions (see Fig.1) :

- monitoring, live display of accepted and rejected data
- data storage, rejected data storage, data replay
- on-line event filtering, histogramming, secondary storage.

Monitoring has to be performed by a specific processor that accumulates for each crystal two spectra during different time intervals, to appreciate temporary deviations from the average behaviour. This can be accomplished either by selecting a crystal to be interactively tested or by running a correlation program which can signal excessive deviations.

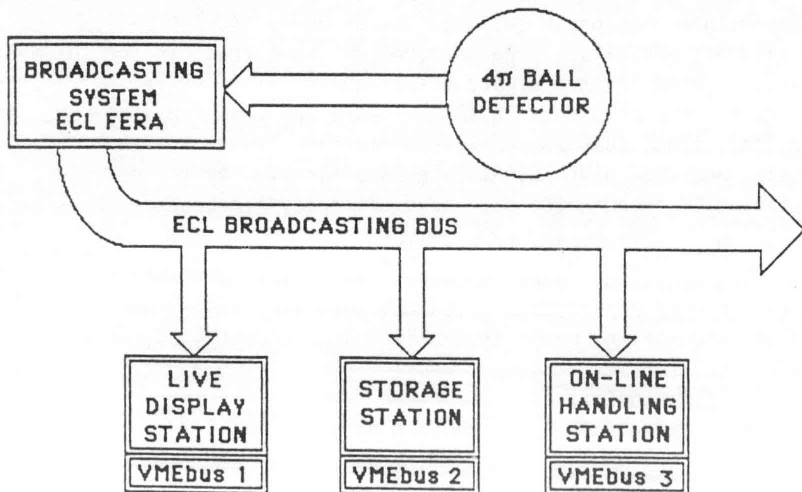

Fig. 1 - General scheme of the system. The broadcasting system includes the first level trigger, the analog to digital converters and the second level trigger. The three receiver stations are three independent multiprocessor systems, each of them can act as a simple or an active listener of the ECL data bus.

Live and rejected data display are performed by dedicated processors running the same program but driven by different action tables and triggered by different data validation signals on the ECL data bus. They easily allow to build high resolution one and two-dimensional histograms, with also colour map drawing facilities.

Data storage can be performed by one or more processors, alternatively enabled to write to one or more storage units: rejected data storage is symmetrical to rejected data display. Moreover, if the preanalysis system is disconnected, a storage process can be reverted to act as a transmitter, allowing the use of the system for the off-line data analysis.

The event filtering is to be performed by means of a parallel alghoritm, running on a number of very high performance processor such as MC 68020. It allows either a drastic reduction of the amount of data to be finally stored or the realization of a fast on-line event handler, sampling data on the ECL bus and making complex calculations to build high resolution histograms bound to the calculation results.

Each receiver can act either as a simple or as an active listener, the difference being its eventual capability to force the preanalysis system to a dead time condition until ready to receive new data.

3. SYSTEM CONTROL

Such a multiprocessor system would usually need a somewhat complicate control process. In our case this task is really simplified, because of the subdivision into non interacting subsystems: all the control actions have to be performed at subsystem level. The whole control process can be splitted into a number of independent sections, each one controlling a particular subsystem, or it can be splitted on different processors, each one running a different section of a control program. The main goal of each control section is to rule the multiprocessor activity on the related receiver subsystem. This can be easily accomplished using the software architecture described later.

The control system has to be implemented on a MacIntosh PC, which acts also as development system, by means of the MacPlinth-MacVee interface [4]) and of the MacSys software environment [5]). In such a configuration a MacIntosh PC can control up to 7 VMEbus crates plus 8 CAMAC crates.

4. SOFTWARE ARCHITECTURE

The software architecture we are using has already been tested on a VAX computer, were we have developed a data analysis system called TERESA [6]), based on three concurrent processes plus a controller one, which also plays the role of user interface.

All the processes have to be realized as table driven programs, using the facilities of the VMEbus in terms of local and global data resource sharing, multiple interrupt generation and handling. A simple message passing protocol is used, while a semaphore and a queue for each processor govern the data handling.

5. USER INTERFACE

The control process has firmly embedded in it the user interface. This last one is a set of functions that allow an easy interaction between the experimenter and the data acquisition system. The ergonomic design of user interfaces is subject to a number of rules regarding the way of selecting commands, options, parameters, etc.

The MacIntosh operating system implements a number of powerful facilities, like menus and windows, that can widely satisfy all the user interface requirements. It also features a well supported graphic kernel, with a number of libraries available that can be used to give a graphic representation of the various parts of the system, simplifying the user intervention and reducing the possibility error occurence.

6. CONCLUSIONS

The basic architecture of the system was already tested, and it proved to work efficiently. Now we are working to fully implement it in a VMEbus environment, and we plan to ultimate a prototype within 1988.

REFERENCES

1) E.Migneco, Proceedings of the Beijing International Symposium on Physics at Tandem, Beijing China May 26-30 1986, 559, World Scientific Publishing
P.Piattelli, Thesis (1985), Catania, unpublished.
2) LeCroy Res. Corp., "CAMAC model 4300 FERA", user's manual.
3) P.Finocchiaro, C.Maiolino and P.Piattelli, Nucl. Instr. and Meth. A, in print.
4) B.G.Taylor / CERN-EP, The MacVee hardware user manual.
5) D.Samyn, CERN-DD-UAl, MacSys a MacIntosh Editor/Assembler/Fortran/ Linker software development system.
6) P.Finocchiaro, C.Maiolino and P.Piattelli, in preparation.

THE CERN HOST INTERFACE AND THE OPTICAL INTERCONNECT

R.A. McLaren, T.J. Berners-Lee, D. Burckhart, R. Divia,
P. Gallno, B. Heurley, K. Hollingworth, D. Jacobs, H. Müller,
C. Parkman, E. van de Bij, A. van Praag
CERN, 1211 Geneva 23, Switzerland

A. Guglielmi, M. Rance
Digital Equipment Corporation GmbH, Freischützstrasse 91,
8000 Munich 81, Germany

T. Almeida, P. Gomes
Laboratorio de Instrumentaçâo e Fisica experimental de
Particulas,
Av. Elias Garcia 14-1, 1000 Lisbon, Portugal

P. Alves
Instituto de Engenharia de Sistemas e Computadores,
Rua José Falcâo 15-6, 4000 Porto, Portugal

ABSTRACT

Interfaces between Digital Equipment Corporation's VAX series computers and VMEbus and FASTBUS have been designed as part of the CERN Host Interface (CHI) project. Both the VMEbus and the FASTBUS interface share a common architecture which includes a powerful MC680x0 central processing unit, large data memories and a link port to connect to different members of the VAX family. Software support allows user software to be split between the VAX and the CHI processors whilst an enhanced VAX/VMS driver reduces operating system overheads. In addition an optical link allows the FASTBUS or VMEbus crate to be up to 1 kilometer from the host computer.

1. INTRODUCTION

The CERN FASTBUS Interface [1], has allowed us to acquire experience with interfacing to FASTBUS [2] and to discover the limitations of the design. A study was initiated to collect the requirements of high energy physics experiments (mainly those planned for CERN's Large Electron Positron accelerator), to review currently available hardware and to discuss various interfacing architectures. The conclusion of the working group, endorsed by the LEP collaborations, was that work should begin on the CERN Host Interface [3], a series of user-programmable interfaces, which would interface FASTBUS and VMEbus [4], to Qbus-, UNIBUS- and VAXBI-based VAX computers from Digital Equipment Corporation (DEC). A more recent requirement was the need to provide a long link between the computers above ground and the equipment in the underground experimental areas; a distance of up to 500 meters.

Experience with the CFI also showed that, for reasons of maintenance and support, it was important to use only DEC hardware and software in the VAXes. Standard DEC input/output registers are therefore used to provide a high performance channel to the VAX.

2. THE FASTBUS IMPLEMENTATION OF THE CHI

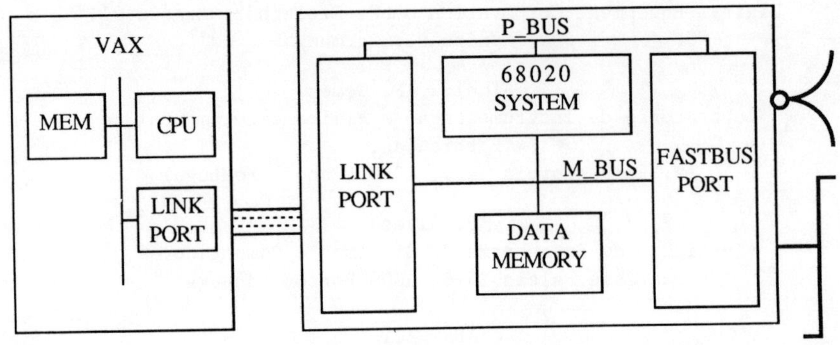

Block diagram of the VAX to FASTBUS interface

The VAX to FASTBUS interface is a single width FASTBUS module with four major components: the microprocessor system, the data memory, the parallel link port which connects to the input/output register in the VAX, and the FASTBUS port which contains the Master and Slave logic for FASTBUS. Two separate address and data buses (the P_bus and the M_bus) interconnect the components. The P_bus allows the microprocessor to access to control registers and to trigger single word data transfers in the link port and FASTBUS port. the M_bus allows Direct Memory Access (DMA) from the ports to the multi-ported data memory and access by the microprocessor to the data memory.

2.1 The Microprocessor System

The microprocessor system is a 16 MHz Motorola MC68020 with 1 Mbyte of local random access memory, 1 Mbyte of erasable programmable read only memory, a real time clock and two serial RS232 ports for communication with a terminal and a host computer during debugging.

2.2 The Data Memory

The 1 Mbyte multi-port data memory can be accessed from the microprocessor, the link port, and both the Master and Slave logic of the FASTBUS port. Arbitration for the data memory is normally performed on a cycle by cycle basis but can also be "locked" to allow unique access to one of the ports which is required, for example, during FASTBUS pipeline transfers.

2.3 The Parallel Link Port

The link to the VAX must be adaptable. Whereas 700 series VAXes and microVAXes have 16 bit full-duplex parallel input/output registers [7], the 8000 series machines have 32 bit half-duplex registers [8]. The link port hardware can multiplex 32 bit words onto 16 or 8 bit wide paths, performing byte or word swap if required. A set of adapter boards then provides the necessary choice of physical interface and drive levels. The on-board microprocessor can read which adapter board is connected, and so configure the hardware appropriately.

The link port may be accessed by microprocessor instructions, or data can be transferred directly to, or from, the data memory or FASTBUS using a 24 bit word counter. A pipeline register allows transfers to proceed at a theoretical bandwidth of 15 Mbytes/sec. In practice, however, the bandwidth is a present limited to 5.5 Mbytes/sec on 8000 series VAXes and 1.2 Mbytes/sec on the microVAX.

2.4 The FASTBUS Port

There are three main components in the FASTBUS port; the Master, the Slave and the Interrupt handler. The Master logic is based on the principle used in the General Purpose Master [9]. Operations on FASTBUS are triggered by accessing "key" addresses in the MC68020 address space, operands for these operations are transferred on the data lines. Any sequence of FASTBUS operations may then be constructed by a number of MOVE instructions. These instructions may also be stretched by allowing the MC68020 cycle to terminate only when the FASTBUS operation is complete. FASTBUS block transfers are handled by a dedicated DMA controller which allows transfers of up to 20 Mbytes/sec. FASTBUS pipeline transfer rates are programmable. In the case of non-zero SS responses, parity errors, or a timeout on the FASTBUS operation, an exception handler is invoked.

In the FASTBUS Slave port, the Control and Status Registers (CSR) for FASTBUS Interrupt Messages are implemented in hardware. Dedicated DMA logic transfers the message to the data memory. When the message is complete, an interrupt is sent to the microprocessor and a flag set. The interrupt service routine resets the flag, which enables subsequent interrupt messages to be received. FASTBUS Service Requests interrupt the MC68020 directly.

All other Slave CSR and DATA registers are emulated by microprocessor software.

3. THE VMEBUS IMPLEMENTATION OF THE CHI

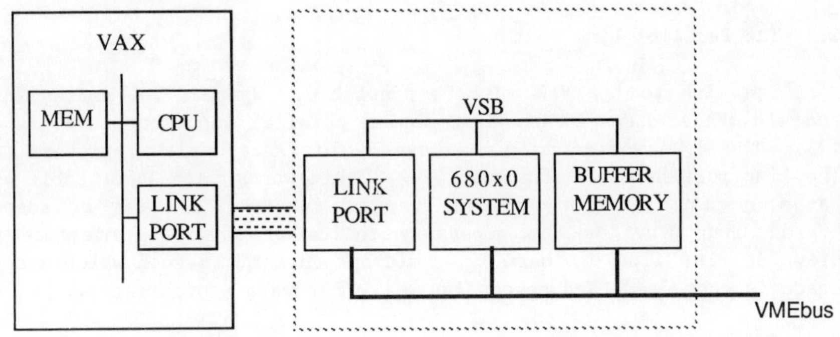

Block diagram of the VAX to VMEbus interface

Two factors influenced the decision to implement the VMEbus interface in a multi-board configuration. Firstly the board size is limited in VMEbus and secondly, VMEbus microprocessor modules and data memory are available commercially. The only module not commercially available was a VMEbus input/output register to connect to the VAX [10]. A detailed specification of this unit was written and the module was designed and built by Creative Electronic Systems of Geneva. This module has similar functionality to the parallel link port on the FASTBUS implementation, providing a link to the input/output register in the VAX via plug-on adaptors and byte and word swap logic. VMEbus logic provides a Master/Slave port with DMA capability. Data can be transferred with standard (A24) or extended (A32) address and D16 or D32 data widths. Block mode data transfer is supported to allow higher bandwidth. A Vertical Subsystem Bus [11] Master port is also implemented.

4. THE CHI OPTICAL CONNECTION

The LEP experiments will have a distance of up to 500 meters between the data acquisition computers and the detectors. This can be bridged in several ways: for example, by a FASTBUS to FASTBUS link, or by providing an optical extension module to the parallel link of the CHI. The latter is being being implemented as part of the CHI project.

The maximum length of the cables linking the VAX input/output register to the FASTBUS or VMEbus based link ports is 15 meters. The optical extension module allows the cable to be replaced, with no loss of bandwidth, by a pair of optical fibres and provides the user, at each end of the link, with a standard DEC input/output register interface.

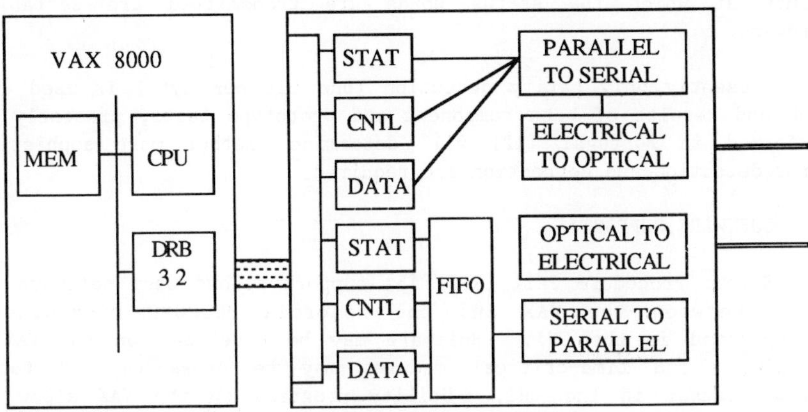

Block diagram of the CHI optical link

The signals generated by the input/output register in the VAX are classified into three groups; control, data and status. These groups request transmission and the winner of the ensuing arbitration sends a byte to the parallel to serial convertor, the status having the highest priority. In the case of a data word, which has four bytes, no further arbitration takes place until the four bytes have been transmitted. To distinguish between data words and control and status words a command byte is transmitted before any control or status word. The serial byte is then converted from an electrical to an optical signal and transmitted over an optical fibre.

In the remote module the serial information is converted back into an electrical signal, and from a serial stream into parallel bytes. The data bytes which form the normal data word are input to the main FIFO memory. Control words and the direction of transfer information, which forms part of the status word, are also input to the main FIFO to ensure that the sending sequence is respected. When information is input to the main FIFO two "tag" bits are simultaneously stored in a secondary FIFO to indicate the type; status, control, or data, of information in the FIFO. At the output of the FIFO the tag bits permit the regeneration of the standard input/output register signals by steering the output of the FIFO to either the status, control or data lines.

Flow control over the link is performed by a system of semaphores. When the receive FIFO is "almost full" it transmits a status words back to the transmitter. The transmitter recognises the status word and stops sending data. The receiver FIFO is then emptied and when the FIFO is "almost empty" the receiver transmits a status word to the transmitter to indicate that it is prepared to accept further data. The points at which the status words are transmitted are settable in hardware.

Presently only parity detection (one bit per byte) is used on the link and results of both component and prototype tests, presently being performed in Portugal [12] will determine whether more sophisticated error detection and correction are required.

5. SOFTWARE SUPPORT

Remote Procedure Call (RPC) [5] support allows user software to be split between the VAX and the Motorola MC68020 microprocessor incorporated in the CHI. Software may be developed on the VAX, for example, and time-critical modules may be cross-compiled for the microprocessor in the CHI. Utility programs on the VAX allow these routines to be loaded into the CHI, controlled, and linked transparently to routines on the VAX. At run time, when a program on one processor calls a subroutine on the other, parameters are passed to the subroutine, the subroutine is executed, and status and data returned to the caller.

5.1 The FASTBUS Routines

The NIM FASTBUS Standard routines [6] have been implemented as a CHI-resident library making use of the "key" address technique. The routines are callable locally or remotely via RPC from application programs residing on the processor or on the host. FASTBUS data may be transferred to and from CHI memory, or the "external buffer" mode of the Standard Routines may be used to transfer it directly to or from the VAX. Service Request and Interrupt Messages, accepted by the CHI FASTBUS port, result in the execution of the corresponding Standard service routines.

The slave software emulating Control and Data registers is independent of the implementation of the FASTBUS Standard routines.

5.2 The VAX to CHI Communication Protocol

A protocol is run over the link to accomplish:

- control of the direction of the (basically half-duplex) link;
- multiplexing of data between different tasks at either end;

- flow control, so that data are never sent unless a receiving buffer exists.

The functionality provided to the user is that of an ISO standard transport service, allowing the creation and deletion of logical task-task links. These links are used by the RPC system, and are also directly available to application programs.

On the MC68020 side, the software distinguishes between transfers of arrays of bytes, words and longwords. Advantage is then taken of the byte swapping ability of the hardware to keep each data type intact after transfer to, and from, the VAX.

6. CONCLUSIONS

The CHI is an example of how the combination of modular hardware, in the form of ports, can be interconnected using a well defined, high speed link. Although conceived with the aim of interfacing VAXes to FASTBUS and VMEbus, the CHI can also provide a VMEbus to FASTBUS connection.

The optical extension module permits the length of the high speed link to be extended from the present 15 meters to 1 kilometer without loss of bandwidth. The optical extension module has a standard Digital Equipment register interface which, coupled with CHI interfaces, can provide optical links from VAX to VAX, FASTBUS to FASTBUS or VMEbus to VMEbus.

User code may run in the interface in a familiar software environment which includes transport connections to tasks on the host machine, and the standard FASTBUS routines. Used in this way, the CHI provides flexibility unavailable with the previous generation of interfaces.

7. ACKNOWLEDGEMENTS

The work described in this paper is part of a joint project between Digital Equipment Corporation and CERN. We would like to thank E. Gerelle, DEC's joint project manager and F. Gagliardi (CERN) for setting up the framework for the project. At the LIP, in Portugal, we are indebted to G. Barreira who has organised the infrastructure required to construct the optical link. At CERN we would like to thank J. Allaby, J.A. Bogaerts, D.M. Sendall and H. Verweij for their support and K.J. Peach, P. Ponting and E. Rimmer, for their advice. Last, but by no means least, we would like to thank that essential group of people, our users.

8. TRADEMARKS

VAX, VMS, DRB32, DRQ-11, DRE-11, VAXBI, UNIBUS and Qbus are trademarks of Digital Equipment Corporation.

9. REFERENCES

[1] Jacobs, D. et al., "CFI user guide", private communication, January 1985

[2] ANSI/IEEE Std 960-1986, "IEEE Standard FASTBUS Modular High-Speed Data Acquisition and Control System"

[3] McLaren, R.A. et al., "The CERN Host Interface". Presented at the Nuclear Science Symposium, San Francisco, 21-23 October 1987

[4] IEEE Standard 1014, "VMEbus: A Standard Specification for a Versatile Backplane Bus"

[5] Berners-Lee, T.J., "Experience with Remote Procedure Call in Data Acquisition and Control", Proceedings of the 5th Conference on Real-Time Computer Applications in Nuclear, Particle and Plasma Physics, San Francisco, May 1987

[6] U.S. NIM Committee, "FASTBUS Standard Routines", U.S. Department of Energy DOE/ER-0324, March 1987

[7] "DRU-11C, DRE-11C, DRQ-11C Alternate Buffer DMA Interfaces", Digital Equipment Corporation, Computer Special Systems, Munich, February 1984

[8] "DRB32 Technical Manual", Digital Equipment Corporation, October 1986

[9] Müller, H., "A General Purpose Master (GPM) and Memory Module (DSM) for Online Applications", IEEE Transactions on Nuclear Science NS-33, no.1, February 1986

[10] Parkman, C.F., "The VMEbus/Host Interface I/O Register Module 'HyperVior' User's Manual", CERN DD/OC, private communication, April 1987

[11] IEC BUS 822 - Parallel Sub System Bus for the IEC BUS 821 (VSB) International Electrotechnical Commission 47B (Central Office) 22

[12] Alves, P., "Transmission Tests on the Optical Link", private communication, October 1987

AN OPAL 32 BIT COINCIDENCE ARRAY INTEGRATED CIRCUIT

M.J.French,F.Slorach
Rutherford Appleton Laboratory
Chilton,Oxon,England OX11 OQX

ABSTRACT

The I.C. is the part of the trigger processor in OPAL that correlates Z/R values generated from the Vertex and Jet detectors. A programable on chip memory array allows the user to program combinations of input data that may represent a valid track. The I.C. then multiplexes out valid combinations as they occur. A detailed description of the design is presented together with an outline of the software methods used.

1 INTRODUCTION

In many particle physics experiments it is necessary to generate triggers by correllating data from separate detectors. The trigger processor is composed of a Jet and Vertex detector that generate Z/R values from detected particles [1]. The coincidence array IC is used to output triggers from this data by comparing these values with data stored within a 32 by 32 bit memory cell array. In this array combinations of value that represent valid tracks are loaded. When a valid combination occurs the coincidence is multiplexed onto an output bus.

1.1 The discrete solution

It is possible to implement this function using RAM look-up tables (Figure1a).

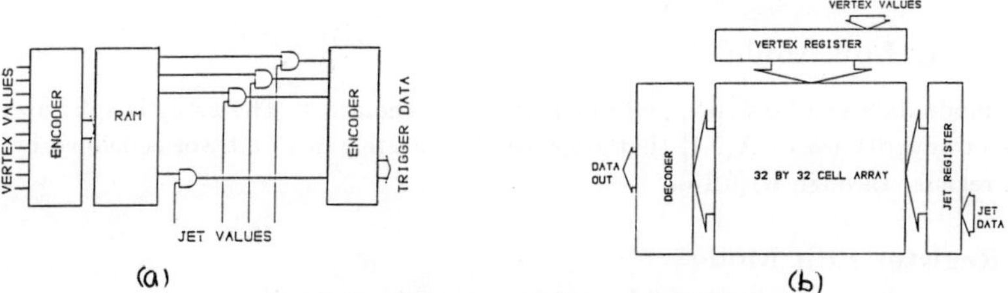

Figure 1: Discrete and Integrated Solution

This method however is complicated by the requirement for the circuit to run with 32 bit resolution, this necessitates the use of more than one RAM, which together with the extra logic required to manage the overlap between them complicates the logic. An equivalent circuit would have to comprise of at least 40 packages. Since this function is required for each segment of the detectors integration represents a very significant reduction in chip count.

2 DESIGN PHILOSOPHY

The IC was designed using the SDA[1] software design suite. This includes several fully integrated tools to assist in each stage of the design for circuits with both standard and custom designed componants. The HSPICE[2] simulator allows the designer to experiment with ideas before the layout begins and to optimise the design during the layout stage. Other utilities are provided to confirm that the layout does not violate the process design rules, and to ensure the electrical connectivity of the layout is consistent with that originally intended. When the layout is complete a final simulation provides the various characteristics of the custom cells which are needed to create a model for the digital SILOS[3] simulator. These are then used to simulate the entire design.

Test vectors were written to test the chip. Fault simulations were then carried out on these vectors which were then modified to acheive high fault cover, these test vectors have been passed on to the foundary to test the fabricated parts and guarantee working devices.

3 FUNCTIONAL DESCRIPTION

It was found to be economically advantageous to split the memory into four 16 by 16 bit quadrants. This improved the aspect ratio of the final design by allowing greater flexibility in the chip placement. Therefore the integrated circuit comprises, four 16 by 16 bit cell arrays, one 32 bit Jet register, two 16 bit Vertex registers, a 32 bit bidirectional data bus and the required data conrol logic. The operation of the circuit is split into four modes:

3.1 Running Mode

This is the normal operational mode of the chip. Data is loaded into the Jet and Vertex registers and combinations that correspond to cells loaded with a '1' are mutiplexed by two address inputs A⟨0:1⟩ onto the first eight bits of the bidirectional bus IO⟨0:7⟩.

3.2 Array Edit Mode

In this mode data can be written into or read out of the array. The array is split up into four quadrants defined by A⟨0:1⟩ that are editted separately in 16-bit words defined by the Vertex register through IO⟨0:15⟩.

3.3 Register Edit Mode

In this mode the Vertex register can be written into or read. It is split into two 16-bit sections defined by A⟨0⟩. This mode is designed to set Vertex bits for 'Array Edit' operations, since in that mode only one quadrant can be editted at a time when one Vertex word is written the 16 bits of the other register are cleared.

[1]The SDA software suite is marketed by the SDA corporation

[2]HSPICE is an analogue circuit simulation program supported by META SOFTWARE

[3]SILOS is a digital simulation program supported by SIMUCAD

3.4 Test Mode

This mode is similar to 'Register Edit Mode' except that full read/write access is allowed for both the Jet and Vertex registers without clearing other registers.

4 MACROCELL DESIGN

The macrocell contains 256 memory cells arranged in 16 columns of 16 corresponding to each Vertex address. Each column may be written to or read from either individually or simultaneously. If two or more columns are being read at the same time, the Or'd function must be read out.

Conventional static RAM (SRAM) cells are based on cross-coupled inverters written to and read from on a common complementary data bus that is addressed using pass transistors (Figure 2a).

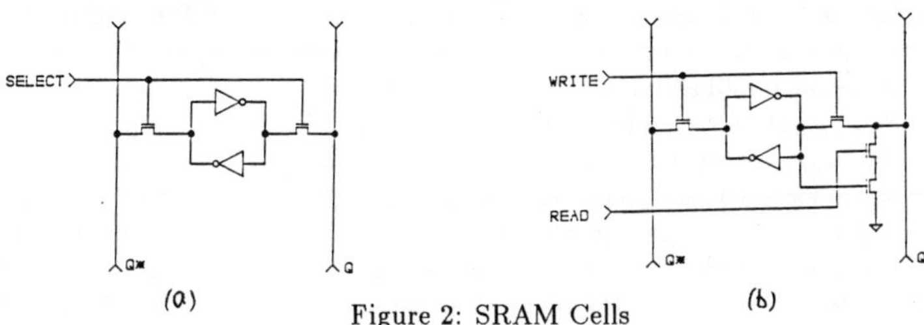

Figure 2: SRAM Cells

This approach is suitable for RAM applications because in this application it is only ever necesary for one cell to be addressed on each bitline at a time. However in this circuit more than one cell may be read at once so a modification to the cell comprising of two extra transistors was designed to facilitate this requirement (Figure 2b). The 'Read' control signal is separated from the 'Write' so that during a read cycle if a '1' is stored in any of the addressed cells the Q bitline will be pulled down whilst Q^* remains high. A read operation must now be preceeded by a bitline precharge. Two P-type devices are included for this purpose (Figure 3).

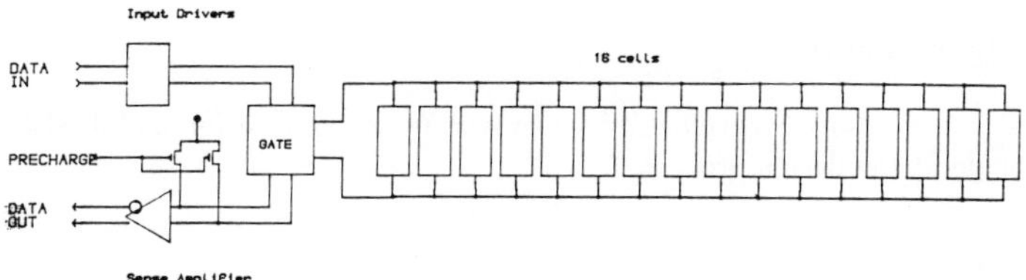

Figure 3: 16-Cell Bitline

4.1 Sense Amplifier

Satisfactory operation of the macrocell depends on the reliable detection of the differential voltages developed by the addressed cells. An inbalanced sense amplifier/latch was designed for this purpose. The heart of this is a cross-coupled pair of N-type devices.

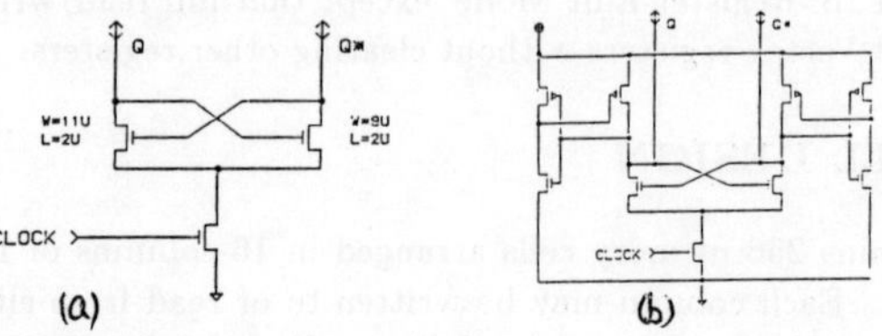

Figure 4: Cross-Coupled Pair/Sense-Amplifier

The positive feedback inherent in this circuit ensures that it latches into one of two states. Asymetry in the design guarantees that the circuit latches in one state if no cell is active (logic state 0), or the other (logic state 1) if any cell is. Small P-type devices, driven by inverters connected to the bitlines are included to restore the high bitline, the resulting circuit for the sense amplifier/latch is shown in Figure 4b.

The correct operation and speed of the sense amplifier is critically dependent upon the judicous choice of transistor sizes. These were evaluated initially by calculating the relative currents in the devices active during the read cyle and applying these to the width/length ratios of the devices (the current an N-type device is proportional its transconductance factor β and hence its width/length[2]). These values were then optimised using SPICE circuit simulation to both obtain the required speed and noise immunity of the bitlines.

5 CHIP LAYOUT

The SDA software utility 'Place and Route' was used to lay out the final design. This allows the user to manually place the custom cells and then automatically place the standard cells and IO ring. Once the placement was completed the layout electrical connectivity was verified and interconnect capacitances extracted to ensure that none of the cells would be overloaded. The dimensions of the final die is 5.8 by 6.8 mm, it contains the equivalent of 10,000 gates. The design is currently being fabricated.

Acknowledgements

We would like to thank S Jarolawsksi for providing the application and J Alsford for his help throughout the design process.

References

[1] S Jaroslawski,'A Fast Track Trigger Processor for OPAL Experiment',Impact Of Digital Microelectronics And Microprocessors On Particle Physics, March (1988).

[2] J A Pretorious,'Latched Domino CMOS Logic',IEEE Journal of solid state circuits, Vol sc-21, No. 4, August (1986).

Application of Bipolar Cell Array Technology to the Development of a Time Digitizer

G. Delavallade [1)], J.J. Jaeger [2)] and J.P. Vanuxem [1)]

Abstract

Application Specific Integrated Circuits (ASIC) are becoming more and more popular, as they offer large scale integration exactly tailored to user requirements. In addition, semi-custom technologies are affordable at moderate cost, which should make them attractive in high energy physics for low-to-medium volume applications. A Logic Cell Array has been designed which integrates the front end part of the LTD, a high performance time digitizer developed for LEP detectors[1]. A description of the basic circuitry is first given, followed by that of the selected ASIC technology. The CAD facility used during the development of this project is reviewed and finally the results are given.

1) CERN, Division EP, Geneva, Switzerland

2) IN2P3, Collège de France, Paris, France

1. THE BASIC CIRCUIT

The circuit to be integrated is entirely digital and consists of the front-end part of the LTD (LEP Time Digitizer)[1], a high performance, FASTBUS[2] compatible, yet low cost time-to-digital converter offering 2 ns resolution, 32 μs dynamic range, multi-hit and multi-event buffering capability, as well as data formatting and fast read-out speeds.

Fig. 1 shows the basic implementation for one channel which uses the standard ECL 10K series of SSI and MSI integrated circuits. The 125 MHz CLOCK and TEST inputs are both common to all the 48 channels implemented in the LTD. All input signals, including the 48 channel input (IN) signals are differential, this ensuring a good noise immunity by the inherent common mode rejection of the differential receivers.

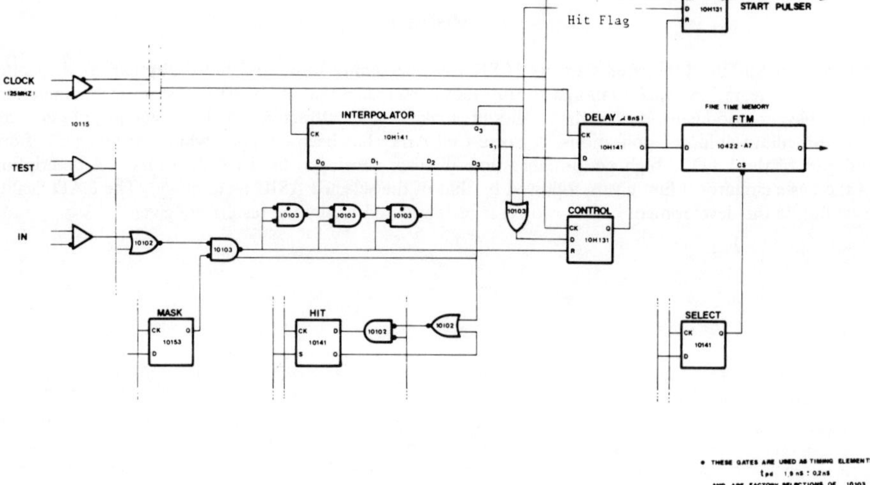

Figure 1: Front End Circuit of the LTD: one channel

The heart of the circuit is a 2 ns time interpolator, which determines in which of 4 possible time bins of 2 ns a hit occured on the wire during a clock period of 8 ns.

The interpolator operates on the following principle: during the data taking interval (the drift time for a drift chamber detector), the CONTROL flip-flop is normally (i.e. in the absence of a signal on the channel input) in the OFF state, its Q output is low, and so is the Mode Select input S1 of the shift register, which keeps the shift register in the "Parallel Entry" mode. At each active transition of the CLOCK signal (i.e. every 8 ns), the state of the channel input "IN" is interrogated and is retained at the Q3 output of the 4th flip-flop of the shift register. Whenever the channel input "IN" is activated by the arrival of a hit signal, D3 becomes asserted, and the immediately following active transition of the clock sets the "Hit Flag" in Q3 to "1" (at time "t") indicating the presence of a signal on the channel input, (or a "Hit"). At the same time t, the fine time (FT) information, which is obtained by successively delaying the negative transition of the channel input signal by 2, 4 and 6 ns, and is presented on the inputs D0, D1 and D2 of the first 3 flip-flops of the shift register, is also stored into these 3 flip-flops. At this stage, the FT information, which designates the time of arrival of a hit (leading edge of signal) relative to the time t-8 ns, is coded as follows:

Q0	Q1	Q2	FT (in ns)
0	0	0	0
0	0	1	2
0	1	1	4
1	1	1	6

The three delays of 2 ns are obtained by means of external ECL 10103 gates which are factory-selected so that their t_{pd} = 1.9 ns ±0.2 ns (the extra 0.1 ns is lost in the unavoidable wiring). Four nanoseconds after the hit information (consisting of the hit flag at Q3 and the FT at Q0, Q1 and Q2) has been entered from the parallel inputs D0 to D3 of the shift register (i.e. at time t + 4 ns), the CONTROL flip-flop is set, asserting S1 and placing automatically the shift register in "Shift" mode. The Hit Flag and its corresponding FT information may then be serially transmitted to the Fine Time Memory (FTM) during the 4 successive active transitions of the clock. Note that an 8 ns delay is necessary between the interpolator and the FTM in order to achieve proper timing; this is implemented by merely inserting a fifth flip-flop in the chain.

The rest of the circuit contains an input multiplexer (symbolically represented by an OR gate on Fig. 1) allowing for a TEST input which is common to all channels and is enabled in test mode, a MASK flip-flop to enable or disable the channel input signal (which either results from the "IN" input signal from the detector in normal data acquisition mode, or from the "TEST" input signal in test mode), a HIT flip-flop which indicates that a signal was present on the wire at least once during the data taking interval (Note that this flip-flop is different from the Hit Flag), a SELECT flip-flop to enable the FTM memory "slice" corresponding to the channel to be enabled (both for Write during data taking and for Read during a subsequent conversion), and a common "Pulser" circuit (not shown on Fig. 1) which is responsible to generate the proper amount (either 5 in normal mode, or 1 in VETO mode) of write cycles to the FTM just after time t.

One of the advantages of this circuit, compared to other time interpolation techniques, is that it is digital, and therefore requires neither a precise and accurate initial component selection (except for the 10103 gates which are used as active delay elements) nor a frequent recalibration to check for possible drifts of component values with time or temperature (the drift figures quoted for ECL gates, and particularly those quoted for the 10103, are very small). Furthermore, the circuit is fully synchronous, and a single clock is required. Finally, such a fully digital approach lends itself very well to integration by means of either full custom or semi-custom technologies available to-day.

The motivations for integrating this circuit are numerous. The first and main one is simply.....SPACE! An early implementation with standard ECL 10K ICs allowed space for only 8 channels on an "extended CAMAC" board, which was built in order to check the "idea" of the proposed TDC technique, and can be seen in Fig. 2. With the use of the standard ECL 10 K and 10 KH families, the maximum number of channels which was then expected to be housed on a standard FASTBUS board was no more than 12. It therefore appeared very clearly that the new TDC technique would have no future without the development of an ASIC containing multiple channels in a single chip.

Figure 2: LTD CAMAC Prototype (8 channels)

The basic circuit was redesigned with this in mind, and a special effort was made to optimize the logic and to incorporate 4 channels in the chip. This second step in the feasibility study is shown by the logic diagram of Fig. 3. Rapid calculation then showed that a semi-custom bipolar technology offering chips with a complexity of about 1000 equivalent gates would do the job, providing that some drastic requirements (such as flip-flops toggling at 250 MHz) could be met. Minimizing the power consumption was also a major concern in the choice of the technology.

2. The ACE Family, and the ACE 900 L Logic Array

2.1 The ACE Family

The ACE (Advanced Customized ECL) family of semi-custom ICs from Philips offers seven different prediffused circuits with a complexity ranging from 200 to 3000 equivalent gates[2], two of these circuits having additional RAM (up to 1280 bits). The family uses a "Subilo-N" process which is characterized by deep-oxide intercomponent isolation in place of the usual p-n junctions.

The advantages of the family are low propagation delays, low power consumption, the possibility to interface with CMOS (5 V), TTL and ECL (either 10 KH or 100 K) circuits and, last but not least, the availability of numerous logic functions in a powerful library[3]. Two versions of the family exist: the "Low Power" and "Turbo". The Low Power version offers an equivalent gate delay of 350 ps at 0.8 mW/gate, while the Turbo version achieves the shorter (200 ps) equivalent gate delay at 1.6 mW/gate. A factor of merit for a given technology may be defined by the product of the equivalent gate delay and the power consumption per equivalent gate. This factor of merit is then 0.28 pJ and 0.4 pJ for the

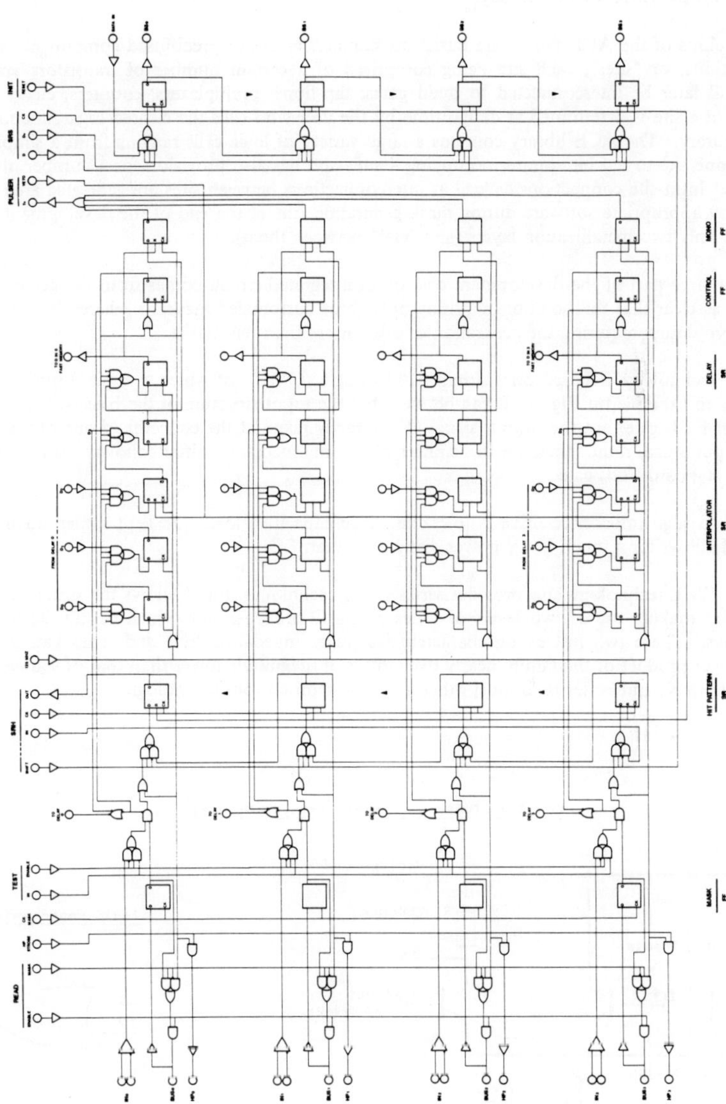

Figure 3: Front End Circuit of the LTD optimized for 4 channels

Low Power and Turbo versions respectively, the Low Power version offering one of the best figures available on the ASIC market to-day.

All chips of the ACE family are based on standard arrays of prediffused components arranged in cell locations, or "sites", each site being comprised of a certain number of transistors and resistors which will later be interconnected to build gates, flip-flops, multiplexers, counters, etc... The functionality of a site is determined at design time by the user who calls the desired logic function or "cell" from a library. The ACE library contains a large variety of logic cells ranging from a simple gate occupying one site to a 4 bit counter occupying 4 sites and has also provision for a number of I/O interface cells. Intra-site connections as well as interconnections between sites are generated after schematic capture by appropriate software during mask-generation (i.e. at the end of the development phase) by means of only two metallization layers (and "vias" between them).

As a large part of the development cost of an integrated circuit comes from the generation of the masks, it is clear that this solution requiring only three customized masks is inherently cheaper than a full-custom design requiring the generation of often more than ten masks.

The technology is based on Current-Mode Logic (CML), of which a basic 2-input OR gate is shown as an example on Fig. 4. It can be seen that the input structure of the basic gate is the same as that of an ECL gate, but the output signal Q is directly taken at the collector of one of the transistors of the input stage, rather than on the emitter of an intermediate emitter follower transistor as in the case of a standard ECL gate.

Such a high impedance drive is tolerable, since capacitive loads are kept rather small inside the chip, and allows for a much lower power dissipation than ECL.

The CML technology also provides series gating techniques. Fig. 5 shows the structure of a 2 to 1 multiplexer making use of two levels of series gating (3 are possible), which permit faster equivalent gate delays. These two interesting characteristics (high impedance drive and series gating) make the speed-power product of the family nearly two orders of magnitude lower than that of standard SSI and MSI ECL 10 K, and better than most other available semi-custom technologies.

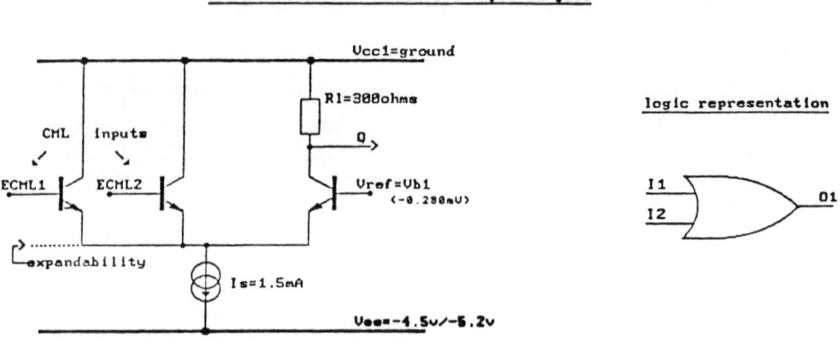

Figure 4: CML (Current Mode Logic): 2-input OR Gate

CML STRUCTURE : INTEGRATION BY SERIES GATING

2 LEVELS ==> 2 to 1 MULTIPLEXER (EX: MXY04)

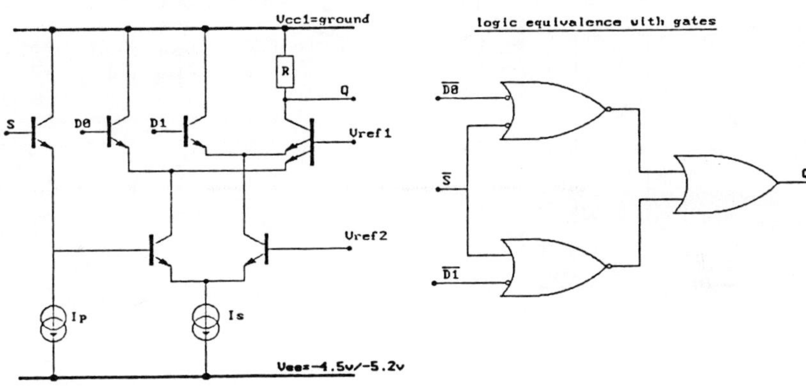

Figure 5: CML: 2-to-1 Multiplexer examplifying Series Gating

2.2 The ACE 900L Logic Array

This type of array was chosen as it fits well with the requirements of the circuit, offering 900 equivalent gates, one of the lowest possible power consumptions in this category and the best price/ performance ratio.

Fig. 6 shows the basic architecture of the array before and after Cell Placement. The centre area contains 36 major sites which are grouped by 4 to allow for space between groups in the two directions for metal interconnections (also called "channels") between sites; each major site is composed of 38 transistors and 16 resistors. On the periphery of the chip are 58 I/O sites, which are used to interface the circuit with TTL or ECL (10 KH or 100 K) external signals, and 22 minor sites, which may also contain logic cells like the major sites, but these cells are of lower complexity due to the smaller size of the sites. The dimensions of the chip are about 6x5 mm^2.

The package which has been selected is a ceramic pin grid array (PGA 64) with 64 pins, of which 58 are dedicated to I/0 interconnections, 2 to V_{EE} (-5.2 V) and 4 to V_{CC} (0 V). This package has good thermal characteristics (Rth_{JA} = 25° C/W maximum), making the use of a heat sink only optional. Various stages of the packaging process include die bonding with silver-filled epoxy resin, automatic ultrasonic gold ball and 32 μm wire bonding, hermetic epoxy lid sealing (lid made of ceramic), and gold plating of the pins (suitable for insertion in a socket). Fig. 7 shows a picture of the chip and Fig. 8 a picture of the selected package.

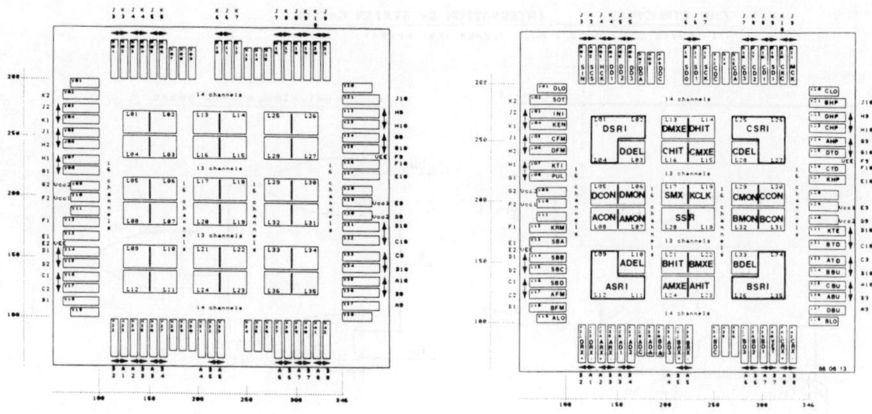

Figure 6: ACE 900 Structure: a) before and b) after Cell Placement

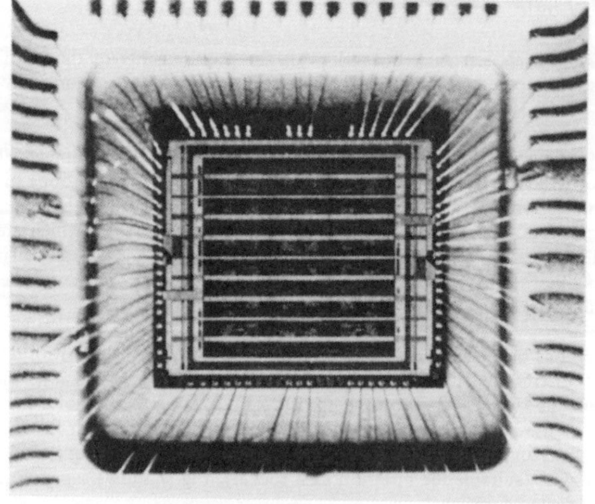

Figure 7: The Chip

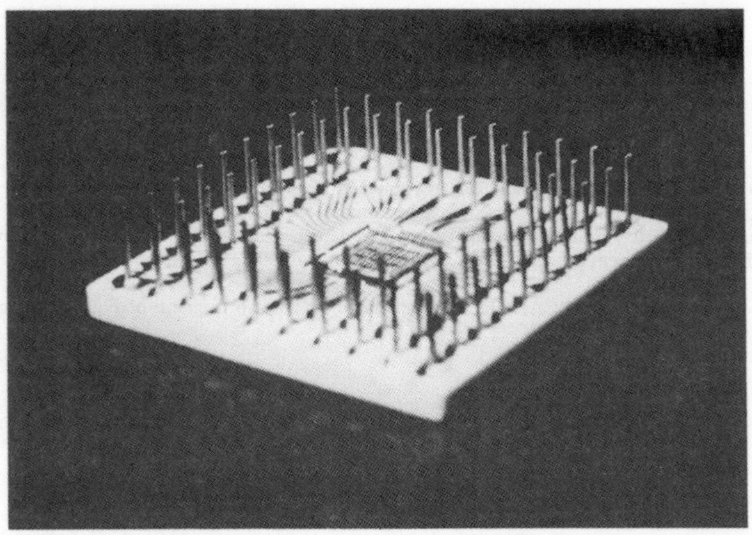

Figure 8: The Selected Package

3. ABOUT THE CAD SYSTEM

This ASIC was developed with a MENTOR GRAPHICS software and hardware system. Philips-RTC design centers offer also to their customers the possibility of using DAISY, VALID or IBM – PC XT/AT (FUTURENET) systems, but MENTOR is the most popular (12 workstations at IMSC Fontenay).

The CAD software consists of an association of MENTOR non-dedicated programs and PHILIPS array-specific programs (ACE or CMOS).

3.1 HARDWARE

MENTOR workstations are APOLLO stand-alone computers (DN 660 or new DN 3000) which are all interconnected by the DOMAIN network.

The main specifications of the DN 660/DN 3000 are listed below:

Central Processor	:	- 32 bits (bit slice or 68020 + coprocessor) - 1 MIPS
Main Memory	:	- 2 Mbytes, expandable
Graphics	:	- 19″ colour, 1024 x 1024 pixels
Peripherals	:	- floppy disk (1.2 Mbytes)

- hard disk (80/140 Mbytes)
- keyboard with function keys
- mouse

3.2 SOFTWARE

APOLLO'S AEGIS O.S. (version 2.0) is a multi-process, multi-window operating system which is well suited to CAD applications. Based on a performant 68020 CPU with up to 4 Mbytes of RAM, the system offers effective multitasking capability and a designer commonly works with 3 to 5 processes in parallel (i.e. concurrent schematic capture, simulation, compiling and peripheral access...).

The different stages of the CAD chain are presented by the two diagrams of Fig. 9. Commenting on these two diagrams:

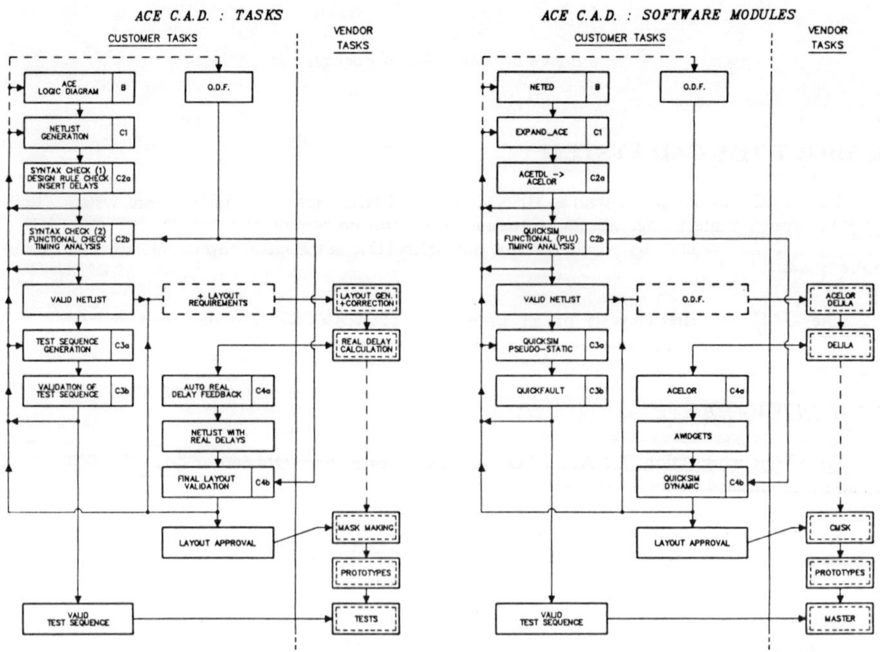

Figure 9: CAD Tasks and Software Modules

3.2.1 Schematic capture:

This first step consists of translating the block diagram of the project into an "implementable diagram", using the proper cells of the appropriate library (i.e. technology). Several iterations are necessary, and direct transcription on the screen is reserved only to experienced ACE designers. So paper work is always useful for this step...., after which schematic capture starts.

- NETED is a graphical editor which enables the drawing of a circuit diagram, the naming of cells, equipotentials, signals, etc...

  Commands are displayed in a menu and can be entered either with the cursor (and mouse), or with function keys, or by typing the command on the keyboard.

  One can work either with a hierarchical design or with a page design. The choice depends on the type of circuit. The only difference is in the translation of the names during compilation of the circuit.
- SYMED is a symbolic editor in complement. It is used for hierarchy and at circuit level for specialised programs.

  NETED and SYMED are MENTOR programs.

3.2.2 Schematic verification - Compilation:

Each graphical editing session ends with a "CHECK SHEET" command, to verify names, pins left open, short-circuits between different wires etc...

- After this syntax verification, "EXPAND for ACE" can be run. It is the compilation of the schematics which stops at the level of ACE cells, and gives a MENTOR netlist as output.
- "EXPAND for SIM" creates a simulation table (down to the gate model level).
- ACETDL translates the MENTOR netlist into a TEGAS netlist.
- ACELOR is an ERC/DRC program. It checks the electrical and design rules:

  - maximum DC fan-outs
  - maximum AC fan-outs
  - expander rules
  - number and type of cells and I/O

  ACELOR also calculates the fan-out delays for each output of each gate, and adds "estimated" delays due to wires. After routing, ACELOR will add the "real" delays extracted from the actual layout.
- AWIDGETS back-annotates the delays computed by ACELOR into the MENTOR netlist as compiled for simulation. ACELOR is a Philips technology-dependent software for all ACE (ECL) arrays of the family.

3.2.3 Simulation

This phase consists of the functional and timing verification of the design.

- This is done by QUICKSIM (Mentor Software). An alphanumeric input stimuli file has to be written and compiled by QUICKSIM. Two types of outputs can be displayed on the screen:

  - waveforms
  - binary maps

A multi-window facility offers the possibility to view also a page of the schematics. Interactive tools enable the monitoring and observation of any node of the circuit. But simulation itself is a batch process.

Of course several iterations are necessary, which can imply schematic modifications, stimuli changes, followed again by a simulation, etc...

3.2.4 Placement and Routing:

A placement file has to be written, which gives a correspondence between the logical names of the internal and I/O cells and their topological positions on the chip.

Graphic tools have been developed by Philips-RTC to perform this task and are now becoming available.

- DELILA is an automatic routing software from Philips. It runs on VAX with a Tektronix (4115 or 4111) high resolution display. In this particular case, 97% of the connections were automatically routed. The others had to be done manually. A lot of routing modifications were necessary for critical pathes of our circuit, due to drastic timing requirements. Routing was done by a member of the RTC support staff. The results of the routing can be seen on Fig. 10.

The data resulting from this step are used for the mask making.

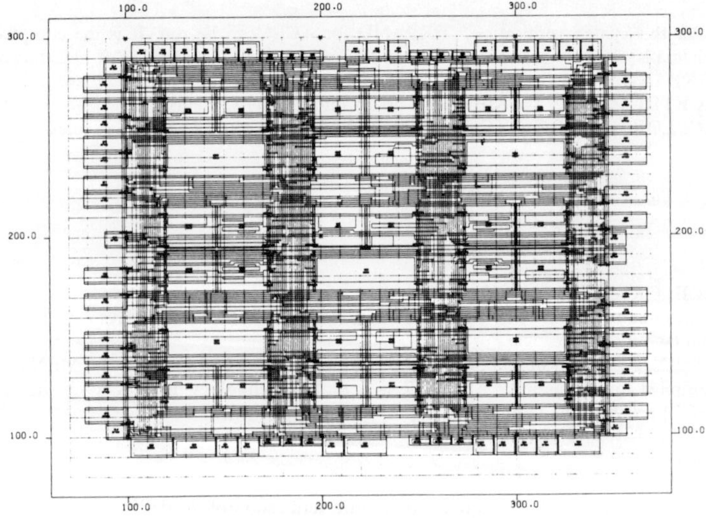

Figure 10: The Routing of the Chip

3.2.5 Delay Extraction

- "DELILA LOAD CHECK" command extracts the wire lengths from the final layout. These values are fed into ACELOR which computes the real delays, which are then fed back into the netlist for simulation by AWIDGETS. Thus the real circuit can be simulated again with QUICKSIM (for typical, worst and best case).

3.2.6 Test sequence

The test pattern file can be written in parallel with the placement/routing work.

The file derives from the already created simulation files, and is meant to test all inputs/outputs of the circuit in a most exhaustive manner.

These patterns will ultimately be applied by the test machine on the pins of the circuit at the end of the manufacturing process. In this particular application, about 1000 test patterns were used.

Mask making and tests were under the responsibility of the manufacturer.

4. RESULTS AND CONCLUSION

The present design makes use of the ACE 900 L chip as follows:

- Major Sites	:	36 used out of 36 available	(100%)
- Minor Sites	:	12 used out of 22 available	(55%)
- I/O Sites	:	58 used out of 58 available (all 58 available I/O pins are used)	(100%)

With such a semi-custom LSI circuit, a gain of a factor 30 in component count and 4 (10 if one excludes the external delay elements from this comparison) in printed circuit board space has been achieved for the front end part of the LTD: this is clearly shown in Fig. 2 and Fig. 11, which give a picture of the same circuit implemented before integration for 8 channels on a CAMAC test module, and after integration for 48 channels on a LTD prototype in FASTBUS. A second major benefit is the reduction of power consumption by more than a factor 5; the typical power consumption of the chip itself (not including the power dissipated in the external pull−down resistors) is 1.3 Watt. Other advantages include an invaluable reliability improvement which is the direct result of the drastic reduction of the number of components, and an overall (component and assembly) cost reduction which increases with quantities ordered, but is already effective for orders as low as 1000 pieces.

The overal development time for this ASIC was 6 weeks, including schematic capture, simulations, routing and test pattern generation, and prototypes were made available within 12 weeks after the development phase.

In conclusion the development of this ASIC has allowed a new time-to-digital conversion technique to be successfully brought to life. Because of their numerous advantages, it is expected that such semi-custom technologies will play a major role in the development of front end electronics for the read-out of future LEP (and post-LEP!) detectors.

Figure 11: LTD FASTBUS Module (48 channels)

5. ACKNOWLEDGEMENTS

This work was jointly supported by the CERN EP-Electronics R & D Group and the Channel Electronics Standardization Group of the DELPHI experiment. We are thankful to the engineers of the Philips International Microelectronics Support Center (IMSC) for their continuous assistance during the development of the project.

6. REFERENCES

1. The LTD: A FASTBUS Time Digitizer for LEP Detectors, G. Delavallade and J.P. Vanuxem, NIM (1985) 596-604, also EP-Electronics Note 85-06 and DELPHI Note 87-68 ELEC-28.

2. IEEE Standard FASTBUS Modular High Speed Data Acquisition and Control System, 1986.

3. ACE (Advanced Customized ECL) Design Manual, Philips, 1985.

4. ACE (Advanced Customized ECL) Cell Library, Philips, 1985.

The Contiguity Processor
A SIMD architecture for a 2nd level track trigger

G. Darbo, INFN, Genoa, Italy
B.W. Heck, CERN, Geneva, Switzerland

ABSTRACT

The contiguity processor (CP), designed for the Delphi time projection chamber (TPC), is a Fastbus 2nd level trigger processor, that makes heavy use of application specific integrated circuits (ASIC). The architecture is based on a bidimensional lattice of 4608 processing elements (PEs). Each of these PEs is connected with its 4 neighbours, and they all execute the same stream of instructions (single instruction multiple data architecture - SIMD). The CP instruction set is tailored to the special requirements of the contiguity algorithm.

1. INTRODUCTION

The contiguity processor (CP) is a special purpose processor that will be used for the 2nd level trigger of the Delphi time projection chamber (TPC). For a detailed description of the Delphi detector and of the TPC we refer to the technical proposal [1]. The general aspects of the contiguity processor as a trigger device have been described elsewhere [2,3]. In this paper we describe more details of its algorithm and internal architecture, which are now compiled in gate arrays.

We can consider the Delphi TPC as a big cylinder (2.8 m both in diameter and length) divided in two halves, each one with 6 sectors of 60° (12 sectors in total). Each of these sectors maps *sensitive cells* on its end-planes directly to the corresponding parallel inputs of one card of the CP. That is, cell $c_{i,j}$ at r_i and ϕ_j is wired to the $IMI_{i,j}$ (image memory input) of the CP. In a TPC sector there are 16 cells in r ($\Delta r = 5\ cm$) and 24 in ϕ ($\Delta\phi = 2.5°$) and therefore 16×24 $IMI_{i,j}$ in a CP card. Each $c_{i,j}$ gives a digital pulse when a charged track crosses its fiducial volume $\Delta r \times \Delta\phi \times \Delta z$ ($\Delta z = 1.4\ m$, one half of the TPC length). The delay, from the BCO (beam cross over) to the time the pulse is seen, is proportional to z.

2. THE ALGORITHM

The processing of an event in the CP is divided into two phases:

- *Data acquisition:* During the drift time in the TPC, i.e. for about 23 μs after BCO, the event in each of the 12 TPC sectors is recorded, formatted and loaded into the 8 image memories (IM) of a CP card. This is done in parallel by the IMIs: $IMI_{i,j} \rightarrow IM^{(n)}_{i,j}$ (where $n = 0,..,7$, $i = 0,...,15$ and $j = 0,...,23$).

- *Processing:* Once the event is in the CP, IMs are sequentially addressed to search for tracks. The processing of each IM takes about 1 μs for a total of 8 μs. The result is the contribution of the CP to the Delphi 2nd level trigger. Then the processing continues for a few hundreds of microseconds more and valuable data are prefiltered for the forthcoming trigger levels.

The full former sequence is executed unless a new warning (WNG) from the trigger supervisor restarts a new event.

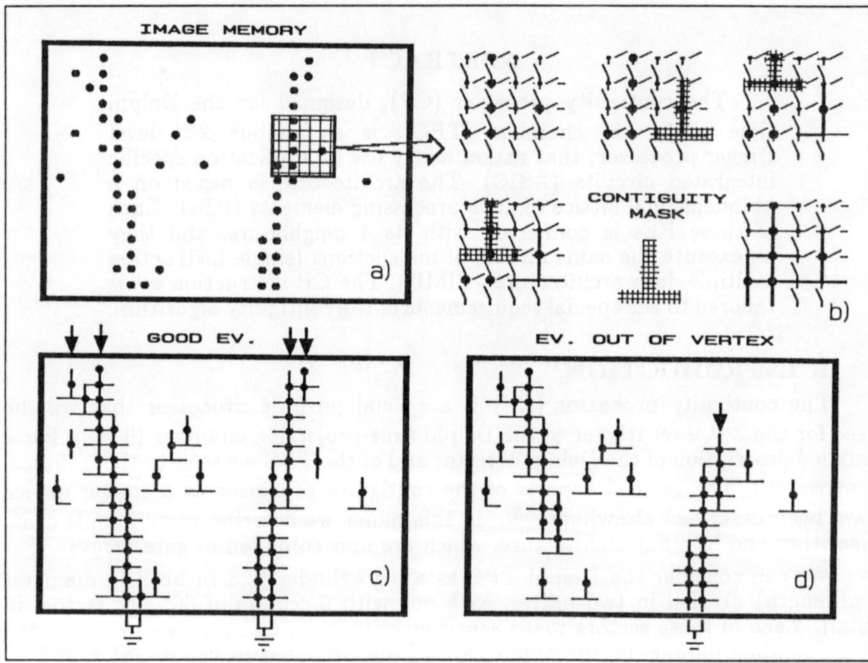

Fig. 1 : **The contiguity mask algorithm.**
In (*a*) a 'good event' on which we apply a contiguity mask (*b*). The result is two connective paths through the whole image (*c*). In (*d*) the two tracks do not come from the vertex region and are therefore not completely contained within one θ-slice.

Let us now see how data are reformatted inside IMs during the acquisition phase. We have seen that the mapping $c_{i,j} \rightarrow IMI_{i,j}$ is a transformation to polar coordinates. If we now consider the TPC z coordinate (TPC drift information) and we enable the recording of the ith row of the nth IM for time intervals $\Delta t_i^{(n)}$ so that corresponding $\Delta z_i^{(n)}$ are fully contained between two succeeding values of θ ($\theta^{(n)}$, $\theta^{(n+1)}$), then each IM will contain a θ-slice of the TPC volume. Spherical coordinates (r, ϕ, θ) are a convenient representation for events with tracks coming from the center of the detector (good events), because particles, with sufficiently high momentum, lay fully inside one of the 8 θ-slices (16 for the two TPC halves), and therefore in a whole IM (see fig. 1).

In the example of fig. 1*a*, we have a slice of a 'good event' as seen in an IM. To understand the contiguity mask algorithm, let us consider a mesh of switches as in fig. 1*b*. We superimpose the event (IM) to the crossing points of the mesh. Initially all the switches are open, then, for each large point (event), we apply the chosen contiguity mask, which causes some switches surronding the 'on' nodes to close. The graphic result is shown in fig. 1*c*. A signal, which is applied to all the switches of the bottom row, ripples through the network up to the top row in correspondence to tracks of a 'good event' (fig. 1*c*), while an 'out of vertex event' (fig. 1*d*) would not trigger. Choosing the proper contiguity mask or using a slightly different shaped one as a function of the IM row, is matter of a compromise in trigger acceptance and background rejection.

3. THE CONTIGUITY PROCESSOR

The contiguity processor is made of two kinds of modules: the contiguity processor slice (CPS) and the contiguity processor controller (CPC), both are Fastbus slave modules. The CPC simply synchronizes the CPS with the Delphi trigger signals and reformats the results from the CPSs before shipping them to the trigger supervisor.

The 12 CPS (one for each TPC sector) are the true core of the processor. Each CPS, a single Fastbus module, has three main components:

- *Program and data memory (PDM):* This 40-bit word memory stores the program used in the trigger. At each BCO the program starts from location 0 and, unless interrupted by early negative trigger response, it continues up to the end.
- *Contiguity array (CA):* This is the CPU of the system. Due to the highly specialized instruction set, we had to make the design of a semicustom IC (gate array). We divided the 16×24 PEs of the CA in 24 gate arrays.
- *Front end buffer (FEB):* This memory is loaded with the results coming from the upper row of the CA and it is read out each time an event passes the 2nd level trigger. The loading of the FEB is under control of the PDM. Its readout is done by a crate processor in Fastbus.

The gate array we have designed for the CA is the HSG7600 from IST (Innovative Silicon Technology), and comes with a 68-pin PLCC package. It uses 2 μm CMOS technology and has a complexity of 6000 gates. Since the end of last year we got 30 samples of those chips and we tested a full size prototype of the CA with 24 chips. Because the PDM and the FEB are still missing, we emulated them with a logic analyser system in which we loaded programs that finally will go into the PDM. The results are read back into the analyser for comparison. We tested the CA also with physics events generated in a special simulation program for the trigger. The tests run successfully up to 18 MHz, which gives enough safety for nominal 10 MHz clock frequency later.

3.1 Processing Elements (PE)

The CA is a bidimensional lattice of PEs, each of them is connected to the 4 nearest neighbours and to a cell ($c_{i,j}$) of the TPC. Connections to the up and down PEs are made by a sigle bidirectional line, while to the left and to the right PEs are made by two separate unidirectional lines (fig. 2*a*). All PEs are controlled by the same instruction stream in the PDM, while their data source is either inside the elements themselves or inside their neighbours or again coming directly from the TPC's input lines (see fig. 2*c*). All that makes of the contiguity processor a SIMD (single instruction multiple data) machine.

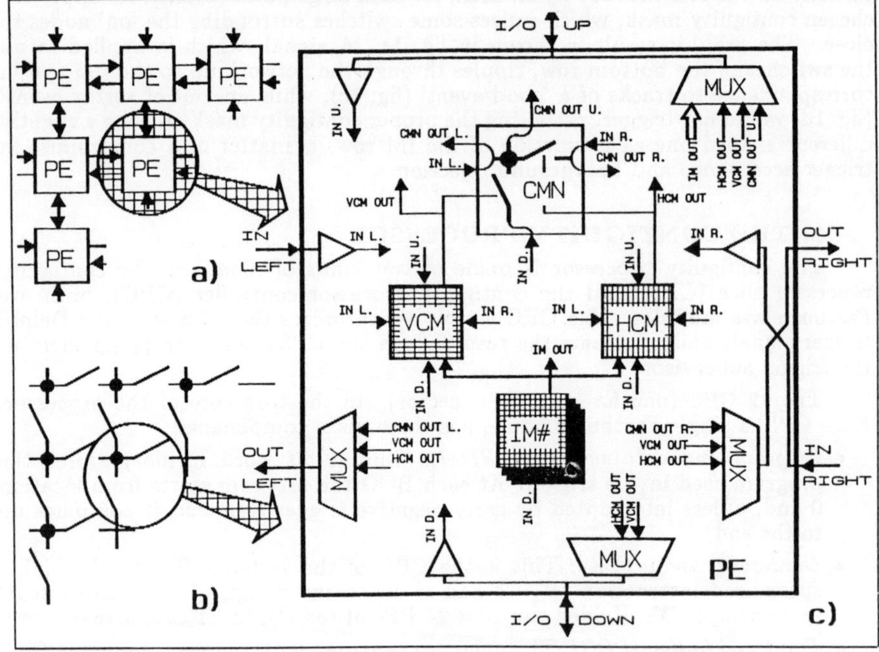

Fig. 2 : **The processing element.**
The contiguity array (*a*) with a blown up processing element (PE) (*c*). A PE includes a node with its two switches (*b*).

3.1.1 Image memories (IM). The 8 1-bit IMs in each PE have the function of recording data from the TPC cell to which they are wired, for being used by the other components in the PE during processing. Many instructions have the IM address (from 0 to 7) in their field. These instructions either use the IM data as an input operand to the connection memories (see paragraph 3.1.2) or the IM is both source and destination of the operation. For example, SHIM(n) (shift $IM^{(n)}$ to the up direction) is used to load simulated trigger data into the CA or to download the real recorded data into the FEB. This is possible because within a single clock cycle $IM^{(n)}_{i,j} \rightarrow IM^{(n)}_{i+1,j}$ $(i = 0,...,14 \quad and \quad j = 0,...,23)$, and the bottom row $(IM^{(n)}_{0,j})$ can be loaded from the PDM while the upper one $(IM^{(n)}_{15,j})$ can be stored into the FEB.

3.1.2 Connection memories (HCM and VCM). Horizontal and vertical connection memories (HCM and VCM) are two 1-bit memories (for each PE), whose values '0' (open) and '1' (close) represent the status of the switch they have to drive (see paragraph 3.1.3). Most instructions operates on HCM and VCM and they allow to generate the required contiguity mask.

HCM and VCM have a completely identical instruction set. Let us consider HCM as an example. The source of operands for HCM is one or more of the following ones: IM (from 0 to 7), up, down, left and right directions, and HCM itself. If the source is an IM the instruction is COH(n) (copy the nth IM to HCM: $IM_{i,j}^{(n)} \rightarrow HCM$), if it is one of the 4 neighbours PEs we have SHHU (shift HCM up), SHHD (shift HCM down), SHHL (shift HCM left) and SHHR (shift HCM right). For each of the previous instructions we can add the content of HCM itself (a kind of feedback), getting a new set of 'copy with feedback' and 'shift with feedback': COHF(n), SHHUF, SHHDF, SHHLF, SHHRF. Moreover some of the previous instructions can be excuted only for selected rows in the CA (contiguity masks can differ from row to row), or more than one can be merged and executed within a single cycle.

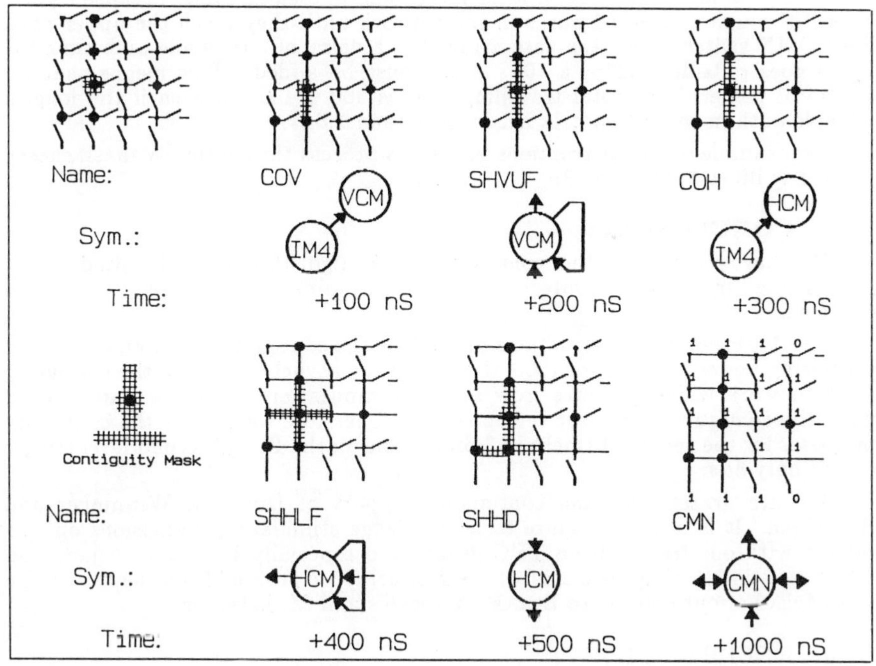

Fig. 3 : Processing an IM.

The same example of fig. 1b is illustrated using the set of instructions in the CA. The contiguity mask is built in five clock cycles, at the same time in all the PEs. Afterwards the signals ripple through the CMN finding the two tracks of fig. 1*a*. The 4×4 nodes shown is what we have in a single gate array, while the time is what we have for the CA as a whole.

The corresponding set for VCM is: COV(n), SHVU, SHVD, SHVL, SHVR; and those with feedback: COVF(n), SHVUF, SHVDF, SHVLF, SHVRF.

3.1.3 Contiguity mask network (CMN). This last layer of the PE has no memory states, but it is a pure combinatory network implementing a horizontal and a vertical switch (see fig. 2). The inputs to CMN come from down, left and right, while the outputs go up, left and right. A '1' propagates to some of the three outputs if there is some signal at the inputs and the switches are in a state allowing propagation. Horizontal signals travel both ways, while vertical ones only from bottom to top. Usually horizontal and vertical switches are controlled by HCM and VCM respectively, generating a connective path defined by the starting image together with the contiguity mask chosen. however any CMN row can also be made 'transparent'. This means that for such a CMN row all the vertical switches are closed while the horizontal ones are open, and the signal travels vertically straight on. Forcing a part of the CMN to be transparent allows either to read or to put stimuli to internal nodes of the network, as well as to bypass full rows.

The CMN has a substantial difference from the other components in the PEs: it is asynchronous, that is the clock has no influence to its output, but only HCM, VCM and the input bottom row define the results at the uppermost row. Moreover CMN outputs are usually valid after more than one clock cycle, because signals 'travel' through connected paths in the network until they reach the upper part. Each CMN switch inside the gate arrays has a delay of $2ns$, while each time the signal goes outside a chip, a $10ns$ delay must be added. This means that for a 'typical pattern' the total travelling time is $300 \div 400$ ns, even if the longest possible path in the full matrix needs more than $1.6\mu s$.

An example of the instructions needed to process the contiguity mask chosen in fig. 1 is illustrated in fig. 3.

4. CONCLUSIONS

We have seen that the contiguity mask algorithm is well suited to be implemented in a SIMD architecture based on a bidimensional array of processing elements.

The basic set of 27 instructions defined for the contiguity array, an equivalent processing power in excess of 1000 MIPS, and the asynchronicity of the contiguity mask network which executes 'row to row computations' in a few nanoseconds make the contiguity processor very powerful. Therefore we can use the contiguity processor for the search of tracks in 3 dimensions in the Delphi second level trigger during only $8\mu s$.

We are grateful for the continuous support of Drs. H. Wenninger and H.J. Hilke. It is also a pleasure to acknowledge stimulating discussions on this subject with our friends from TPC group, and especially Drs. J.V. Allaby and J.E. Augustin. During the design of the gate array we profited from the knowledge and advises of our colleagues Drs. F. Bourgeois and M. Letheren.

5. REFERENCES

[1] DELPHI Technical Proposal, CERN/LEPC 83-3, LEPC/P 2, 17 May 1983.

[2] G. Darbo and B.W. Heck, The TPC trigger for the DELPHI experiment, IEEE Trans. Nucl. Sci. **34,1** (Feb 1987) 227-231.

[3] G. Darbo and B.W. Heck, A three-dimensional ripple trigger for the DELPHI Time Projection Chamber, Nucl. Instr. and Methods **A257** (1987) 567-575.

The Transputer and Occam

David May

Current technology allows a high performance computer to be implemented ion a single chip. One example is the Inmos IMS T800 transputer [Homewood et al 1987], which incorporates 2 processors (central processor and floating point processor), 4Kbytes of RAM (random access read/write memory) and a communications system on a single chip about 1 sq.cm. in area. The floating point performance of this device is over 1 Mflop (1 Mflop = 1 million floating point operations per second).

The communications system of the transputer allows real time systems and concurrent computers to be constructed as a network of transputers. The emergence of such concurrent computers has necessitated the development of new programming languages and algorithms. One example is the occam language which has been designed to enable applications to be expressed in a form suitable for execution on a variety of parallel architectures.

1 VLSI architecture

VLSI technology has significant implications for computer architecture:

1 processors are cheap; they should be used even to save interconnect or memory

2 locality of operation is essential

The need for locality arises because of the high cost in area and delay of moving data between chips. It is therefore important that that the various functional units of a processor (registers, arithmetic logic units and control logic) are combined on a single chip in order that data can be moved rapidly between them.

The inevitable conclusion is that an architecture based on a network of simple computers operating independently and communicating data as needed is ideally matched to the technology. However, such a radical change of architecture cannot be hidden from the programmer; new languages and algorithms are needed to exploit such machines.

2 Occam

The occam programming language [Inmos 1988] enables an application to be described as a collection of processes which operate concurrently and communicate through channels. In such a description, each occam process describes the behavior of one component of the implementation, and each channel describes a connection between components.

The design of occam allows the components and their connections to be implemented in many different ways. This allows the choice of implementation technique to be chosen to suit available technology, to optimise performance, or to minimise cost.

Occam has proved useful in many application areas. It can be efficiently implemented on almost any computer and is being used in applications ranging from single processor embedded control systems to concurrent supercomputers. Occam programs can also be directly compiled and implemented as specialised VLSI devices.

2.1 Locality

Almost every operation performed by a process involves access to a variable, and so it is desirable to provide each processing element with local memory in the same VLSI device.

The speed of communication between electronic devices is optimised by the use of one directional signal wires, each connecting only two devices. This provides local communication between pairs of devices.

Occam can express the locality of processing, in that each process has local variables; it can express locality of communication in that each channel connects only two processes.

2.2 The occam primitives

Occam programs are built from three primitive processes:

v := e   assign expression e to variable v
c ! e   output expression e to channel c
c ? v   input variable v from channel c

The primitive processes are combined to form constructs:

SEQ sequence
IF conditional
WHILE loop

PAR parallel
ALT alternative

A construct is itself a process, and may be used as a component of another construct; in other words, occam is a hierarchical block-structured language.

Conventional sequential programs can be expressed with variables and assignments, combined in sequential and conditional and loop constructs. Concurrent programs make use of channels, inputs and outputs, combined using parallel and alternative constructs. A very simple example of an occam program is the buffer process below.

```
WHILE TRUE
  VAR ch:
  SEQ
    in ? ch
    out ! ch
```

Indentation is used to indicate program structure. The buffer consists of an endless loop, first setting the variable **ch** to a value from the channel **in**, and then outputting the value of **ch** to the channel **out**. The variable **ch** is declared by **VAR ch**.

2.3 The Parallel Construct

The components of a parallel construct may not share access to variables, and communicate only through channels. Each channel provides one way communication between two components; one component may only output to the channel and the other may only input from it. These rules are checked by the compiler.

The parallel construct allows its component processes to be executed concurrently. Two examples of the parallel construct are shown below.

```
CHAN c:
PAR
  WHILE TRUE
    VAR ch:
    SEQ
      in ? ch
      c ! ch
  WHILE TRUE
    VAR ch:
    SEQ
      c ? ch
      out ! ch
```

```
VAR ch1:
VAR ch2:
SEQ
  in ? ch1
  WHILE TRUE
    SEQ
      PAR
        in ? ch2
        out ! ch1
      PAR
        in ? ch1
        out ! ch2
```

The first consists of two concurrent versions of the previous example, joined by a channel to form a "double buffer". The second is perhaps a more conventional version. As 'black boxes', each with an input and an output channel, the behavior of these two programs is identical; only their internals differ.

2.4 Synchronised communication

Communication is synchronised, and takes place only when both processes are ready. Synchronised communication prevents accidental loss of data arising from programming errors. In an unsynchronised scheme, failure to acknowledge data often results in a program which is sensitive to scheduling and timing effects.

Synchronised communication requires that one process must wait for the other. However, a process which requires to continue processing whilst communicating can easily be written:

```
PAR
  c ! x
  P
```

2.5 The Alternative Construct

In occam programs, it is sometimes necessary for a process to input from any one of several other concurrent processes. Consequently, occam includes an alternative construct similar to that of CSP [Hoare 1978]. As in CSP, each component of the alternative starts with a guard - an input, possibly accompanied by a boolean expression. A simple example of the alternative is shown below; this is a 'multiplexor'

```
WHILE TRUE
  ALT
    in1 ? ch
      out ! ch
    in2 ? ch
      out ! ch
```

The multiplexor inputs from either channel **in1** or channel **in2**. It then outputs the value it has just input.

2.6 Channels and hierarchical decomposition

An important feature of occam is the ability to successively decompose a process into concurrent component processes. This is the main reason for the use of named communication channels in occam. Once a named channel is established between two processes, neither process need have any knowledge of the internal details of the other. Indeed, the internal structure of each process can change during execution of the program.

The parallel construct, together with named channels provides for decomposition of an application into a hierarchy of communicating processes, enabling occam to be applied to large scale applications.

2.7 Future developments

The initial version of occam was a very simple language including only integer data types and one dimensional arrays. In view of the need to provide a concurrent language suitable for a wide range of numerical applications, a derivative of occam including real arithmetic has been developed; this language is known as occam2.

Current implementations of occam assume explict control over the physical allocation of processes to processors and channels to interprocessor links. Allocation of physical resources is determined by the compiler. This allows an extremely efficient implementation of concurrency and communication. However, it would be natural to remove these restrictions - especially if further hardware support could be used to make dynamic resource allocation fast. The result would be a language supporting concurrency and recursion, together with recursively defined data types. This would make occam more suitable for concurrent *symbolic* application areas.

3 The transputer concept

VLSI technology allows a large number of identical devices to be manufactured cheaply. For this reason, it is attractive to implement an occam program using a number of identical components, each programmed with the appropriate occam process. A transputer eg. [Homewood et al 1988] is such a component.

A transputer is a single VLSI device with memory, processor and communications links for direct connection to other transputers. Concurrent systems can be constructed from a collection of transputers which operate concurrently and communicate through links.

A transputer system consists of a number of interconnected transputers, each executing an occam process and communicating with other transputers. As a process executed by a transputer may itself consist of a number of concurrent processes the transputer has to support the occam programming model internally. Within a transputer concurrent processing is implemented by sharing the processor time between the concurrent processes.

The most effective implementation of simple programs by a programmable computer is provided by a sequential processor. Consequently, the transputer processor is fairly conventional, except that additional hardware and microcode support the occam model of concurrent processing.

3.1 Sequential Processing

The design of the transputer processor exploits the availability of fast on-chip memory by having only a small number of registers; six registers are used in the execution of a sequential process. The small number of registers, together with the simplicity of the instruction set enables the processor to have relatively simple (and fast) data-paths and control logic.

3.2 Support for concurrency

The processor provides efficient support for the occam model of concurrency and communication. It has a microcoded scheduler which enables any number of concurrent processes to be executed together, sharing the processor time. This removes the need for a software kernel. The processor does not need to support the dynamic allocation of storage as the occam compiler is able to perform the allocation of space to concurrent processes.

At any time, a concurrent process may be

active	-	being executed
	-	on a list waiting to be executed
inactive	-	ready to input
	-	ready to output
	-	waiting until a specified time

The scheduler operates in such a way that inactive processes do not consume any processor time. The active processes waiting to be executed are held on a list. This is a linked list of process workspaces, implemented using two registers, one of which points to the first process on the list, the other to the last.

A process is executed until it is unable to proceed because it is waiting to input or output, or waiting for the timer. Whenever a process is unable to proceed, its instruction pointer is saved in its workspace and the next process is taken from the list. Actual process switch times are very small as little state needs to be saved; it is not necessary to save the evaluation stack on rescheduling.

3.3 Communications

Communication between processes is achieved by means of channels. Occam communication is point-to-point, synchronised and unbuffered. As a result, a channel needs no process queue, no message queue and no message buffer.

A channel between two processes executing on the same transputer is implemented by a single word in memory; a channel between processes executing on different transputers is implemented by point-to-point links. The processor provides a number of operations to support message passing, the most important being 'input message' and 'output message'

A process performs an input or output by loading the evaluation stack with a pointer to a message, the address of a channel, and a count of the number of bytes to be transferred, and then executing an 'input message' or an 'output message' instruction.

The 'input message' and 'output message' instructions use the address of the channel to determine whether the channel is internal or external. This means that the same instruction sequence can be used for both internal and external channels, allowing a process to be written and compiled without knowledge of where its channels are connected.

3.4 Internal channel communication

At any time, an internal channel (a single word in memory) either holds the identity of a process, or holds the special value 'empty'. The channel is initialised to 'empty' before it is used.

When a message is passed using the channel, the identity of the first process to become ready is stored in the channel, and the processor starts to execute the next process from the scheduling list. When the second process to use the channel becomes ready, the message is copied, the waiting process is added to the scheduling list, and the channel reset to its initial state. It does not matter whether the inputting or the outputting process becomes ready first.

3.5 External channel communication

When a message is passed via an external channel the processor delegates to an autonomous link interface the job of transferring the message and deschedules the process. When the message has been transferred the link interface causes the processor to reschedule the waiting process. This allows the processor to continue the execution of other processes whilst the external message transfer is taking place.

3.6 Inter-transputer links

A link between two transputers is implemented by two signal wires, one in each direction. These two signal wires of the link can be used to provide two occam channels, one in each direction. This requires a simple protocol. Each signal line carries data and control information.

The link protocol provides the synchronised communication of occam. Each message is transmitted as a sequence of single byte communications, requiring only the presence of a single byte buffer in the receiving transputer to ensure that no information is lost. Each byte is transmitted as a start bit followed by a one bit followed by the eight data bits followed by a stop bit. After transmitting a data byte, the sender waits until an acknowledge is received; this consists of a start bit followed by a zero bit. The acknowledge signifies both that a process was able to receive the acknowledged byte, and that the receiving link is able to receive another byte. The sending link reschedules the sending process only after the acknowledge for the final byte of the message has been received.

Data bytes and acknowledges are multiplexed down each signal line. An acknowledge is transmitted as soon as reception of a data byte starts (if there is room to buffer another one). Consequently transmission may be continuous, with no delays between data bytes.

3.7 Summary

The transputer demonstrates that the concurrent processing features of occam can be efficiently implemented by a small, simple and fast processor. The time taken for a process to start and terminate is about 1.5 microseconds - a small enough overhead to allow processes consisting of only a few statements. An inter-process communication has a fixed overhead of about 1.5 microseconds, and also requires time to transfer the message. Messages are transferred at up to 40 million bytes / second on-chip, and up to 1.5 million bytes / second through a link.

Experience with occam and the transputer has shown that many applications naturally decompose into a large number of fairly simple processes. Once an application has been described in occam, a variety of implementations are possible. In particular, the use of occam together with the transputer enables the designer to exploit the peformance and economics of VLSI technnolgy.

The transputer therefore has two important uses. Firstly it provides a new system 'building block' which enables occam to be used as a design formalism. In this role, occam serves both as a system description language and a programming language. Secondly, occam and the transputer can be used for prototyping highly concurrent systems in which the individual processes are ultimately intended to be implemented by dedicated hardware.

4 Future developments

The development of the next level of static RAM technology (256Kbit) is well advanced, and by 1990 it is expected that 1Mbit static RAMS will be in production. By this time 4Mbit dynamic RAMs will be available. This means that a transputer could be equipped with 100Kbytes of on-chip RAM and a processor at least twice as fast as the present one. Of course there is also the possibility of considerably more on-chip processing and communication capability than is currently being achieved.

With scaling to smaller geometries comes an increase in speed. Present transputers achieve processor cycle times and memory cycle times of about 50ns. The 1990 technology is expected to reduce this to about 20ns. Notice that the switching time scales linearly but the number of devices scales quadratically; another important reason for the increased use of concurrency.

4.1 Processing Node Architecture

The possibilty of wide (256 bit pages) access to the internal memory makes allows several processors to share the same (internal) memory. This architecture would allow, for example, a communications processor to transfer application specific data structures through communication links whilst the other processor(s) perform the computation associated with the node.

The processor itself can be provided with additional microcoded operation at very little cost. Support for dynamic storage allocation to permit recursive, concurrent programs and their associated user defined data types is a logical step.

4.2 Communications architecture

One important use of the extra silicon area available within a processing node is the provision of hardware for global message routing. As a message arrives (or is generated locally), its destination address is decoded and, if necssary, the message is forwarded via the appropriate communications link.

Another possibility is the provision of hardware support for global memory addressing - even in a distributed memory multiprocessor. Non-local addresses can be decoded in hardware, and read and write operations translated into messages to remote processing nodes. These are routed in the same way as normal messages.

The result of these developments is that the programmer will not need to adapt his programs to a particular system structure, nor will reconfiguration be needed. However, it will still be important to preserve locality in mapping algortihms onto the processor array. This is an important area for future language and compiler development.

4.3 The input-output system

Future computer input and output will require specialised concurrent processors. These will perform operations such as image analysis, graphics, animation, speech and public key encryption. These functions can be implemented as either fixed configurations of standard components such as transputers - or as specialised VLSI systems.

The construction of tools for the design and programming of systems constructed from standard components and specialised silicon devices is the subject of current research. Recent work has shown that a communicating process language such as occam is suitable as a source language for silicon compilation [May et al 1987]. This approach allows a wide range of concurrent algorithms to be implemented directly in VLSI, and allows reliable migration of algorithms from software into specialised hardware as performance requirements increase. It also allows a wide range of program development and verification tools to be used to for VLSI designs.

These tools are essential to realise the potential of the technology: *a complete concurrent computer including input and output on a single chip.*

5 Conclusion

A general purpose concurrent computer will contain a processing array with many (between 10 and 1,000,000) similar processing elements. These will incorporate automatic message routing providing low-latency communication between all the processing elements. The elements will be connected in a fixed network.

The general purpose processing array will be surrounded by input and output systems also containing many processing elements. These will be more specialised, consisting of general purpose devices connected in specialised configurations or alternatively of specialised devices.

The concept of *communicating processes* can provide an architectural basis for both general purpose and specialised parts of the computer. Languages based on communicating processes, such as occam, serve as the system language for these concurrent computers. Experience gained from the application of these machines will form the basis for higher level user languages.

6 References

C A R Hoare: Communicating Sequential Processes, Communications of the ACM Vol. 21, 8 (August 1978) p. 666.

Inmos Limited: Occam2 Reference Manual. Prentice-Hall International 1988.

Inmos Limited: Transputer Reference Manual. Prentice-Hall International 1988.

D May, R Shepherd, C Keane: Communicating Process Architecture, in Future Parallel Computers, ed P Treleaven and M Vanneschi, Springer-Verlag 1987.

M Homewood, D May, D Shepherd, R Shepherd: The IMS T800 transputer, IEEE Micro Volume 7 Number 5, (October 1987).

Recent Experience with Transputer based Processor Farms

Jeremy M. Carter
High energy physics group
Royal Holloway and Bedford New College, TW20 OEX UK

Ian Glendinning
Concurrent computation group
The University of Southampton, SO9 5NH UK

ABSTRACT

The processor farm has recently become recognised as a cost-effective architecture for the rapid analysis and simulation of high energy physics event data. The INMOS transputer is a processor which forms a natural building block for such farms, since it is provided with communications links on-chip. This paper reports experience gained in porting the ALEPH detector simulation program, GALEPH, to Meiko transputer Fortran. Execution times for transputers are compared with those for a VAX 8200. A useful by-product of this work has been the implementation of a substantial portion of the CERN program library in transputer Fortran.

1. INTRODUCTION

The development of increasingly complex detectors for high energy particle physics experiments, culminating in those for LEP and HERA, has brought with it the necessity to process ever larger volumes of data. For instance, the ALEPH data acquisition system is expected to read out 100-200 kbyte events at a rate of around 1Hz, each of which will require about 30s (IBM 168 seconds, roughly equivalent to VAX 8600 seconds) to process. Generating sufficient Monte Carlo events to compare with this experimental data is an even more daunting task. The ALEPH detector simulation program, GALEPH, can take up to 300s to perform a full event simulation.

Faced with such severe demands for processing power, physicists have been forced to consider more novel, and cost-effective, architectures than the traditional mini and mainframe computers. One architecture which is particularly well suited to event simulation and analysis is that of the processor farm.

2. PROCESSOR FARMS

A processor farm consists of a set of processors running identical code, completely independently, but each operating on different data. Typically, there is a single stream of input data 'packets', which are distributed to any processor which is ready for more data. Output packets are merged into a single output stream on a first come first served basis (figure 1). This scheme is clearly ideal for the processing of high energy physics events, since they are all independent of one another. It should be noted that the order of input packets is not in general preserved in this simple scheme, although it is possible for that to be arranged, if desired, at the cost of a small loss in performance.

The INMOS transputer represents a natural building block for various multiprocessor systems, of which the processor farm is one instance, because of the presence of on-chip communication links. A practical realisation of a transputer based processor farm is shown in figure 2.

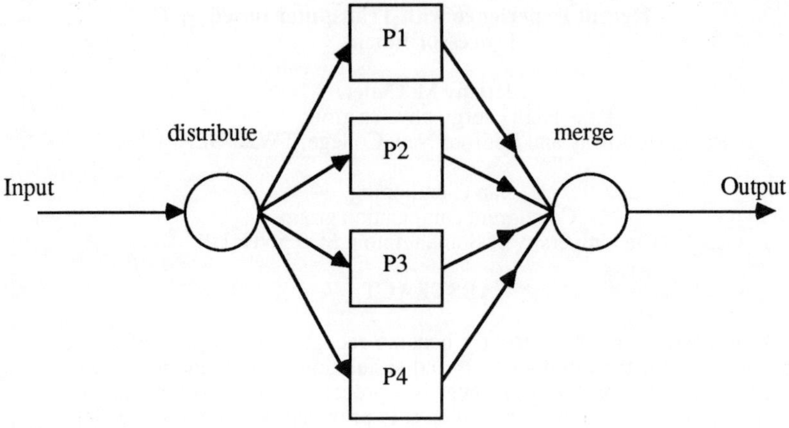

Figure 1

The farm elements (workers) are arranged in a linear chain, connected at one end to a 'farmer' processor, which interfaces to a host system, such as a micro VAX, which provides access to peripherals. Each farm element runs two processes concurrently, (transputers have hardware support for concurrency) one of which performs the task which has been farmed out, while the other is responsible for routing input packets to the first available worker, and output packets back to the host. On the face of it, it might be thought that, say, a tree would perform better than a linear chain, but apart from having less latency when starting and stopping, this is not the case. It has been shown [1)] that, in the steady state, a linear chain is optimal, in that the rate at which packets are processed grows essentially linearly with the number of processors, until the link bandwidth of the farmer is saturated. In the case of the T800 transputer, this is at around 2 Mbytes/s, which is comparable to the bandwidth of a good tape drive.

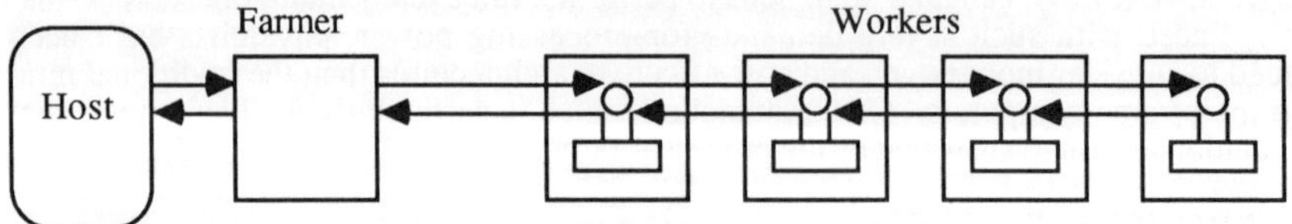

Figure 2

The farmer and routing processes together constitute what is known as a 'harness'. The application programmer need not know about its workings - it is merely linked with the application program. Farm harnesses for Meiko transputer systems have been available for some time, while a harness which is compatible with INMOS software is presently being developed at Southampton [2)], as part of an extra mural research project, funded by the SERC/DTI transputer initiative.

3. GALEPH

GALEPH is the ALEPH experiment's detector simulation program, and is coded in Fortran 77. Figure 3 represents a simplified picture of the flow of data between program modules, and their relation to the main library packages (shaded grey).

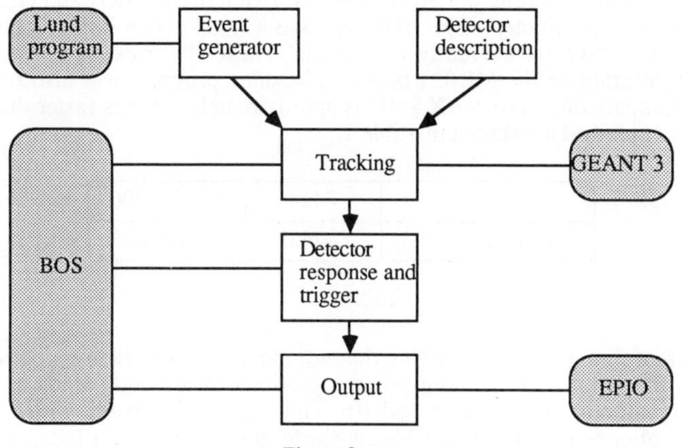

Figure 3

GEANT 3 builds the detector geometry from information stored in a detector description file. Particle four-vectors from an event generator, such as the Lund Monte Carlo, are tracked, and the subdetector responses and trigger are simulated. Events may optionally be output to a mass storage device, using EPIO. Memory management is performed by the ZEBRA package, within GEANT3, and by the BOS package within the general framework of GALEPH.

4. IMPLEMENTATION DETAILS

The implementation of GALEPH required the porting of elements of the CERN program library (CERNLIB) and, in fact, this constituted the major part of the work. The library consists of a nucleus of basic utility and numerical routines, known collectively as KERNLIB, together with other, more specialised, routines and packages.

Two kinds of problem were encountered in porting the code - non-standard language features and machine dependencies. The bulk of the machine dependency was contained in KERNLIB. In particular, bit, byte and character handling routines had to be written. This task was made relatively simple by the support in Meiko Fortran for non-standard bit manipulation routines, although bit formats differed in some cases. Furthermore, many CERNLIB routines make use of the LOCF function in KERNLIB, which returns the address of a variable. An equivalent function (IADDRESS) was kindly provided by Meiko [3)] specifically for this purpose. Most of the remaining work involved modification to code relying on implementation specific use of character variables. The syntax of hexadecimal constants, which do not form part of ANSI standard Fortran, also required modification.

The hardware used consisted of a Meiko M10 crate, containing an MK014 local host board and an MK021 mass store board. This equipment was supplied by the Rutherford Appleton Laboratory Transputer Loan Pool[4)]. The MK021 was populated with 8 Mbytes of memory and a 20 MHz T414 transputer. The host system was an IBM PC clone, containing an MK026 link adapter card. The size of the program required an 8 Mbyte board for the linking stage (the executable image is some 3.5 Mbytes).We note in passing that the current version of the linker is one-pass, requiring either careful ordering of routines presented to it, or multiple inclusions of the same code.

5. RESULTS

We have successfully ported KERNLIB, HBOOK, ZEBRA and GEANT 3 from the CERN program library. The BOS package has also been implemented, although references to EPIO are currently dummied. These libraries were used to run the GALEPH program on the MK021 board. The same program was also run on a VAX 8200 for comparison. (The VAX 8600 is approximately 4 times faster than the 8200). The timings obtained are shown in Table 1.

	T414	VAX 8200
seconds/event	209.0	29.0

Table 1

Lack of time precluded running the program on a T800 floating point transputer. However, the number of floating point operations in the code leads us to expect a speedup by a factor of between 6 and 10. This may be compared with a value of 8.7 previously obtained for the Lund Monte Carlo generator [3], run independently from GALEPH.

6. CONCLUSIONS

We have run a large high energy physics Fortran program on a transputer system. In doing so, we have implemented several extensively used librarys. The program has not yet been run on more than one transputer; however, this step is a straightforward exercise, as our previous experience with transputer based processor farms[5] has shown. Benchmark timings show that the T800 transputer has a processing power which is comparable to that of a VAX 8200, indicating that a T800 based processor farm would represent an extremely cost-effective source of compute power for high energy physics simulation.

REFERENCES

1) Pritchard, D.J., "Mathematical Models of Distributed Computation", in proc. 7th occam users group meeting (September 1987) Grenoble, ed. T. Muntean.

2) Surridge, M., work in progress at: Transputer Support Centre, Chilworth Research Centre, Southampton SO1 7NP, UK.

3) Cownie, J., (private communication), Meiko Ltd., Whitefriars, Lewins Mead, Bristol, UK.

4) We would particularly like to thank Mike Jane and Fran Childs of the RAL Transputer Loan Pool for support and co-operation.

5) Glendinning, I. and Hey, A., "Transputer Arrays as Fortran Farms for Particle Physics", Comp. Phys. Comm. 45, 367-371 (1987).

A TRANSPUTER BASED SECOND-LEVEL CALORIMETER TRIGGER FOR THE ZEUS EXPERIMENT

H. Boterenbrood, S. C. Goble, S. J. de Jong, G. N. M. Kieft, G. Poletiek, K. Schutte, J. C. Vermeulen, A. J. de Waard and L.W. Wiggers

NIKHEF-H, P.O. Box 41882, 1009 DB Amsterdam, the Netherlands

ABSTRACT

The second-level calorimeter trigger system for the ZEUS experiment at HERA at DESY will be discussed. The system will process the data of the uranium calorimeter. In the system each of the VME-crates with front-end electronics is equipped with a VME-board with 2 Inmos transputers. The transputers read-out the front-end electronics via the VME-bus and execute the second-level trigger algorithm. All communication between the crates in the system and with external systems, as needed for obtaining a second-level trigger decision and - after a positive decision - for the transport of the data, proceeds via transputer links. In this contribution the architecture of the system, the two-transputer VME-card and the principle of the trigger algorithm will be described.

1. SHORT OVERVIEW OF THE ZEUS DATA-ACQUISITION AND TRIGGER SYSTEMS

The ZEUS collaboration is building a detector for measuring the reaction products of electron-proton interactions, to be produced in the HERA accelerator at DESY. This detector has to be ready to take data at the start of accelerator operation in 1990. The number of background events is expected to be high compared to the number of genuine electron-proton interactions at the interaction point. In successive steps the event rate has to be brought back to 1 kHz after the first-, 100 Hz after the second- and 2-5 Hz after the third-level trigger. While at the first level in hardware an evaluation is made of the event from a reduced set of data, the full data is shifted through local digital or analog pipelines. The first-level trigger decision is broadcasted within 5 μs after the interaction, a time longer than the bunch crossing interval of HERA of 96 ns. The pipelines are necessary for minimizing the dead-time. After a positive decision the data in the pipelines corresponding to the bunch-crossing for which a positive decision was taken, is stored in local buffers. The trigger algorithms at the first level will be simple and defined in hardware. At the second level more data and time are available, allowing for a more detailed and accurate analysis. Because of the incoming rate of 1 kHz a decision has to be produced every millisecond. The second-level trigger system for the calorimeter is the subject of this paper. In this system (and also in the global second-level trigger and in the event builder) Inmos transputers 1) will be applied.

2. THE INMOS TRANSPUTER

The Inmos transputer has been designed as a building block for concurrent systems. Important properties, for our type of application, are: support of high bandwidth communication between transputers via four fast bidirectional links under DMA control (maximum effective transfer rate of 1.77 MByte/s); a powerful 32-bits CPU; support of multi-tasking and inter-process communication in micro-code; a minimal need for external components; the possibility for booting via a link (no ROM's are needed); the straightforward support for handling of parallel processes in the occam programming language on one or more transputers; the possibility of programming in conventional languages (PASCAL, C, FORTRAN-77).

3. THE SECOND-LEVEL CALORIMETER TRIGGER SYSTEM

3.1 Introduction

The calorimeter consists of three parts: a forward part (FCAL), a barrel (BCAL) and a rear part (RCAL), with 460, 452 and 448 towers respectively. Most towers consist of 6 cells. Every cell is read-out at both sides by photo-multipliers. The total number of read-out channels is 12864. After a positive first-level trigger decision 4 - 6 samples of the data stored in analog pipelines, are digitized by ADC's on VME-cards 2), processed locally by an on-board digital signal processor (DSP) and put into on-board memory. A card with two ADC's and one DSP handles the data from one tower. The data processed by the card can be read-out via the VME-bus, for further analysis. Each of the 80 VME read-out crates will be equipped with a two-transputer card in development at NIKHEF-H, for both the second-level trigger processing and for the read-out of the calorimeter data. Each crate is associated with a well-defined region of the calorimeter.

3.2. The Two-Transputer VME-Card

A block scheme of the card is presented in fig. 1. Each transputer (a T800) has access to the VME-bus, has a local memory of at least 1 MByte, and has access to a dual-port (static) memory of at least 128 KByte. One transputer has direct access to the dual-port memory, the other transputer only via the local bus. It is envisaged that the latter transputer will transport data from the VME-bus to the dual-port memory, while the first transputer concurrently will execute computing-intensive tasks on basis of data previously transported. Each transputer may cause a signal on the event pin of the other transputer. In this way the synchronization needed for data transfers between the transputers can be achieved. The dual-port memory may be accessed directly from the VME-bus via the local bus. Since this transport proceeds via the local bus of the card and each transputer has its own local memory, both transputers can execute in the meantime other tasks without being disturbed by the ongoing data transport. Each transputer may generate external logical signals, as well as receive external signals via its event pin. Each transputer has a special register from which the source of the event signal can be determined. The logical signals may also reset other transputers, so a transputer may be rebooted under control of another transputer. The VME-bus interface has been designed to accommodate various VME-cycles. The transputer itself can only generate two types of bus cycles: 32-bit read or write cycles and 8-bit write cycles. The address is always 32-bits wide. The VME-bus supports also other cycles, e.g. a 24-bit address, 16-bit data cycle. The bus interface automatically converts a transputer cycle into the appropriate VME-cycle. The type of VME-cycle is selected via the address output by the transputer: the memory map of the transputer is divided in several regions corresponding to different types of VME-cycles. This limitation of the addressing range is not serious, as the total addressing range of a transputer amounts to 4 GByte. In the present design the transfer rate during read-out by a transputer via VME is about 4 Mbyte/s. The read-out of the dual-port memory by another module on the VME-bus will be faster.

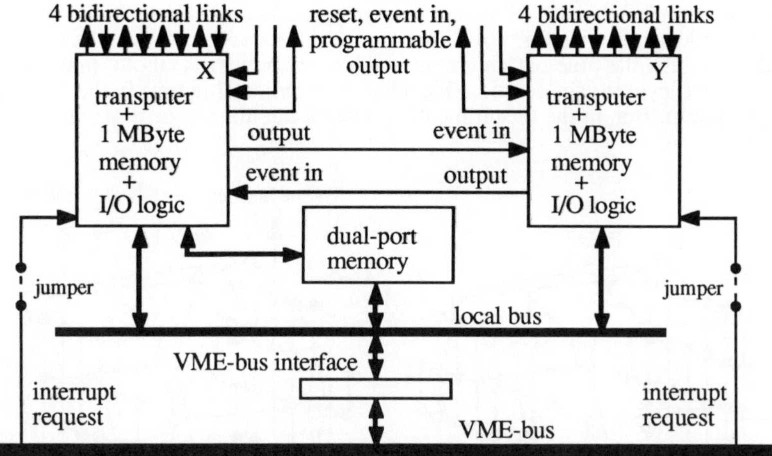

Fig.1. Block scheme of the two-transputer VME-card

3.3 Processing by the Transputers

First, data exclusively for the trigger processing is read. These should be the six sums of energies obtained from the left and right photo-multipliers per tower, to be calculated by the DSP. Those data are read from the ADC-DSP cards via the VME-bus and put in the dual-port memory by transputer Y. The amount of data is only 114 16-bit words per crate, read in 57 VME-cycles. The data is subsequently processed by transputer X, which is connected to its geographical neighbors (i.e. the transputers that are associated with the most nearby regions of the calorimeter) via its links. A possible configuration is shown in fig. 2a. The trigger algorithm, running in the X transputers has to determine:

a) general quantities, like total energy deposited and total transverse energy, already determined at the first-level trigger level, but with higher accuracy,
b) the presence of jets of particles, with their energies,
c) the presence of single isolated particles (electrons).

For the recognition of hadronic jets the amount of deposited energy is related to the area of energy deposition. The radius of the area should obey a predefined, approximate inverse linear relation. Also the depth profile of the energy deposition is considered. Single isolated showers of electromagnetic origin will be tested on having a small radius and on having a large fraction of their energy deposited in the top layer of the calorimeter. An algorithm with this strategy has been found to function satisfactorily on a single processor. Work has started on the realization of a multi-transputer version of this algorithm.

After local processing - which may involve communication between neighboring transputers - the results of the trigger algorithm are transferred over the network towards one of the two master processors. At the transputers on the way results of the local trigger algorithms are merged. Finally the results of the trigger master processors are combined and sent to the global second-level trigger. The processing, including transport, will take less than 1 ms. In the meantime other events can already enter the system. While transputer X is processing, transputer Y will transmit data of a previous event that satisfied the overall second-level trigger criteria, via a separate network to a read-out transputer (see fig. 2b). From there the data are sent via links to the event builder. In total, the time before arrival at the event builder is about 6-10 ms after a positive first-level trigger decision, including 1 ms processing time in the second-level trigger system. But, in the meantime other events can already be passing through the system.

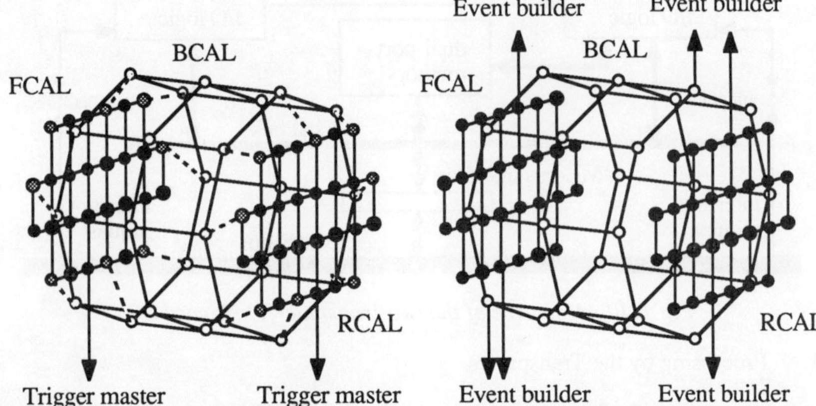

Fig. 2a. Association of the X transputers with regions of the calorimeter plus the interconnections for the second-level processing. The grey nodes in the FCAL and RCAL parts indicate transputers that are connected with transputers in the BCAL part

Fig. 2b. Connections between the Y transputers for the read-out of the calorimeter. No connections exist between the FCAL, BCAL and the RCAL parts

The tasks of the hardware and software of the transputer system can be summarized as: (i) read-out and control of the ADC-DSP cards, execution of the trigger algorithm and transport of the results and of data to the event builder, (ii) monitoring of the incoming data, the data throughput, the functioning of the hardware and software items in the system; communication and presentation of results and setting of alarm conditions, (iii) execution of control tasks: e.g. downloading of DSP programs and constants; controlling the state of the system (initialization, data-acquisition, system checking), (iv) system checking (not during normal operation): testing of correct wiring of the system and of correct functioning of all hardware components, testing of correct functioning of the system with simulated or "real" data, presentation of the test results and fault diagnosis

It is intended to construct the hardware and software of the system such that incorrect functioning of individual items does not lead to a break-down of the complete system.

References:

1. M. Homewood et al., "The IMS T800 transputer", IEEE Micro, 7, No.5, 10 - 26 (October 1987)
2. J. W. Dawson and J. S. Berg, "VME data acquisition card for the ZEUS calorimeter", unpublished (1987)

THE FERMILAB ADVANCED COMPUTER PROGRAM MULTI-ARRAY PROCESSOR SYSTEM (ACPMAPS) A SITE ORIENTED SUPERCOMPUTER FOR THEORETICAL PHYSICS*

T. Nash, H. Areti, R. Atac, J. Biel, A. Cook, J. Deppe, M. Edel, M. Fischler, I. Gaines, D. Husby, T. Pham, and T. Zmuda
Advanced Computer Program
Fermi National Accelerator Laboratory+
Batavia, IL 60510 USA

E. Eichten, G. Hockney, P. Mackenzie, H. B. Thacker and D. Toussaint*
Theoretical Physics Group
Fermi National Accelerator Laboratory+
Batavia, IL 60510 USA

ABSTRACT

The ACP Multi-Array Processor System (ACPMAPS) is a highly cost effective, local memory parallel computer designed for floating point intensive grid based problems. The processing nodes of the system are single board array processors based on the FORTRAN and C programmable Weitek XL chip set. The nodes are connected by a network of very high bandwidth 16 port crossbar switches. The architecture is designed to achieve the highest possible cost effectiveness while maintaining a high level of programmability. The primary application of the machine at Fermilab will be lattice gauge theory. The hardware is supported by a transparent site oriented software system called CANOPY which shields theorist users from the underlying node structure.

1. INTRODUCTION

The ACP Multi-Array Processor System (ACPMAPS) is a highly cost effective, local memory parallel computer designed for floating point intensive grid based problems. The project is a joint effort of Fermilab's Advanced Computer Program (ACP) and Theoretical Physics Group. Processing nodes of the system are single board array processors based on the FORTRAN and C programmable Weitek XL chip set. They are connected by a network of very high bandwidth 16 port crossbar switches. The architecture is designed to achieve the

---

*Talk given by Thomas Nash at the Adriatico Conference on the Impact of Digital Microelectronics and Microprocessors on Particle Physics, International Centre for Theoretical Physics, Trieste, Italy, March 28-30, 1988.

+Fermilab is operated by Universities Research Association, Inc., under contract with the U.S. Department of Energy.

*On leave from Department of Physics, University of California, San Diego, La Jolla, California 92093 USA.

highest possible cost effectiveness while maintaining a high level of programmability. At Fermilab the primary application of the machine will be lattice gauge theory.

To obtain some estimates of the computing needs of lattice quantum chromodynamics one can consider the calculation of the deconfining temperature in SU(3) gauge theory without quarks,[1] which is one of the most solid four dimensional calculations done so far. Something like 500,000 MFlops - hours (peak, ~70% delivered) were used on a Star ST-100 array processor. This calculation required a lattice spacing of less than 0.1 fermi and a volume of close to (2 fermi)^3, resulting in lattices with spatial sizes of up to 19^3. It is virtually certain that calculations with quarks will require even larger lattices than this for comparable accuracy. Lattices with space-time sizes 32^4 to 64^4, requiring 1-20 GBytes of data memory, are a reasonable guess. Calculations of hadron masses in the approximation of ignoring dynamical quarks have not yet achieved a reasonable understanding of calculational errors, even on Cray-sized supercomputers. Although algorithms for the inclusion of dynamical quark effects have made tremendous progress in the last few years, at present they still seem to require at least two orders of magnitude more computer time than comparable calculations without quarks. It is thus clear that large increases in combined CPU power *and* algorithmic power are still required for even simple QCD calculations.

The aim for the Fermilab ACP Multi-Array Processor System (ACPMAPS) is the delivery of such large amounts of memory and CPU power at the lowest possible cost, without compromising the programmability required for rapid algorithm development which is just as important as raw computing power in achieving the goals of lattice gauge theory. For other large-scale computer projects aimed at lattice gauge theory, see Reference 2.

A 16 node system will be built in the summer. Two switch prototypes are working and under going a rigorous testing program now. The Floating Point Array Processor (FPAP) node design is being simulated. PC board lay out is underway and a prototype planned for May. Fermilab intends to proceed to a 256 node (5 GFlop for about $750K) system as soon as the 16 node system is operational. Maximum system size is 2048 nodes. The system is being designed in the ACP tradition to be commercialized and available to other institutions.

2. ARCHITECTURE OVERVIEW

A block diagram of the system is shown in Figure 1. The individual single board Floating Point Array Processors have peak performance of 20 MFlops. They each contain 8 MBytes of data and 2 MBytes of program memory. The FPAPs are plugged into a crate whose backplane is a 16 fold bidirectional high speed cross bar. The nodes can speak with each other in pairs at a full 20 MBytes/sec simultaneously. The architecture of ACPMAPS is a hypercube network of such crossbar switch crates each supporting 8-16 FPAPs. In a typical configuration 8 array processor nodes will be plugged into each switch crate along with up to 8 I/O modules that interconnect crates in a hypercube (or better). The switches handle intercrate routing automatically. The system therefore does not operate with all nodes in lock step like an SIMD machine as is the case in other projects of this type (Columbia, IBM GF11 and APE). It also does not strongly favor local communication (as existing hypercubes do). It thus allows for any conceivable new lattice algorithm unconstrained by synchronous or local communication requirements. Despite its algorithmic flexibility the system ranks as the best (or nearly so, we won't argue) in terms of cost effectiveness of MFlops/$.

The system is controlled by a host microVAX (or a mainframe VAX) operating part time. The host starts a control program running on a control node, which is identical to all the other nodes of the system except in software. This node controls the lattice wide parts of the program and starts subprocesses on the individual nodes.

As noted, the nodes operate completely asynchronously. The use of MIMD (multiple instruction, multiple data) rather than SIMD (single instruction, multiple data) architecture is one of the most important features of the system. There are many advantages to this type of architecture. It is very flexible: it can handle problems which are awkward or impossible in SIMD such as heat bath and incomplete LU decomposition algorithms and random lattice problems. The allowed sizes and shapes of the lattices are independent of the details of the hardware. The node structure of the machine can be made invisible in much or all of the high-level user code, resulting in improved programmability. This also results in improved fault tolerance, since the system can be reconfigured with one fewer node when one node is down, without requiring changes in user software or the setting aside of any spare nodes. Complications which have to be faced with MIMD include the potential for synchronization conflicts, which can't occur with SIMD. This requires care in designing and understanding the communications system. In addition, a nontrivial system software design effect is required to ensure that overheads associated with the communications software are kept to acceptable levels.

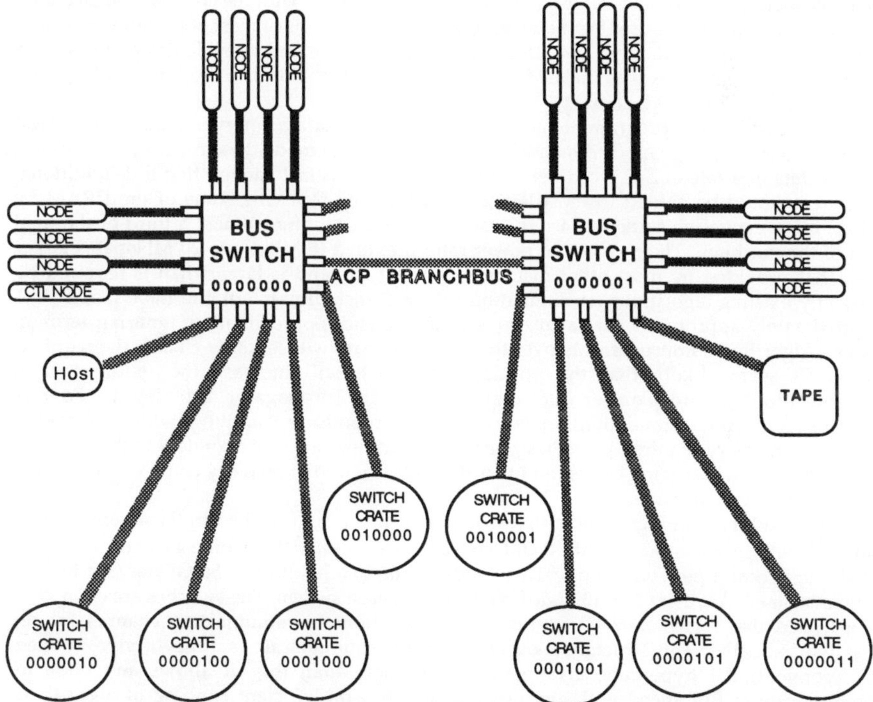

Figure 1. ACP Multi Array Processor System 256 node configuration.

A major new package of software (CANOPY) is being developed for this system. Theorist users will think only in terms of sites and fields on sites. The system will automatically allocate sites to nodes and handle all site to site communication whether on the same node or another. Thus users will not have to know details of the hardware for effective use of the system. Routines that are used heavily will be microcoded. The skeleton of all programs will be written in FORTRAN or C using a series of special subroutine calls that will make the programs particularly readable for lattice gauge theorists and others with site oriented algorithms. In this way, despite ease of use and flexibility, the system will approach 10 MFlops/$3000 node in FORTRAN or C.*

3. HARDWARE

The nodes are single board floating point array processor using the Weitek XL chip set which contains a 32 bit, 20 MFlop (peak) floating point unit, an integer processor, 32 floating point and 32 integer registers, and an instruction sequencer. The chip set as a whole is programmable in FORTRAN and C, at some sacrifice in performance. Thus, these modules incorporate the functions of a high level language programmable single board computer and a high performance floating point array processor. No external CPU is required as a controller for these stand alone floating point engines.

The FPAP modules (Figure 2) contain the XL chip set, the data and code memory, and the interface logic and input and output queues for communicating with the crossbar switch crates. One floating point unit is used per node, in contrast to the designs of most of the SIMD machines aimed at lattice gauge theory. In addition to being a flexible and sensible design for a wide variety of problems, this was dictated by the desire to be able to use the Weitek FORTRAN and C compilers for the XL chip set.

The 2 MBytes of program memory and 8 MBytes of data memory is made from 1 Mbit 80 nsec access time page mode dynamic RAM chips. In page mode these memory systems can deliver data at a rate of one word per 100 nsec. This rate is fast enough that little additional efficiency would be gained in most lattice algorithms by replacing some of the DRAM by faster, more expensive static RAM. The memory chips constitute almost a third of the total cost of the machine. The memory to power ratio provided (8 MBytes to 20 MFlops) is larger than that provided by most other machines of this type, and is larger than is required by presently existing algorithms for simulating full QCD including internal fermion loops. It is approximately appropriate for calculations in the valence approximation, ignoring fermion loops. Algorithmic improvements over the next few years will certainly change the required ratio. It seems likely that the possibilities which will increase the required ratio (preconditioning and Fourier acceleration of quark propagator calculation, Fourier acceleration of gauge simulation) are currently more promising than those which reduce the amount of memory required per CPU cycle (such as adding nonlocal operators to the action to reduce finite lattice spacing errors) and that the large amount of memory could easily become crucial in the years to come.

The nodes are plugged into a network of switch crates (described in the accompanying paper[3]) whose backplanes handle full sixteen port crossbar switching at bandwidths of 20 MBytes/second per connection. This yields a total bandwidth of 2.56 GBytes/sec for 256 node machine. A cluster of 8-12 nodes is attached to each switch. The switches are connected in a hypercube, which may be augmented by additional communication channels along heavily used paths. This structure allows the nodes to communicate as if they were connected in a conventional hypercube arrangement, but more than this, it allows any node to communicate at full speed with *any* other node, allowing efficient running of algorithms requiring nonlocal communications. The switch crates allow any node to access any other node's data memory without needing to know where the other node is located on the network. With the current switch crate hardware, systems of up to 2048 nodes are possible before this transparent nonlocal communication feature is lost. The switch crates are based on SN74AS8840 sixteen input crossbar switch chips from Texas Instruments. They will also be

*Memory prices are fluctuating widely at this writing. FPAP costs given here are based on Fall 1987 DRAM prices.

used in a variety of high performance experimental particle physics applications of the ACP Multiprocessor System,[3] including the "Level-3" programmable trigger for the Collider Detector at Fermilab (CDF).

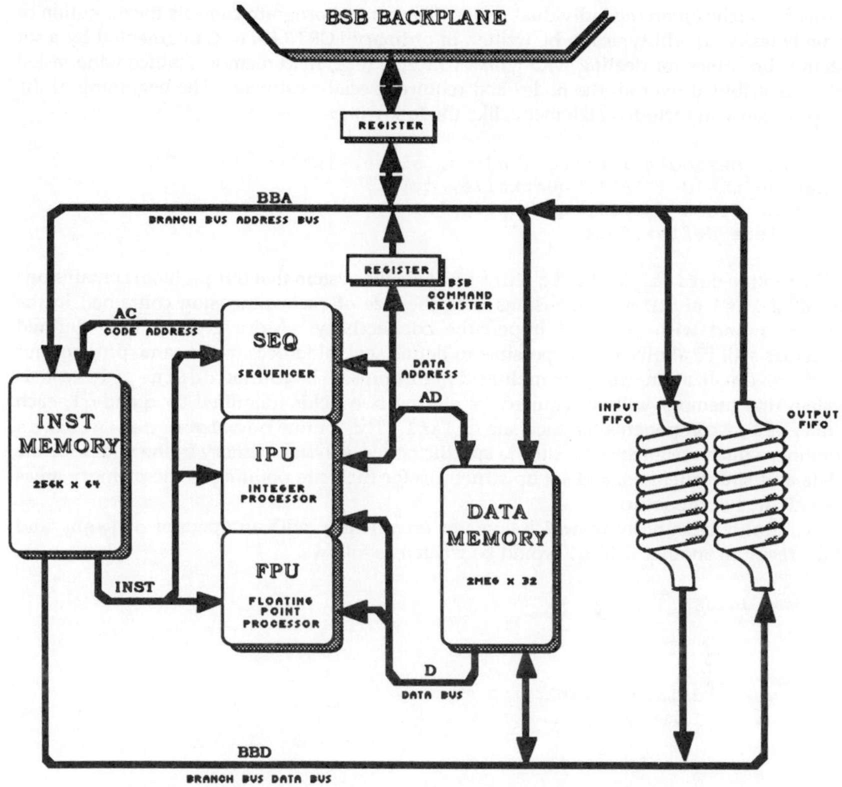

Figure 2. Schematic Design of the Floating Point Array Processor (FPAP)

Low cost video technology tape drives will be used for check pointing long calculations and for archiving of gauge fields and propagators. These tape drives cost a few thousand dollars and can store 2 GBytes of data on an 8 mm video tape cartridge costing $5-$10. One drive will be attached to every switch crate, enabling all of memory to be stored in under five minutes.

4. SOFTWARE

Lattice gauge theories are part of a large class of grid-based problems derived from discretization of a set of differential equations which are very suitable for a parallel architecture like this one. The natural breakdown of the problem is to assign a certain subset of the sites in the space or spacetime to each node, which stores the data for the field variables defined on the sites assigned to it in its local memory and does calculations for its sites. The system software,[4] known as CANOPY, has been designed to shield the user as much as possible from the hardware dependent node structure of the parallel architecture. The user thinks in terms of sites not nodes.

User programs are divided conceptually into two pieces: the control program, which is called from a microVAX host or mainframe VAX and runs on the control node, and site subroutines, which run on the individual nodes. The control program controls the execution of lattice-wide tasks. It will typically be written in ordinary FORTRAN or C augmented by a set of system subroutines for dealing with global concepts (e.g., field memory, lattice wide tasks) which are distributed over all the nodes and require special treatment. The beginning of the control program will included statements like the following:

```
call define_periodic_lattice ( ndims, sizes, lat1 )
call define_field ( lat1, quarksize, q )
call define_field ( lat1, quarksize, q1 )
call complete_definitions
```

The routine `define_periodic_lattice` tells the system that our problem contains one lattice called `lat1` of `ndims` dimensions with the size of each dimension contained in the array `sizes` and with standard hypercube connectivity. More general user defined connectivities will be allowed. It is possible to define several lattices in the same program for block spin renomalization group or multigrid algorithms. The routine `define_field` tells the system that memory will be required for storing two fields identified by `q` and `q1`, each with `quarksize` components for each site of `lat1`. The routine `complete_definitions` calls routines which assign specific sites to specific nodes, allocate memory in the nodes for the field data and site structures, and set up structures for each site pointing to the memory areas of adjacent sites of the lattice.

A control node subroutine which operates on a field `q` with an operator `dslash_` and stores the result in another field `q1` would be written as follows.

```
subroutine dslash ( q, q1 )
        .
        .
        .
call do_task ( dslash_, lat1,
                pass$, q, 1,
                pass$, q1,1,
                end$)
return
end
```

The system subroutine `do_task` passes to all the nodes a pointer to a subroutine `dslash_` which operates on a single site and a pointer to a list of sites on which to operate, which may be the entire lattice `lat1` or some previously defined set of sites such as `red_sites`. A system routine on the node, invisible to the user, calls `dslash_` for all the sites in the set of sites which have been assigned to the node. `do_task` may also be used to pass (`pass$`) to the nodes parameters required by the site subroutine (like the field identifiers `q` and `q1`) and to integrate (`integrate$`) data returned from the individual nodes.

The site subroutines access and replace data from global fields with subroutines like:

```
call get_field  ( q,  site, qtemp )
call put_field  ( q1, site, qtemp )
```

which determine if the desired data is already resident on the node and open a channel to the communications hardware if necessary.

Most site subroutines can be written in FORTRAN or C. CPU intensive kernels such as SU(3) matrix multiplication and essential routines like `dslash_` will be microcoded for maximum efficiency. We expect that lattice gauge algorithms prepared in this way will run at 8-10 MFlops per node.

5. PHYSICS PROGRAM

The main interest of the Fermilab lattice group is the application of lattice gauge theory to QCD and "beyond the standard model" phenomenology. During the initial phase of running the machine we will pursue a program which as much as possible serves the multiple purposes of machine shakedown, algorithm development, careful error analysis for heavily studied quantities like the hadron spectrum, and production of new physics results.

Enormous progress has been made in the last years on algorithms for inclusion of dynamical quark loops in QCD calculations. Many (possibly most) groups with large machines have begun attempts to calculate hadron masses with the new algorithms. Since there do not yet exist definitive spectrum results in the approximation of ignoring quark loops, this may seem somewhat premature. It is dictated to some extent by the availability of machines with low memory to CPU power, which cannot easily go to lattices with larger volumes and smaller lattice spacing. With our machine, it will make more sense to do a careful study of the statistical, finite volume, and finite lattice spacing errors in the valence approximation before going on to the inclusion of internal quark loops.

In the valence approximation, CPU time is dominated by the efficiency of the quark matrix inversion algorithm (as opposed to the product of the efficiencies of the quark matrix inversion and the simulation algorithms in dynamical fermion algorithms). The relative efficiencies of these algorithms is very dependent on lattice size, the quark mass, etc. Our first algorithm project will be the testing on large lattices of the most promising of these methods (conjugate gradient, minimal residual, . . .) and various methods of preconditioning (Fourier acceleration, incomplete LU decomposition).

During the initial period of running, we will be doing careful analysis of errors in the valence approximation on a variety of lattice spacings and volumes. We intend to have as complete a set of analysis programs as possible for quantities which can be calculated from the basic set of quark propagators to run as a standard package on all data sets produced. These will include meson and baryon heavy quark potentials, the light hadron spectrum and decay constants, the kaon bag parameter, and the properties of D and B mesons in the 1/M expansion, including masses, spin splittings, decay constants and bag parameters.

6. REFERENCES

1] Gottlieb, S.A. , et al., Phys. Rev. Lett. 55 1958 (1985); Christ, N.H. and Terrano, A.E., Phys. Rev. Lett. 56 111 (1986).

2] Christ, N.H. and Terrano, A. E., IEEE Trans. Comp. C-33(4) 344 (1984); Seitz, C.L., Commun. ACM 28 22 (1985); Beteem, J., Denneau, M. and Weingarten, D., J. Stat. Phys., 43 1171 (1986); Albanese, M., et al., The APE Computer, ROM2F/87/005 (1987).

3] Gaines, I., Areti, H., Atac, R., Biel, J., Cook, A., Deppe, J., Edel, M., Fischler, M., Hance, R., Husby, D., Nash, T., Pham, T. and Zmuda, T., "Multi-Processor Developments in the United States for Future High Energy Physics Experiments and Accelerators" presented at the Adriatico Conference on the Impact of Digital Microelectronics and Microprocessors on Particle Physics, International Centre for Theoretical Physics, Trieste, Italy, March 28-30, 1988.

4] "CANOPY – ACP MAPS Software - Specifications Document", Fermilab Advanced Computer Program internal document. To be released. Availability of ACP publications may be determined by reference to the file at the HEPNET location, FNACP::ACPDOCS_ROOT:[DOCS]DOCLIST.DOC

Integrated Microsystems as a Driving Force in Modern Detector Designs

Helmuth Spieler
Lawrence Berkeley Laboratory, Berkeley CA 94720, U.S.A.

1. Introduction

Numerous SSC and LHC reports [1] have defined performance requirements and set the scale for detectors at the next generation of high-energy particle accelerators. Although the design of these new detectors is far from set, it is clear that they must provide a formidable combination of performance parameters that go significantly beyond incremental improvements in existing techniques. These new detectors require new techniques and new technology. However, new technology by itself is no panacea; many promising new developments have not yet matured to the point where one can assess their practical viability. On the other hand, there also exist new techniques that are well developed and understood, but have not made it to the marketplace - either because the market does not exist, or has not yet been perceived. This paper will focus on applying *existing* technology to integrated microsystems in radiation detectors.

2. Integrated Microsystems

The integrated microsystems considered here are semiconductor devices that include

1. Detecting elements
2. Front-end electronics
3. Event preselection.

Semiconductor vertex and tracking devices have been chosen as prototypical examples, since they offer unique features coupled with a particularly demanding mix of performance requirements - low noise, high speed, low power, fine segmentation, and radiation hardness.

There are several reasons for taking this arduous approach:

1. The high interaction rates and the event complexity drive detector designs towards greater segmentation. This leads to a large number of data channels ($>10^8$ for a typical vertex detector, for example), which in turn dictates local event preselection.
2. The increased quantity of data imposes a need to simplify the structure of the raw data in order to facilitate analysis, for example through non-projective tracking devices that provide unambiguous position information.
3. For high-resolution tracking/vertex detectors semiconductor microsystems offer an unequalled combination of essential or desirable characteristics.
4. Complex semiconductor microsystems are required to obtain the radiation resistance needed in SSC detectors.

The last statement may cause some surprise, as the use of semiconductor devices is generally considered to be the cause, rather than the cure for radiation damage problems. A brief discussion will clarify this point.

The dominant radiation background inside an SSC detector is due to albedo neutrons emanating from the calorimeter [2]. Current estimates assume a roughly uniform fluence of 10^{13} cm^{-2} per year within the detector cavity. In a semiconductor detector the primary damage mechanism is displacement of lattice constituents, which leads to an increase in leakage current and associated shot noise. The leakage current is proportional to the area of a detector element. Hence, the area of a detector can be chosen to provide the required level of radiation resistance [3]. For typical detector parameters at the SSC the maximum allowable area per detector element is about $5 \cdot 10^{-3}$ cm^2, which corresponds to $2 \cdot 10^6$ elements per m^2. Following these considerations, a viable design for a pixel device for use as a vertex detector at the SSC could still provide a signal-to-noise ratio >100 after nearly 20 years of operation at a luminosity of 10^{33} $cm^{-2}s^{-1}$. This device would include about $2 \cdot 10^8$ elements [4]. These arguments hold for any semiconductor detector, also in the calorimeter, for example.

Clearly, any detector system for the SSC or LHC will include a significantly larger number of data channels than any existing system. Dealing with them is a major challenge; a first step towards resolving this problem is the inclusion of some form of sparse readout circuitry within the detector.

3. Sparse Readout

Several groups have designed monolithically integrated readout systems for silicon strip detectors with output multiplexing [5,6,7]. A chip designed at LBL takes this one step farther by including sparse data scan circuitry on-chip to control the readout process [8]. This IC (called the SVX chip) has been designed for a proposed silicon strip vertex detector at CDF. Fig. 1 shows a block diagram of the device. The chip includes 128 channels of electronics (one per detector strip), connected to bonding pads on a 50 micron pitch. Each channel consists of a charge-sensitive preamplifier followed by a series of switched-capacitor stages that, by an appropriate sequence of clock pulses, perform pulse shaping (e.g. a correlated double sample), threshold discrimination, and analog storage. Thresholds are externally set through the test input of the preamplifier, and can be updated if necessary. The discriminator outputs are fed to the sparse readout circuitry that automatically scans the detector channels for hits; this proceeds at <4 ns per channel. Readout can be confined to hit information alone (digital hit address) or can include analog signals. Up to 128 chips, i.e. >16000 channels, can be daisy-chained to a common readout bus.

The chip is implemented in a conventional double-metal 3-micron CMOS process; prototypes have been fabricated through MOSIS, a silicon foundry "brokerage" [9]. The die is about 6.3 mm on each side and includes over 11000 transistors. Even with this level of circuit complexity the power dissipation is <200 mW per chip. The measured equivalent input noise is 250 e + 38 e/pF (for an internally set peaking time of 500 ns). Fig. 2 shows a readout chip bonded to a

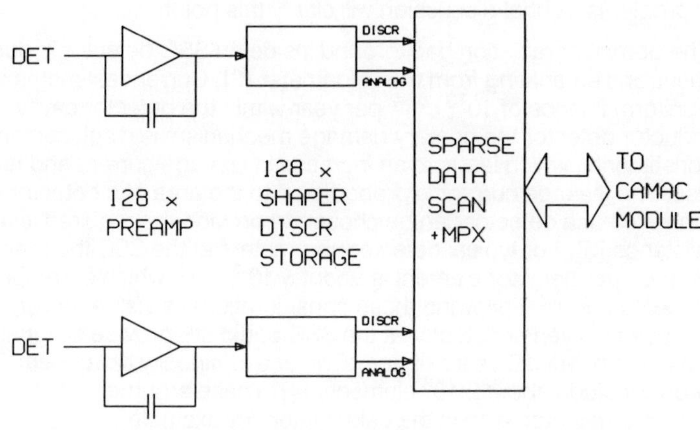

Fig. 1 Block diagram of SVX strip detector readout IC

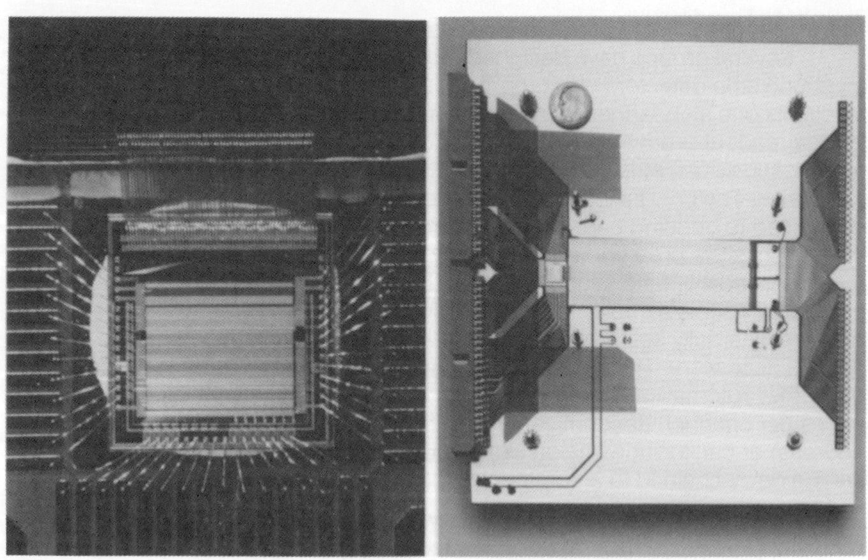

Fig. 2 SVX readout IC bonded to strip detector

strip detector as used in a beam-test at Fermilab. Several iterations of this device have been fabricated and tested; a pre-production prototype is currently being evaluated.

This development demonstrates the feasibility of combining low-noise front-ends with digital circuitry on the same chip in order to perform rudimentary signal preselection. With this technique the **number of output data channels is determined not by detector segmentation, but by event multiplicity**.

4. Strips vs. Pixels

To a great extent the popularity of strip detectors is due to the feasibility of readout systems. In fixed target experiments one can accommodate front-end systems constructed of discrete components; detectors for colliders have only become possible through the availability of monolithic readout ICs. In many respects technical expedience determines the choice of detector. However, even current experiments have demonstrated the need for truly two-dimensional, i.e. pixelized, detectors [10]. The arguments for pixel devices become stronger at future high-luminosity accelerators:

- Track Separation
  For 1 TeV/c QCD jets (multiplicity 10...30) pixels of 100 micron size will unambiguously resolve every track in >90% of all events. For comparison, short strips with dimensions 1 cm x 25 microns will resolve all tracks in only 70% of all events [4].
- Remove ambiguities inherent to all projective geometries
  For a multiplicity of 20 a crossed-strip configuration will simulate nearly 400 "phantom" hits. These can be recognized only through track reconstruction, which incurs substantial computing effort.

Small pixels also provide significant technical benefits, besides the increased radiation resistance of the detector discussed above. The small capacitance allows a high signal-to-noise ratio with low overall power dissipation. This results in simplified signal processing circuitry. More important, it increases the radiation resistance of the front-end electronics by providing substantial performance reserves, *i.e.* the system can tolerate greater degradation by radiation.

5. Smart Pixel Devices

The only pixel devices used in high energy pixels are CCDs [11]. However, these suffer from unacceptably long readout times in SSC applications. Random access pixel devices have been used in infra-red imaging [12]. Although these devices allow addressing of individual pixels, they do not include the hit recognition circuitry that is needed for the efficient readout of a sparse data field.

The following discussion outlines a conceptual design for a pixel device that would fulfill all requirements for SSC operation at $L = 10^{33}$ $cm^{-2}s^{-1}$. This is the result of a substantial effort within the Advanced Detector R&D Group at LBL. The data are based on extensive device measurements and simulations (the SVX strip detector readout IC described above is a spin-off from this work).

The architecture of this pixel device is inherently matched to a sparse data structure in high-background environments. The pixels provide prompt signals when they are struck to allow fast time-sliced pattern recording with on-chip circuitry. On receipt of an event trigger on-chip circuitry selects the pixels to be read out based on the pattern information stored for the time slice of interest (for a detailed discussion see [4]). The proposed device has the following configuration:

- o A two-dimensional array of detector elements ("pixels") with a readout cell associated with each pixel
- o Each readout cell includes a preamplifier, pulse shaper, discriminator and analog storage. *This rather complex circuitry is needed to obtain the necessary radiation resistance.*
- o Event buffers and sparse readout circuitry on the periphery of the chip.

The design of a pixel device can easily become utopian if some general considerations are not observed.

- o The signal flow and the circuitry must be kept simple. All fast circuitry must be on-chip to minimize power consumption and propagation delays.
- o Data is only read out on receipt of a first level trigger.
- o Fast event buffering (every 16 ns in step with the SSC bunch crossings) must be by digital hit pattern. *Power considerations preclude large numbers of ADCs or the additional high gain-bandwidth analog circuitry needed for fast analog time-slicing.*

The LBL design eliminates the need for multichannel high-speed analog time-slicing. Analog information can be stored on the struck pixel if both the pixel address and the time of hit are recorded. This is possible since the small pixels yield a low double-hit probability. Consider a detector at $r_p = 10$ cm from the beam axis. At $L = 10^{33}$ cm^{-2}s^{-1} the hit rate is $10^8/r_p^2$ cm^{-2}s^{-1} [1]; hence, the rate on a pixel of 100x100 micron^2 will be 100 s^{-1}. If a first level trigger is generated within 10 µs, the double hit probability will be $< 10^{-3}$. This scheme replaces time-slicing and its rather high power requirements with low-power distributed memory.

5.1 Architecture

The detector consists of an array of pixels, each with its own amplifier/discriminator cell. Every cell has two output lines, which can be enabled either individually or simultaneously. The detector can operate in two modes:

a) Fast pattern mode

For fast pattern recording the outputs of all pixels are bussed to form a strip-like configuration and provide x and y coordinates for each hit (Fig. 3, top). This is the normal mode of operation. Hit pattern information is clocked into digital buffer registers in step with the bunch period of the accelerator (16 ns for the cur-

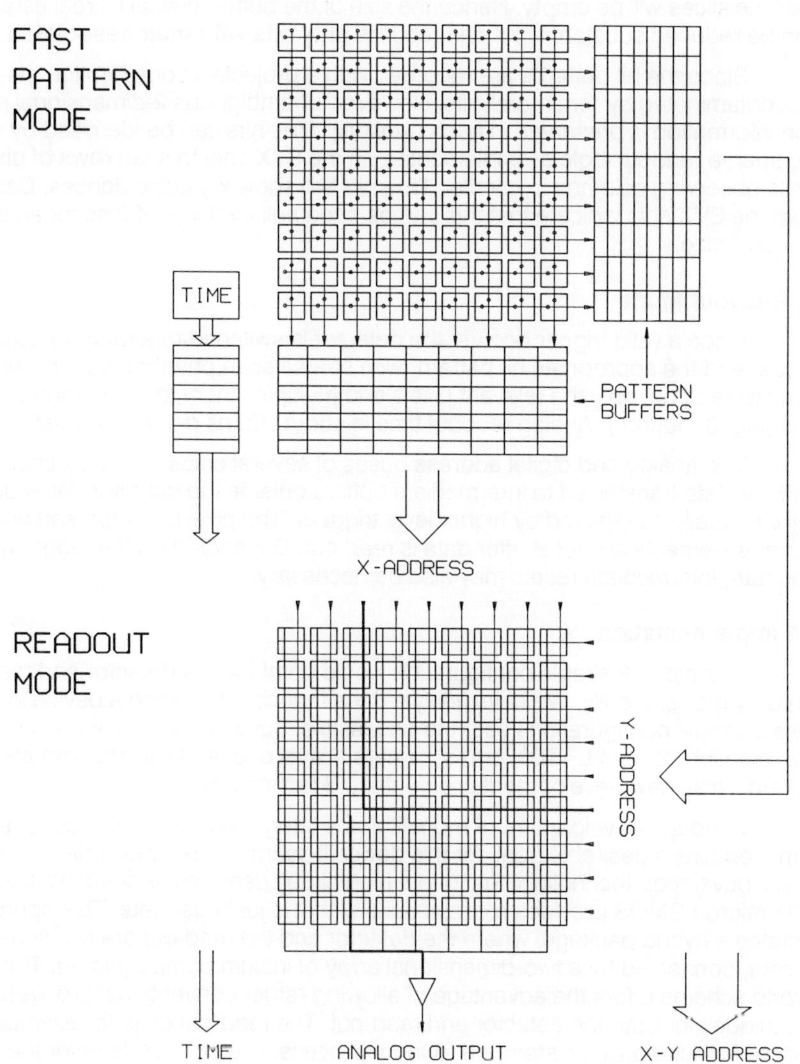

Fig. 3 Readout modes of "smart" pixel device

rent SSC design). The buffers are made deep enough to accommodate the time necessary to generate a trigger. Since the hit rate per pixel is only 100 s^{-1}, most of the time slices will be empty. Hence the size of the buffer (critical IC real estate) can be reduced substantially if one records **only hits with their associated time**.

Since the hit patterns are recorded with a projective configuration they will be contaminated by "phantom" hits. However, unambiguous 2-dimensional position information is preserved in each pixel cell. True hits can be identified by utilizing sparse readout logic as implemented on the SVX chip to scan rows of pixels while observing columns in parallel. True hits will show x-y coincidences. Data from the SVX chip indicate that this can be accomplished in <500 ns for an array of 1 cm^2 size.

b) Readout Mode

Once a valid trigger occurs, the detector is switched to a random access mode, and the appropriate hit pattern (with sparse scan phantom suppression) is used to read out only the relevant pixels and multiplex them to a common output bus (Fig. 3, bottom). Analog readout time is about 200 ns per struck pixel.

The analog and digital address buses of several chips are daisy-chained and the data transferred to intermediate buffers outside the detector. Most data will eventually be rejected by higher level triggers. The pixel preamps and analog storage elements are reset after data is read out. Depending on the trigger type and rate, intermediate resets may also be necessary.

5.2 Implementation

The most desirable configuration would combine the detector and readout circuitry monolithically. We have analyzed the fabrication of such a device in some detail. Simple configurations seem to be feasible, but a circuit designed to sustain the radiation dose at $L=10^{33}$ $cm^{-2}s^{-1}$ incurs a level of complexity that requires considerable new development in an untested technology.

Although development of monolithically integrated detector/readout systems remains a desirable goal, for this device it seems more appropriate to utilize a well-developed technology that offers high circuit densities. Indeed, we find that 1.25 micron CMOS provides a circuit density that is just adequate. This approach dictates a hybrid package, where the detector and the read-out are two separate wafers, connected by a two-dimensional array of indium bumps (Fig. 4). The hybrid scheme offers the advantage of allowing rather conventional processing techniques for both the detector and read-out. The read-out chip, for example, can be fabricated with a standard foundry process. In a collider detector the pixels would not be square, but elongated along the beam direction. A size of 50 x 200 μm^2 seems quite feasible, and 25 x 400 μm^2 should be possible. The latter would provide <10 micron position resolution in phi direction.

5.3 Predicted Performance

Potential device performance was evaluated by a combination of measurements on test devices (transistor and amplifier arrays) and computer simulations. Electrical data were determined for a 3 micron CMOS process; only capacitance was scaled in extrapolating to a 1.25 micron feature size (available from several foundries). Power dissipation is of crucial importance in this device. Measured device parameters have been used to optimize transistors and circuitry for minimum power dissipation commensurate with the required level of performance [13]. Our data predict (assuming a total input capacitance of 100 fF):

Noise:	65 electrons rms
Rise-time:	100 ns
Power per Pixel:	10 μW

The predicted noise yields a signal-to-noise ratio of nearly 400 for minimum ionizing particles traversing a 300 micron thick detector. Timing accuracy is several ns, determined primarily by the required circuit simplicity. The total power dissipation including the sparse readout logic and the output drivers is 120 mW/cm^2 (for 10^4 pixels/cm^2). If a "10 Mrad" radiation hard process [14] is used for the readout, the predicted lifetime at a distance of 10 cm from the beam axis is about 20 years; after this time shot noise from the detector would reduce the signal-to-noise ratio to 100. An SSC vertex detector will require 100 to 200 million pixels; the corresponding power dissipation would be 1...2 kW (comparable to an optimized strip detector system with inferior performance). Power dissipation in this design is determined primarily by the required radiation hardness; it could be reduced for other applications.

5.4 Pixel Device Summary

The pixel device described above demonstrates how the application of contemporary microelectronics technology can shape detector architecture to provide a unique combination of characteristics that have not been achieved in any other type of detector:

- **Unambiguous track identification** both in **space and time**
- **High position resolution**
- **Data reduction**: The quantity and rate of data are not determined by the interaction rate, but by the first level trigger rate and the event multiplicity
- **Radiation resistance** adequate for extended operation near the interaction region of future high-luminosity colliders.

All of this can be obtained through application of **existing technology**. Clearly, many technical details must be developed, but viable techniques and steps toward the resolution of crifical issues have been identified. The implementation of this device depends less on technology than on a commitment by the high energy physics community and the funding agencies.

6. Microsystems R&D - Required Effort

Is semiconductor microsystems technology within reach of the HEP community? Does semiconductor device research and development require facilities and manpower beyond the scale of more traditional HEP activities? The possible directions in semiconductor device R&D span a wide range of required expertise and resources. Areas include

- o Gate array design
- o Full custom IC design
- o Detectors
- o Detector/readout systems
- o Device and Process R&D

Gate array and full custom ICs (usually termed collectively as Application Specific ICs - ASICs) became generally accessible to small groups with the proliferation of foundries (the latest VLSI Systems Design survey lists 45 Si and 11 GaAs foundries [15]). Gate arrays provide standard cells whose interconnections must be defined. Conceptually, the design is quite similar to working with discrete logic circuits on PC boards. Individual vendors provide software that greatly facilitates this process.

Full custom design offers significantly grater flexibility. In recent years full custom layout has evolved to a highly formalized procedure that requires only basic knowledge of IC fabrication processes. This has put IC design within the reach of layout engineers with only a rudimentary understanding of electronics. However, the main reason for adopting a full custom path is the need for device and circuit optimization in advanced designs. This requires a more thorough understanding of device physics and circuit fundamentals than does conventional

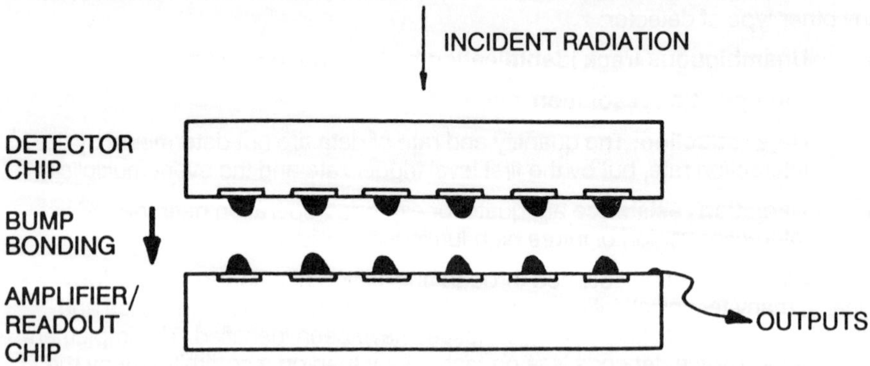

Fig. 4 Bump bonding of a detector and readout chip

electronic design, which adopts predetermined transistors and circuit blocks. Since ICs cannot be "fixed" after they have been fabricated, the design relies on extensive circuit simulation, which in turn must be performed with skepticism and judgement if one is to obtain reliable results. In this respect high quality IC design requires more intellectual effort and discipline than the "cut and try" style prevalent in traditional detector and electronic design.

On the other hand, the required facilities for IC design are quite modest; at LBL public domain software (SPICE for circuit analysis and MAGIC for IC layout) are used [16]. A MicroVAX II is used as a dedicated graphics workstation. Layout data files are transmitted by telephone-modem to the foundry. Testing and characterization of the chip requires substantial effort, due in part to the complexity of the circuits that one typically integrates, but also because problems often cannot be traced directly.

Quite a few groups in the US and Canada are now engaged in IC design for high energy physics. The following is a (partial) list of institutions and activities:

1. Univ. Hawaii/UC Santa Cruz/SLAC: Strip detector readout, pixel devices
2. SLAC: Signal processing/Analog Storage
3. Univ. of Pennsylvania/BNL/IMEC: Drift Chamber Readout (bipolar, CMOS)
4. UC Santa Cruz: Strip Detector Readout (bipolar)
5. TRIUMF: GaAs CCDs, Gate Array ASICs for Fastbus
6. Lawrence Berkeley Laboratory: Analog/Digital signal processing systems (CMOS), e.g. SVX Strip Detector Readout; New Devices (e.g. Pixel Detector System); Process & Device R&D (detector/readout integration)

Detector design and evaluation can be performed in collaboration with commercial detector manufacturers. An important area that requires additional effort is the investigation of radiation effects. This work is very much in line with traditional HEP detector research.

More complex devices, however, require a research effort that is appropriate to multidisciplinary laboratories. The pixel device discussed above combines advanced circuit design, detector development, and packaging technology. Device and process R&D are important components of this effort. An example for this is a new detector fabrication process developed at LBL by Steve Holland [18]. In contrast to conventional semiconductor detector fabrication, this process is fully compatible with conventional IC processing; high-quality small geometry MOSFETs have been fabricated on the same wafer with detectors without compromising the performance of either. Furthermore, and perhaps more important, this process actively counteracts the effects of contaminants and reduces the sensitivity to variations in the fabrication environment. Since the cost of semiconductor detectors is dominated not by materials, but by fabrication, the assurance of process repeatability and high fabrication yields is a critical contribution to the economic viability of the large-scale detector systems needed for high energy physics.

7. Conclusion

Integrated front-end circuitry in detectors provides a multitude of opportunities. It is essential for many large-scale detectors, but even small scale detector systems, for example in x-ray imaging, can benefit from compact detector-readout assemblies that provide efficient data collection. More importantly, however, integrated microsystems allow new detector architectures with unprecedented performance. Integrated pixel devices, for example, can provide a combination of signal-to-noise ratio, position resolution, speed, and readout efficiency that surpasses by far any other type of detector. Integrated semiconductor detector systems are still in their infancy and a sustained effort will be necessary to exploit their potential, but integrated microsystems are already a major driving force in modern detector design.

Acknowledgements: Substantial contributions to this work by were made by D. Nygren, L. Bosisio, S. Holland, and S. Kleinfelder. This work was supported by the U.S. Department of Energy under contract No. DE-AC03-76SF00098.

References

1. for example: *Report of the Task Force on Detector R&D for the Superconducting Super Collider*, SSC Central Design Group, Lawrence Berkeley Laboratory, 1986, SSC-SR-1021
2. D.E. Groom (*ed.*) *Radiation in the SSC Interaction Regions*, SSC Central Design Group, Lawrence Berkeley Laboratory, 1988, SSC-SR-1033
3. D. Nygren and H. Spieler, *Semiconductor Detector Size and Radiation Hardness*, in *Report of the Task Force on Radiation Effects at the SSC*, SSC Central Design Group, Lawrence Berkeley Laboratory, 1988
4. D. Nygren and H. Spieler, to be published
5. J.T. Walker *et al.*, *Development of High Density Readout for Silicon Strip Detectors*, Nucl. Instr. and Meth. **226** (1984) 200
6. G. Lutz *et al.*, *Low Noise Monolithic CMOS Front End Electronics*, Nucl. Instr. and Meth. **A263** (1988) 163
7. P.P. Allport *et al.*, *Results of Silicon Strip Detector Readout using the CMOS Low Power Microplex*, IEEE Trans. Nucl. Sci., **NS-35/1** (1988)
8. S.A. Kleinfelder *et al.*, *A Flexible 128 Channel Silicon Strip Detector Instrumentation Integrated Circuit with Sparse Data Readout*, IEEE Trans. Nucl. Sci., **NS-35/1** (1988)
9. L.K. Quinones, *MOSIS - VLSI Prototyping Service*, IEEE Trans. Nucl. Sci., **NS-35/1** (1988)
10. P. Weilhammer, *Experience with Si Detectors in NA32*, in *Proceedings of the Workshop on New Solid State Devices for High Energy Physics*, Berkeley, 1985, LBL-22778
11. S.J. Watts, *CCD Vertex Detectors*, Nucl. Instr. and Meth. **A265** (1988) 99
12. S. Gaalema, *Low Noise Random-Access Readout Technique for large PIN Detector Arrays*, IEEE Trans. Nucl. Sci. **NS-32** (1985) 417
13. H. Spieler, *Design Considerations for Monolithic Charge Amplifiers*, to be published
14. K.H. Lee *et al.*, *Radiation Hard 1.0um CMOS Technology*, IEEE Trans. Nucl.Sci. **NS-34** (1987) 1460
15. *Survey of Semiconductor Foundries*, VLSI Systems Design, **8**, August (1987) 60
16. *VLSI Tools*, Computer Science Division, EECS Dept., University of California at Berkeley, 1986
17. S. Holland, to be published in Nucl. Instr. and Methods

VLSI STRUCTURES FOR TRACK FINDING

Mauro Dell'Orso

Dipartimento di Fisica - Università di Pisa
Piazza Torricelli 2 - 56100 Pisa - Italy

Luciano Ristori

INFN Sezione di Pisa
Via Vecchia Livornese 582a - 56010 S.Piero a Grado
Pisa - Italy

ABSTRACT

We discuss the architecture of a device based on the concept of *Associative Memory* and capable of solving the track finding problem in a few microseconds even for very high multiplicity events. This "machine" may be implemented as a large array of custom VLSI chips. All the chips are equal and each of them stores a number of "patterns". All the patterns in all the chips are compared in parallel to the data coming from the detector while the detector is being read out.

INTRODUCTION

The quality of results from present and future High Energy Physics experiments depends to some extent on the implementation of fast and efficient track finding algorithms. The detection of *heavy flavor* production, for example, depends on the reconstruction of secondary vertices generated by the decay of long lived particles, which in turn requires the reconstruction of the majority of the tracks in every event.

Particularly appealing is the possibility of having detailed tracking information available at trigger level even for high multiplicity events. This information could be used to select events based on impact parameter or secondary vertices. If we could do this in a sufficiently short time we would significantly enrich the sample of events containing heavy flavors.

Typical events feature up to several tens of tracks each of them traversing a few position sensitive detector layers. Each layer detects many hits and we must correctly correlate hits belonging to the same track on different layers before we can compute the parameters of the track. This task is typically time consuming: it is usually solved using "constraint equations" which apply to hits from the same track and going through a large number of different hit combinations using a "trial and error" approach.

We propose here to use modern VLSI technology to build a device capable of solving the pattern recognition problem in a time span of a few microseconds even for the most complicate events.

THE DETECTOR

In this discussion we will assume that our detector consists of a number of layers, each layer being segmented into a number of *bins*. When charged particles cross the detector they *hit* one bin per layer. No particular assumption is made on the shape of trajectories: they could be straight or curved. Also the detector layers need not to be parallel nor flat. This abstraction is meant to represent a whole class of real detectors (drift chambers, Silicon microstrip detectors etc.). In the real world the coordinate of each hit will actually be the result of some computation performed on "raw" data: it could be the center of gravity of a cluster or a charge division interpolation or a drift-time to space conversion depending on the particular class of detector we are considering. We assume that all these operations are performed upstream and that the resulting coordinates are "binned" in some way before being transmitted to our device.

THE PATTERN BANK

For each event we know which bins have been hit and from this information we want to reconstruct the trajectories of all the particles. We call this process *track finding*.

The problem of track finding can be solved, at least conceptually, by a "brute force" approach. We consider all the possible tracks that go through our detector. Each track generates a *hit pattern*. Since the detector has a finite spatial resolution (bin size), many different tracks generate the same hit pattern. The number of different hit patterns generated by all the tracks is finite and it is possible to store all of them in a sufficiently large memory. The collection of all these patterns defines both the space of the tracks we are looking for and how they appear in the detector: we will refer to this collection as the *pattern bank*.

For each event, a number of tracks traverse the detector and a particular configuration of hits is thus generated: we will refer to this configuration as the *event*. A conceptually simple way to perform the track finding algorithm is to scan the *pattern bank* and compare each pattern to the *event*. A track candidate is found whenever all the hits in the pattern are present in the event. Going through the totality of the patterns in the *bank* yields a number of track candidates.

The number of different patterns to be stored in the bank depends on the detector granularity and geometry, and on the characteristics of the tracks we want to detect. As an example we will consider the situation shown in fig.1: the detector consists of four parallel planes and each plane is segmented into n bins. We consider all straight tracks crossing all four planes. We want to estimate the number of different patterns (N_p) that can be generated by a single track.

A fairly good approximation is

$$N_p \approx 3*n^2 \qquad (1)$$

By selecting one bin in plane 1 and one bin in plane 4 we define a *road*: there are n^2 different roads. From fig.2 it should be obvious that all the tracks belonging to a road generate three different patterns corresponding to the three *subroads* delimited by dotted lines.

Expression (1) can be generalized as follows:

$$N_p \approx (m-1)\ n^2 \qquad (2)$$

where

N_p = number of patterns
m = number of detector planes
n = number of bins/plane

The main problem with this approach is that the number of patterns to store in the bank for a practical situation may be very large. For example, if we consider 4 planes with 256 bins/plane we obtain:

$$N_p \approx 3*256^2 \approx 2*10^5$$

To deal with such a large number of patterns we need a lot of memory and we expect the process of matching all the patterns sequentially to be very time consuming.

ASSOCIATIVE MEMORY

The pattern matching algorithm can be easily implemented on a parallel architecture because different patterns can be compared to the event independently and in any order; in particular, any number of comparisons can be performed in parallel provided that this is allowed by the hardware.

If our main goal is speed, we can try to push the degree of parallelism to the limit and compare *all* the patterns to the event at once. To do this we need a special type of *associative memory* to store the pattern bank: each cell of the memory is big enough to hold one pattern and has enough logic built in to compare its contents to the event. A possible architecture for this device is shown in fig 3 Each row represents one cell and is designed to store one pattern. Each cell is structured into a number of words, one word per detector layer (four of them are shown). Each word stores the address of one hit on the corresponding layer. All the words in a cell define a pattern by specifying one hit per layer. The Data Bus connects all the words in the same layer, this bus is used to load the pattern data into the memory cells during the initialization phase. During normal operation, for every event, the coordinates of all the hits in each layer are transmitted one after the other on the corresponding Data Bus; all the words continuously compare their contents to what is on the bus and if a match is found the corresponding flip-flop (FF) is set. After all the hits have been transmitted, any cell that has all the flip-flops set is a track candidate because all the hits that define that pattern are present in the event. The addresses of all the track candidates are transmitted sequentially on the Output Bus. It is important to note that each hit needs to be fed into the associative memory only once. When a hit coordinate is placed on the bus all the flip-flops in all the cells containing that hit will be set simultaneously. Feeding a whole event into the associative memory will take a time proportional to the number of hits and will usually be performed in parallel with the detector readout. The actual pattern recognition process takes place within the associative memory during the time needed to extract the information from the detector. By the time all the hits have been read out, all the track information is also available.

VLSI

Typical applications require a number of cells of the order of 100k or more. The amount of logic involved rules out the possibility of using standard components and requires the development of appropriate ASIC's (Application Specific Integrated Circuits).

Using the present VLSI technology, large numbers of logic gates can be integrated on a single chip. The structure we are proposing is basically a large array of very simple cells made of a memory bit and a comparator. A significant effort can be devoted to the design and optimisation of this cell in terms of Silicon area and this basic building block can then be assembled into arrays of different size. In this way we believe we can design a CMOS chip with 256 cells of 32 bits (4 planes x 8 bits). We could then implement a system with 100k cells using only 400 chips.

A characteristic of VLSI technology is that chips can be manufactured in large quantities at relatively low cost; the input/output architecture of our chip should then be designed to facilitate the assembling of many of them into arrays of arbitrary size. One possible solution is described in the following sections.

THE AM CHIP

Fig. 4 shows the Associative Memory (AM) chip and its input/output signals.

Data Bus: is used to input pattern data and event data. The number of bits depends on the number of layers and bins/layer in the detector.

Address Bus: is used to input pattern addresses while loading the memory and to output pattern addresses when reading out track candidates. The number of bits depends on the number of cells in the chip.

Control Bus: specifies the function to be performed (Reset, Read, Write, Compare, etc..).

Output Ready (OR): signals the presence of at least one track candidate ready to be output.

Select (SEL): directs the chip to output the first track candidate in the output queue on the Address Bus.

The main functions that can be performed are:

Write: Patterns are loaded into the memory through the Data Bus at the address specified by the Address Bus.

Compare: hit coordinates are input through the Data Bus, one coordinate per clock cycle.

Readout: addresses of track candidates are output on the Address Bus, one track per clock cycle.

LARGE AM ARRAYS

Fig 5 shows how a number of chips (16 in this case) can be connected to implement a larger Associative Memory. The logic in the shaded area is what is needed to expand the memory by a factor 16: we will call it "glue".

During readout the OR signals are priority encoded and only the leftmost chip is enabled by SEL. As soon as all track candidates from one chip have been read out, the OR from that chip goes away and SEL is automatically transferred to the next chip with available data. When all the OR's are off then also the global OROUT signal is turned off. The output from the priority encoder identifies the chip being read out and it is used to form the high order bits on the address bus, low order bits being the pattern address within the chip. During write operations, the high order bits are input from the bus and fed to the decoder to select only the appropriate chip.

In this way we obtain a device which contains 16 times the number of patterns but has the same I/O lines and protocol as a single chip with 4 additional address lines. It is then straightforward to replicate this structure hierarchically to expand our associative memory to an arbitrary size. Fig 6 shows how to implement a very large array of AM chips. The overall structure is tree-like the leaves being AM chips and the nodes being "glue" chips. Each level in the tree deals with a number of address lines (4 in our case). Each link from one level to the next represents on OR line going to the right and a SEL line going to the left; other control signals and the Data Bus are broadcast to all the AM chips in parallel.

TWO-STAGE TRACK FINDING

In some case, when the number of detector bins is large, the size of the associative memory required to solve completely the track finding problem might be too big. Tracking systems with thousands of channels per layer are often required by modern HEP experiments leading to an order of magnitude of many million patterns, a number which is certainly out of reach of present technology. What would be the use of associative memory in this case? If the number of bins is too large we should treat the detector as if it had lower resolution by logically ORing adjacent bins (which is easily done by simply ignoring a number of low order bits in hit addresses) and perform a first stage of track finding in the associative memory at this reduced level of spatial definition. We can then pass track candidates to a second stage which will refine the definition of the track down to the detector limit. The key point here is that this second stage can be implemented with more conventional techniques (e.g. a sequential processor) since virtually all of the combinatorial problem which makes track finding so time consuming has already been solved. This happens because the number of patterns in the associative memory is orders of magnitude larger than the number of tracks in one event and therefore the probability of having two tracks "hitting" the same pattern is negligibly small.

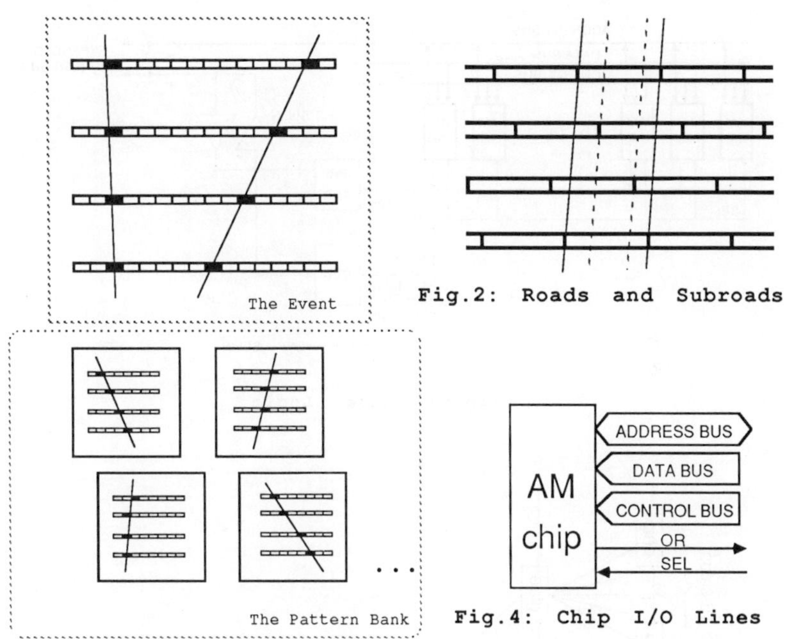

Fig.2: Roads and Subroads

Fig.4: Chip I/O Lines

Fig.1: Pattern Matching

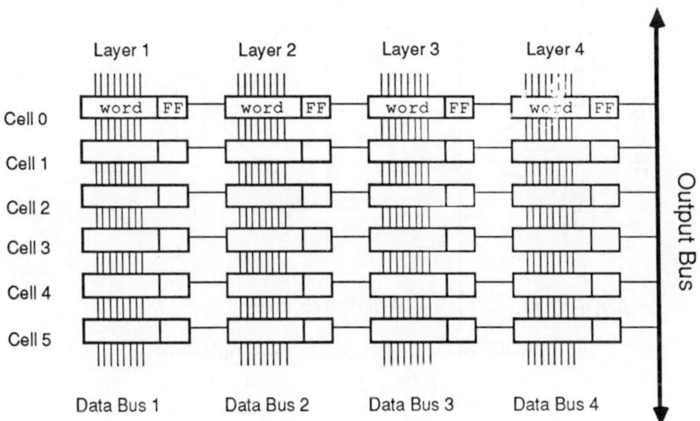

Fig.3: Associative Memory

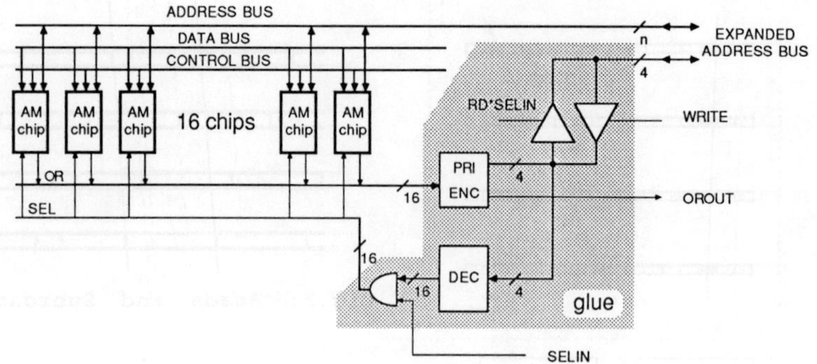

Fig.5: "Glue" Logic

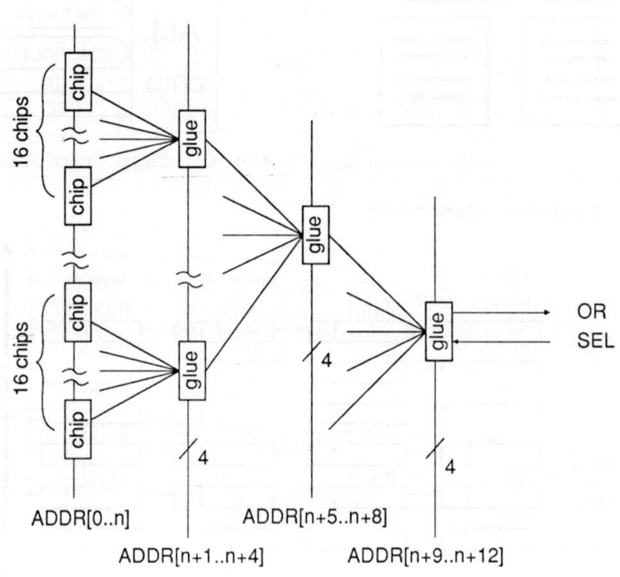

Fig.6: Structure of a Large AM Array

The UA2 Data Acquisition System

The UA2 Online Group, CERN - EP Division
G. Blaylock, K. Einsweiler, G. Fumagalli, S. Hellman,
L. Mapelli, C. Petridou, L. Rasmussen, S. Stapnes

ABSTRACT

The UA2 detector has recently undergone an extensive upgrading program in preparation for running with the new ACOL anti-proton source at the CERN SPS Collider. This upgrade required a corresponding improvement in the performance of the data acquisition and trigger systems. In particular, a new front-end system, including three levels of event selection in a hierarchical trigger system, and two stages of event processing and buffering, has been constructed. In this contribution, we describe the overall structure of the new data acquisition system, placing particular emphasis on the configuration of the front-end along with some of the general principles which influenced its design. In addition, a short discussion of the performance of this system during the initial 1987 Collider run is included.

1. INTRODUCTION

The new ACOL anti-proton source for the CERN SPS Collider is intended to increase the luminosity by a factor of ten relative to the values experienced during the last UA2 Collider run in 1985. Furthermore, the UA2 detector itself underwent a corresponding upgrade to improve the quality of the missing energy and electron measurements, resulting in an increase in the event size by a factor of roughly three. This places a large additional burden on the trigger and data acquisition systems, and a completely new design was required to satisfy these requirements. It was necessary to introduce two additional levels of software triggering to supplement the previous single level hardware trigger. It was also vital to provide multi-event buffering along with intelligent read-out and data formatting.

Several general goals influenced the design of the new system. A first principle was to reduce the number of component types and to implement the system in a uniform, high speed bus structure. For this reason, the extensive processing power is based in FASTBUS and organized around two processor types : a fast trigger processor (XOP), and a general purpose FASTBUS Master using the M68020 microprocessor (ALEPH Event Builder - AEB). A second principle was that the system should, to a large extent, detect and localize any errors which might occur. This requires that all of the participants in the data acquisition system be intelligent enough to support a reliable set of protocols for acquiring events. Finally, it was very important that this collection of processors appeared to be closely integrated with the main data acquisition computer. This integration should include such features as code management, databases/parameters, error reporting, run control, and performance monitoring. This is achieved in practice by connecting all of the processors to Ethernet as well as FASTBUS, providing a uniform trigger and data acquisition environment which is tightly coupled to the host VAX and programmed using extensive cross-software.

2. HARDWARE OVERVIEW

The UA2 upgrade placed several important restrictions on the data acquisition system. While most of the new digitization electronics was constructed in FASTBUS, a substantial amount remained embedded in REMUS branches, requiring an integration of these two systems. In addition, none of this electronics was capable of buffering events, i.e., it could not digitize a new event until the previous event had been safely transferred. The result was a system of digitization electronics which required a time of 1-10 msec for read-out, and which could not accept new events during the read-out period. This long read-out period was not compatible with the expected primary trigger rate of 100 Hz, and a second, more refined, trigger decision was needed. In order to solve the read-out problem, this trigger was required to reach its decision in a time of about 1 msec. Such a trigger could not be sophisticated enough to provide the full rejection needed before recording the data, and so the final design contains a trigger consisting of the following three levels:

- The first level (LVL1) is implemented in a mixture of analog and digital hardware. It makes a decision in less than 1.5 microsec, which is less than the SPS bunch crossing interval of 3.9 microsec, and therefore generates no deadtime.

- The second level (LVL2) is implemented in a special purpose trigger processor, known as XOP (eXtended Online Processor), which uses a dedicated fast input stream to reach its decision. This decision, including the digitization of the fast input stream, requires less than 1 msec. Since the digitization electronics is unbuffered, this decision must generate deadtime.

- The third level (LVL3) is implemented in a pool of general purpose FASTBUS processors, and uses the complete event and up to several hundred msec to reach a final trigger decision. As long as the pool of processors is sufficiently large, this decision generates no deadtime.

In parallel with this trigger organization, the data acquisition system itself requires a minimum of two stages to provide the necessary performance. The initial layer operates at the detector or crate level and provides multi-event buffering, read-out control and data formatting. The processors in this layer are known as Read-Out Processors (ROPs). The second layer is responsible for the event building and interface to the third level trigger, and consists of the Master Event Builder (MEB) and the pool of third level trigger processors (LVL3).

Before proceeding to a more detailed discussion of the system as a whole, it is useful to provide a short description of the important modules. The first such 'module' is the second level trigger processor (XOP), described in detail elsewhere in this conference. This is a highly parallel processor, running a microcycle of 100 nsec, that has a measured performance on the standard UA2 calorimeter clustering algorithm equivalent to about 30 VAX/780. This processor system also contains a FASTBUS Master interface, a large NIM I/O Register, LeCroy ECLine input and output busses, and a VME-based 68000 control processor with an Ethernet interface.

The second important module is the REMUS/FASTBUS interface. This device looks like a standard Read/Write Branch Driver (RWBD) on the REMUS side. On the FASTBUS side, the configuration depends on whether the REMUS operation is a read or a write. For a REMUS read, an autonomous scan is initiated from either the front panel or via FASTBUS, and the resulting data is transferred into an 8K 16-bit word buffer. This allows the read-out of many (slow) REMUS branches to proceed in parallel. For a REMUS write, a FASTBUS data cycle is mapped into a corresponding REMUS write cycle. Eight of these modules were used in the 1987 system configuration; this will be increased to nine for 1988.

The third important module is the FASTBUS/VAX interface. For this, we use the CFI (CERN FASTBUS Interface), which is capable of autonomously executing lists of operations. The elements in a list can be either FASTBUS operations or references to user subroutines which are then executed on a 68000 processor in the module. These user subroutines implement features such as a FASTBUS T-pin scan to locate and process Service Requests.

Finally, the most important module in the list is the ALEPH Event Builder (AEB, Ref. 1). This module consists of two FASTBUS cards, one containing a 20 MHz 68020/68881 processor, a Cheapernet (ThinWire Ethernet) interface, 1.0 MByte of RAM, 0.5 MByte of EPROM, serial and parallel ports, and a FASTBUS Master/Slave interface to a crate segment. The second card contains 0.5 MByte of dual-port RAM in Data Space, 2 KByte of dual-port RAM in CSR Space, and a FASTBUS Slave interface to a cable segment. Several additional FASTBUS features are also supported : Service Request can be generated on the cable port and detected as an interrupt on the crate port. The full sixteen FASTBUS Interrupt Receiver blocks are implemented on the crate port. The 68020 can be interrupted from the cable port by writing to a special CSR. Both ports are capable of generating/checking Parity to insure reliable data transfers. This module is the heart of the new system, and plays the role of Read-Out Processor, Master Event Builder and LVL3 Processor. There were eleven such processors used during the 1987 run; nine of them were ROPs, one was both the MEB and a LVL3 processor, and one was simply a LVL3 processor. For 1988, the number of LVL3 processors will be increased by two.

These building blocks are incorporated into the UA2 system, appearing in a schematic form in Fig. 1. The simplest way to explain this diagram is to provide a condensed summary of the sequence of events occurring during data acquisition. For the case in which the event is rejected by the second level trigger, the digitization electronics is cleared without initiating any activity in the other processors. For the case of an accepted event, the sequence divides into two components, one taking place in real time (generating deadtime), and the other involving data which exists in buffered form. The real time sequence is:

- The LVL2 trigger mask determines a read-out pattern for the detectors in the experiment. This information is sent to the ROPs and MEB via a FASTBUS Interrupt Message, containing the Event ID, Trigger Mask and Read-Out Masks.
- The ROPs transfer data from the digitization electronics to their internal buffers. This takes from 1 to 10 msec, depending on the event complexity and the detector read-out pattern.
- The ROPs signal the End of Read-Out to the LVL2 trigger via an ECL signal, at which point the system can accept the next event.

The buffered sequence is more complex:

- Raw events are formatted by the ROPs. This process can be very complex, requiring up to 50 msec for one detector in which the data is clustered to provide a compression of 2-3 in event length.
- When finished, the ROP notifies the MEB using a FASTBUS Service Request. When all of the required ROPs have notified the MEB, it builds a description of the event and assigns a free LVL3 processor sending the event description as a FASTBUS Interrupt Message.
- The LVL3 processor builds the actual event in a 'SuperEvent' buffer, and then applies the trigger algorithm to the completed event. If the event is not accepted, it is purged from the buffer.
- When a buffer is full, the CFI is notified using a FASTBUS Service Request. The CFI transfers the 'SuperEvent' buffer to the VAX, where it is unpacked, recorded on tape and then made available for online monitoring. This monitoring may take place on the main VAX/785 or, after distribution over DECNet, on one of three additional microVAX GPX-II workstations which also provide the graphics support for online displays.

The data acquisition sequence has two important features which improve its reliability. The first feature is the extensive use of message-based protocols. A sixteen longword Event Descriptor is created and distributed in FASTBUS by LVL2, passes through all of the ROPs and the MEB, and is then used by LVL3 and the CFI. This descriptor contains information about the Event ID, Trigger/Status Masks, Detector Masks, and Processor Masks, as well as the length and address of the corresponding data block. This information allows each processor in the event stream to know what responses are expected at each step, giving immediate detection of sequence errors, pointer errors, dead or misbehaving processors, etc. It also leads to a very modular system, so that the 'LVL3 Processor' and 'Master Event Builder' software can be executed (transparently) on the same hardware processor, allowing more efficient use of the available processing power.

The second important feature is the use of hand-shake protocols for buffer management and flow control. This means that a slave will not release a buffer until it receives an acknowledge message from a master, thereby controlling the data flow when the system becomes congested. In addition, Parity checking is performed on all FASTBUS cable segments, and any detected errors result in an automatic retry before issuing the acknowledge. Again, this provides prompt detection of errors, limited mainly by the inadequate error detection provided by FASTBUS (a single Parity bit for 32 data bits, nothing for control bits).

3. SOFTWARE FEATURES

In this section, a brief description of the software used in the AEB is presented. For information on the LVL2 software, the reader is referred to the preceding talk by S. Stapnes. The AEB software consists of several major components. These components can be separated into cross-software tools, system software, and communications software.

The cross-software tools chosen by UA2 consist of a suite of cross-software products developed at CERN for either VMS or UNIX hosts. These products are based on the CERN Universal Format for Object Modules (CUFOM), and include: an assembler, a C and a PASCAL compiler, and a linker/pusher. Much of the user software is written in FORTRAN-77 using the RTF68K (Real-Time FORTRAN for the 68K) compiler developed by H. Von der Schmitt.

The system software is stored in EPROMs located on the AEB and consists of : a real-time kernel, a debugger, a terminal interface, and Ethernet support. The kernel is the Motorola RMS68K Real-Time Multi-Tasking Kernel, which supports semaphores, event queues, task synchronization, asynchronous service routines, and other useful services. The CERN-written debugger RMON runs as a system task and supports assembly/disassembly, task breakpoints, program tracing, formatting and modification of system tables, etc. The CERN-written MIOS terminal interface supports simultaneous access to the terminal by many tasks, each one owning a 'virtual screen' for output. The Ethernet hardware is supported by a multi-user driver hidden under an RPC (Remote Procedure Call) communications layer.

The Remote Procedure Call communications layer consists of two components (Ref. 2). The first is a compiler which takes procedure descriptions (in an ADA-like syntax), and generates PASCAL "stubs" to support remote procedure calls. The second component is the Run Time System which provides the facilities necessary for two "stubs" on different processors to exchange messages (in our case, over Ethernet) and thereby execute a remote procedure call. This software is vital for the integration of the system, and its application is discussed at length in the next section.

4. SYSTEM INTEGRATION TOPICS

In this section, we present a brief enumeration of some of the features of the system contributing to the integration of the front-end with the Host.

- Program Development and Code Management in a centralized environment. This has the advantage that commercial tools can be used (e.g., MMS/CMS from DEC), that there is a reliable common file base, and that the definition of system-wide data structures and parameters is simple.

- Remote Reset and Program Loading. The ability to reset the remote processors from Ethernet has proven to be very useful. In addition, downloading/starting tasks using file lists and executable code from the Host makes the remote processor initialization almost transparent.

- Task Level Error Reporting to a central message logging facility. This provides a unified format/structure for error reporting at all levels of the system, from LVL2 to the main VAX.

- Remote Access to Central Databases. These databases typically contain detector parameters, trigger parameters, and system configurations (module addresses, etc.). There are many advantages to keeping all of this information in a common place. Each processor in the system can then make remote accesses during either processor initialization, or during run initialization. As an example, each ROP in the system typically needs information about module addresses, module parameters (thresholds...), and data formatting (the mapping from electronics channels to detector channels...).

- Distributed Run Control. This has the advantage that a run is always initiated from the Host. The remote processors access run parameters according to the run type, then perform there preparation for the run. Finally, when all processors have signaled their readiness, the trigger is activated.

- Remote Statistics Accumulation. The remote processors accumulate information on their performance (CPU usage, buffer usage, error statistics, etc.). These statistics blocks are periodically transferred to the Host for monitoring, and provide a record of the system state throughout the run.

- System Integrity Checks performed by remote processors under Host control. The lengthy task of module testing/initialization and system testing can be performed quickly by dividing the problem among the remote processors, which operate in parallel.

All of these features rely heavily on Remote Procedure Calls over Ethernet. This technique encouraged a modular approach to the system design which gives flexibility in distributing the workload in a multi-processor system.

5. SYSTEM PERFORMANCE

The system described above was completed in August 1987, and the first Collider run took place in November/December 1987. During this run, the peak luminosity never exceeded the previous maximum values of 3.5×10^{29} $cm^{-2}sec^{-1}$, providing an integrated luminosity of 46 nb^{-1}. This corresponded to roughly 6.8×10^{6} triggers recorded on 500 magnetic tapes. During this period, the data acquisition and trigger systems worked reasonably well. Since the luminosity was considerably less than the design value, several special runs were made to test the performance of the system under simulated full rate conditions. Again, no particular problems were found, and the system operated with a livetime of about 80% at a primary trigger rate of 150 Hz.

6. REFERENCES

1. A. Marchioro et al. IEEE Trans. Nucl. Science 34(1986), 133-136
2. T. J. Berners-Lee, CERN/DD, "RPC Users Manual"

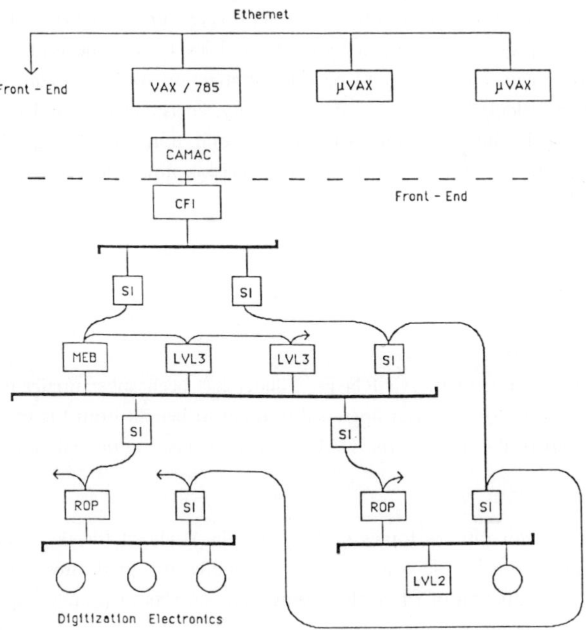

Figure 1. Schematic diagram of the UA2 system

The XOP trigger processor integrated into the UA2 Date Acquisition System

P.Baehler,N.Bosco,K.Einsweiler,C.Ljuslin,S.Stapnes
Presented by S.Stapnes

CERN, CH-1211, GENEVA 23

Abstract

XOP processors are being used by the LEP L3 and the UA2 experiments at CERN in their second level trigger. The present paper describes the integration of the XOP into the UA2 trigger system. In UA2 the XOP is used in the 2nd level trigger, supervising the 1st level trigger and initiating the readout processors in the system. The 2nd level trigger decision, the main task of the XOP, comprises gain correction and pedestal subtraction of 1200 channels of calorimeter data, cluster finding, calculation of the center of gravity, energy leakage and radius of the clusters, calculations of the invariant mass of clusters, total transverse energy deposited in the calorimeter and missing transverse momentum. The decision time is less than 1 msec.

INTRODUCTION

The antiproton source of the CERN $\bar{p}p$ collider has been substantially upgraded. At the same time the UA2 detector has been upgraded in order to benefit from the foreseen increase in luminosity. As a part of the upgrade the DAQ system has been improved and a multilevel trigger introduced.

The most important physics topics in UA2 are the study of the decay of the W and Z intermediate vector bosons into lepton pairs, search for the top quark, investigation of multijet production and the production of jets with missing transverse momentum. The rates for the different triggers at full luminosity can be estimated from Monte Carlo studies and extrapolations from the old UA2 experiment. In the 1987 run the collider reached less than 10% of it's nominal peak luminosity but nevertheless the run gave valuable experience with the new detector.

The improved UA2 detector still concentrates on detection and measurement of electrons, jets and missing transverse momentum. The most important part for the trigger and early data selection is the calorimeter (fig.1).

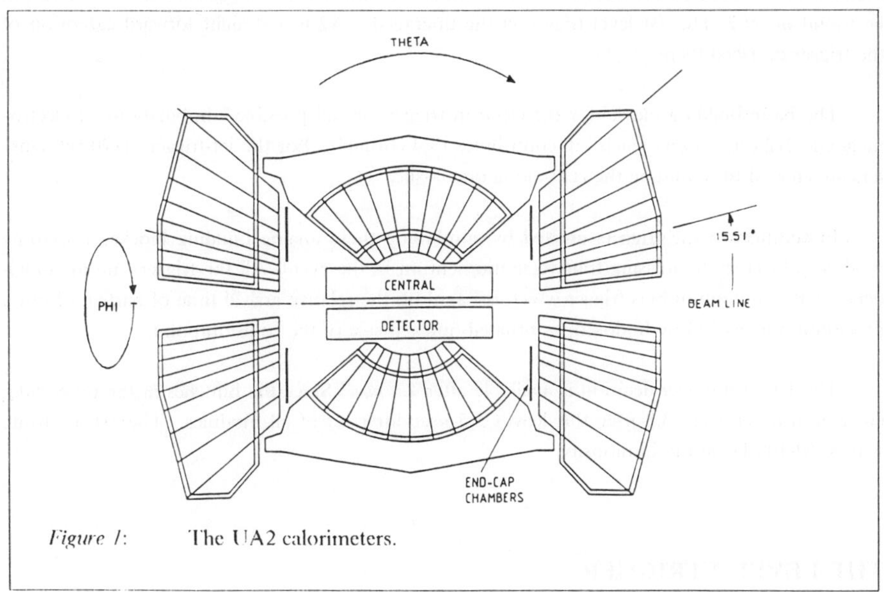

Figure 1: The UA2 calorimeters.

The central calorimeter consists of 240 cells, each covering 15° in azimuthal and 10° in polar angle (the angles are indicated in fig.1). The endcap calorimeter cells cover 15° in azimuthal and between 6.5° and 2.7° in polar angle. Longitudinally the calorimeter is divided into an electromagnetic compartment and a hadronic compartment (which again is subdivided into two compartments in the central calorimeter), and covers down to an angle 5.6° with the beam axis. The cells down to $\theta = 15.51°$, 120 in each endcap, are included in the trigger. The light from each cell compartment is collected with 2 wavelength shifting light guides into two separate photo multipliers (PM's). A full description of the detector can be found in ref. 1.

THE LEVEL 1 TRIGGER

The trigger algorithms are dictated by the properties of the detector and by the physics events that are to be detected. A detailed description of the 1st level trigger for the old UA2 can be found in ref.2. The 1st level trigger in the upgraded UA2 is a straight forward extension of the trigger decribed there.

The basic building blocks for the electron triggers are all possible combinations of electromagnetic 2*2 cell clusters (each cell contributes to 4 clusters). For the jet-triggers 1*20 cell clusters are created by summing the cells along one ϕ-slice.

In addition to the triggers created by combining these analog building blocks, a separate module calculates the missing transverse momentum in the event. All the triggers are in coincidence with a minimum bias trigger which is based on the relative arrival time of particles from a $\bar{p}p$ collision measured by hodoscopes situated on each side of the collision point.

The 1st level trigger decision is available after 1.3 μsec. Since the bunches in the $\bar{p}p$ collider cross each other every 3.8 μsec this leaves 2.5 μsec for reset of all modules. The 1st level rate will be 50-100 Hz at full luminosity.

THE LEVEL 2 TRIGGER

The 2nd level trigger is a completely new feature in UA2. With a primary trigger rate of around 50-100 Hz and a readout time between 2 and 10 msec, a reduction to less than 10 Hz with a rejection time less than 1 msec, is necessary in order to keep the dead time below 20%.

The XOP processor (ref.3) is a micro programmable 16 bit integer arithmetic processor especially targeted at trigger applications. The basic instruction execution time is 100 ns, but due to it's relatively long instruction word (192 bits) and the usage of parallelism and pipelining the overall performance is in fact much higher than the cycle time alone would indicate. The parallelism covers arithmetic operations, data address calculations, data accessing, loop control, condition checking and next instruction evaluation. Certain instructions can be performed in vector mode at a rate of 20 MIPS. The data memory bandwidth is 40 Mbytes/sec. The XOP ECLine interface allows a data input/output rate of 40 Mbytes/sec. The input port is equipped with a preprocessor which can perform in-line gain correction and pedestal subtraction without influencing the maximum data transfer rate. The XOP Fastbus Master interface supports all the different Fastbus data transfer modes. The pipeline block transfer allows a data transfer rate of up to 32 Mbytes/sec. The XOP NIM in/output ports have two 16 bit input registers and four 16 bit output registers.

The XOP system configaration (fig.2) includes a VME system with a M68000 processor ,a memory board and the VME/XOP interface. The EPROM resident monitor on the VME system is an extension of the CERN supported monitor Monica and incorporates additional commands for the XOP processor. In UA2 the VME system also includes a FILTABYTE 25.1 board which allows downloading of programs and control of the XOP over Ethernet from the data acquisition computer.

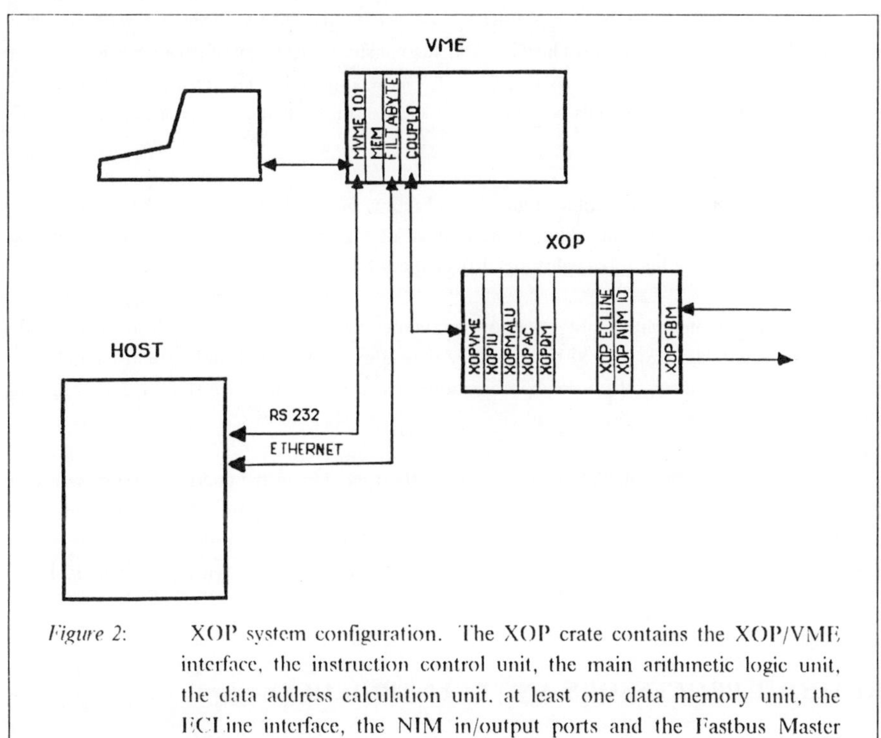

Figure 2: XOP system configuration. The XOP crate contains the XOP/VME interface, the instruction control unit, the main arithmetic logic unit, the data address calculation unit. at least one data memory unit, the ECLine interface, the NIM in/output ports and the Fastbus Master Interface.

A 1st level trigger will start a fast digitization of the calorimeter information. The compartment energies (analogue sum of the two PM's per compartment) are multiplexed through a LeCroy FERA ADC system to the XOP ECLine interface. The transfer speed is limited by the FERA driver to 105 nsec per word. The data coming in over the ECLine is gain corrected and pedestal subracted before being stored in the XOP memory. It takes 165 μsec, including the ADC convertion time, to transfer the 1200 10 bit words into the XOP memory.

The XOP will in the meantime execute a short initialization routine and wait for an end-of-conversion pulse from the multiplex controller. The pulse will set a bit in the ECLine interface status register. After finding this bit set the XOP will read NIM input port 1 which is reserved for the level 1 triggermask. The three compartment energies in the central calorimeter and the two compartment energies in the endcap calorimeters will be added into a cell energy array. Reading, addition and writing take 100 ns per word taking full advantage of the instruction pipelining. It takes 170 μsec to organize the data for the trigger algorithm.

After these initial operations the cells are collected into clusters. Since a memory read, compare, loop control, condition checking and next instruction evaluation can be done within one instruction it takes 100 nsec to reject a cell below threshold. The processing time for an event with 10 clusters and 10 cells per cluster is 250 μsec (more than 20 times faster than a VAX 780).

Next step is to calculate center of gravity and radius of the cluster. This is done in cell-units in the θ and ϕ projections. In addition the leakage of the cluster, which is the ratio of cluster energy deposited in the hadronic calorimeter over the total cluster energy, is calculated.

At this point one can identify electron candidates using the transverse cluster energy, the cluster leakage which is compared to an energy dependent leakage cut and the cluster radius or calculate invariant masses of cluster combinations (electron-candidate clusters, the two leading clusters or all clusters).

The precision of the calculations is limited by the precision in the energy measurement (use the analogue sum of the two PM's through a 10 bit ADC in the trigger, while for the full readout, the two PM signals are digitized by two separate 12 bit ADC's) and by the 16 bit integer constraint. With suitable cuts, the 2nd level trigger will bring the rate down to less than 10 Hz.

READOUT PROCESSORS AND EVENTBUILDERS

From the 2nd level triggermask a readoutmask is constructed which is the logical OR of the readoutmasks for the bits set in the triggermask. The readoutmask defines which subdetectors should be read out. The XOP will send Fastbus interrupt messages to the readout processors in the system (Aleph Event Builders, ref.4) containing the eventnumber, statusword, triggermask, readoutmask etc... . Only the readout processors which are responsible for the readout of the subdetectors flagged in the readoutmask will receive a message. The same message is also sent to the Master Eventbuilder. The readout, dataflow and DAQ system in general are described in a separate contribution to this conference.

In the level 3 eventbuilders all the data is available and a more sophisticated trigger algorithm can be executed. A repetition of the 2nd level algorithm with floating point calculations and improved energy measurements will already give a factor 2-3 rejection. Since the event processing does not interfere with the readout of new events and several processors can work in parallel a processing time above 100 msec can easily be accommodated.

HOSTCOMPUTER-XOP LINK

An Ethernet cable provides the link between the host computer (a VAX 11/785) and the M68000 processor which controls the XOP. Start, stop, reset, loading of code and data etc.. are done by Remote Procedure Calls of subroutines in the M68000 processor which will access the XOP through the XOP/VME interface. This allows loading and control of the XOP from a standard data acquisition program and testing and debugging of the XOP code from the host computer. All the parameters describing the Fastbus system and detector readout are read from a common data-base and loaded into the XOP at runstart.

In the opposite direction the XOP can interrupt the M68000 processor and force a Remote Procedure Call of an errorhandler-routine on the VAX.

CONCLUSION

An XOP trigger processor has been integrated into the UA2 data acquisition system and successfully operated in the 1987 run. Only minor software modifications are necessary before our main data taking runs in 1988 and 1989.

REFERENCES

Ref.1 C.N.Booth, The UA2 experiment at ACOL, CERN−EP/87−78, April 13th 1987

Ref.2 V.Hungerbuhler, UA2 trigger and Data Acquisition, Topical Conference on the application of microprocessors to High Energy Physics Experiments, 4−6 May 1981, CERN 81−07 p.46

Ref.3 P.Baehler et al., Computing in High Energy Physics Conferance, Amsterdam 25−28 June 85, CERN−DD / 85−15

Ref.4 R.Benetta et al., Proceedings of the Fastbus Software Workshop, 23−24 September 1985, CERN 85−15 p.131

SU(3) LATTICE GAUGE THEORY CALCULATIONS ON T800 TRANSPUTER ARRAYS

Jaap Hoek

Dept of Electronics and Computer Science

University of Southampton, Southampton SO9 5NH, UK

and

Rutherford Appleton Laboratory

Chilton, Didcot, Oxon OX11 0QX, UK

The transputer series introduces a new concept in programming. Because processes are scheduled and descheduled fast by the transputer hardware a program can be written on one processor (for example for debugging or on a personal workstation) and then run efficiently on a large number of transputers. The scale-up in performance can be linear for a substantial number of processors, of course depending on the problem. With the advent of the T800 the transputer has become very interesting for scientific applications. In contrast to the T414 transputer, the T800 has floating point hardware, and the performance of one T800 is of the order of 2-3 Vax 11-780's in pure computational power (about 1.5 Mflops). A block diagram of the T800 is given in fig 1.

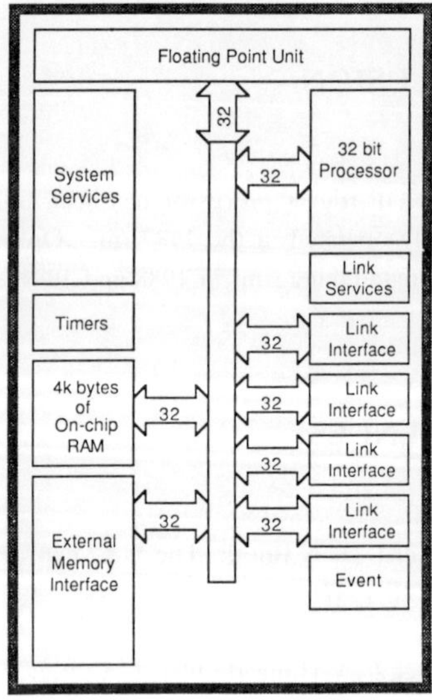

Fig 1. Block diagram of an INMOS T800 transputer.

Another aspect of the transputer is that they are very easy to connect into larger networks, since all the communication hardware is on the transputer chip. So a twisted pair of wires will suffice, although for larger networks electronic switches have now been developed. Communication on 4 bidirectional links can take place concurrently with computation. The University of Southampton takes part in an ESPRIT project to develop a low cost supercomputer (the Reconfigurable Transputer Array) [1].

This consists of 18 transputers in an electronically reconfigurable node, and these nodes can be connected with other nodes to form larger (and still completely reconfigurable) networks. As part of the evaluation of scientific applications on transputer arrays we have been investigating SU(3) lattice gauge theory.

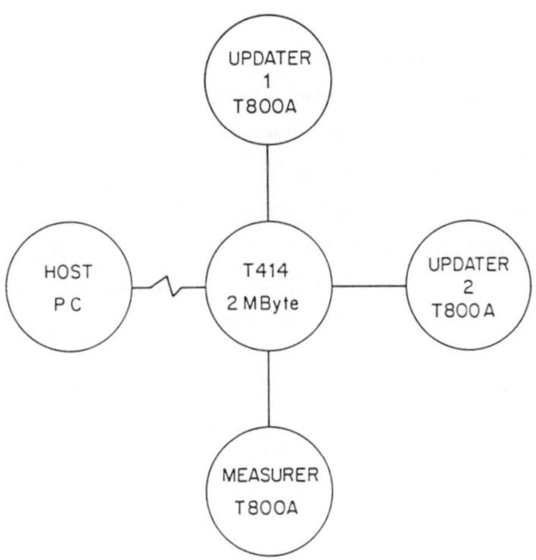

Fig 2. Typical production configuration

The lattice size used in our study (8^4) is small compared to what can be treated on conventional supercomputers at the moment. It was chosen as part of a larger research program on optimized renormalisation group transformations [2], and thus of scientific value. The lattice size that will be used on the 1000-transputer machine in Edinburgh will be much larger. Lattice gauge theoretical calculations can run efficiently on large sets of transputers since the problems can be split up geometrically. In our study we only had a few of the first revision T800's available. The configuration on which most of the work has been performed is depicted in fig 2.

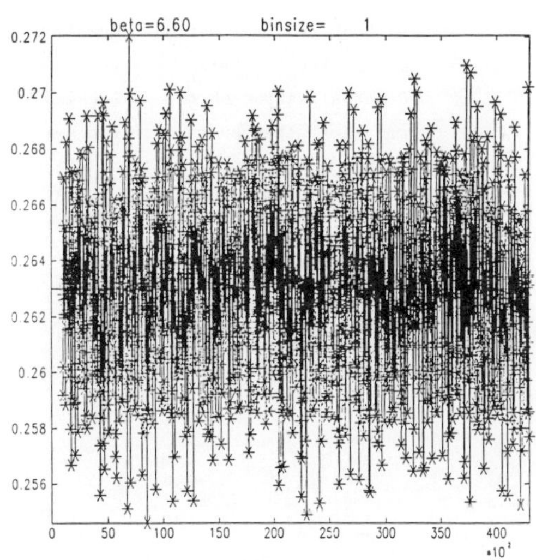

Fig 3. Flucatuations of the 2x2 Wilson loop on an 8^4 lattice at β=6.6.

Two of the T800's are used for link updating whereas the third is used to rform perform measurements on the configurations as they are produced. The measurement consists of measuring the values of planar and non-planar Wilson loops in various representations up to size 2x2. As examples of the physics results we show in fig 3 the fluctuations of the 2x2 Wilson loop on the 8^4 lattice at a coupling value β = 6.60, and in fig 4 the errors if this data is binned in bins of size 116.

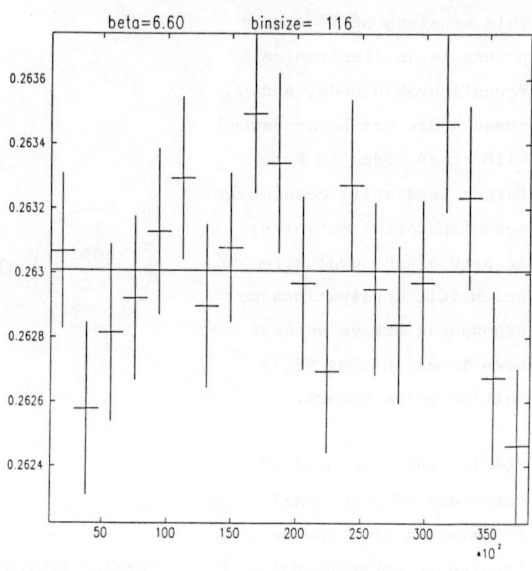

Fig 4. Data of fig 3 binned over 116 sweeps.

The performance figures for linkupdates are:

	Without renormalisation of the SU(3) matrix [Renormalisation of the matrices after every update]	Relative performance (scaled to top speed)	Remarks
T414	44680 μs [46370 μs]	69	single processor top speed .15 Mflops
T8A	4445 μs [4500 μs]	80	single processor top speed 1.3 Mflops
T8C	3990 μs [4040 μs]	77	single processor top speed 1.5 Mflops
CRAY XMP	22 μs	100	single processor, 64 bit precision top speed 210 Mflops
Cyber 205	40.9 μs	56	two pipes, 32 bit precision top speed 200 Mflops

Note that the SU(3) code on the Cyber 205 [3] and on the CRAY XMP [4] are highly optimized, in both cases with help from the manufacturers. The transputer code we used is not in such a state. It is written in OCCAM but for example does not have floating point versions of the intrinsic functions (the ones used on the T8 are recompiled T4 versions). Also the 256 kbyte external memory modules used on the T8's are certainly not the fastest ones around. It is probably that with optimisation a relative performance figure like that of the CRAY XMP can be obtained and therefore a system with about 140 transputers would perform on this problem like a single processor CRAY XMP. This can be tested out in the near future on the Edinburgh machine which upon completion will have 1000 transputers.

To illustrate the efficiency of overlapped IO and computations we have used a version of the program in which the IO and computation were not overlapped. This increased the update time by 34%.

References

[1] J.G. Harp, in Proceedings of the 4th Annual ESPRIT conference, Elsevier Science Publishers, 1987.

[2] C.B. Lang and M. Salmhofer, Optimization of Renormalization Group Transformations in Lattice Gauge Theory, preprint UNIGRAZ-UTP-04187 (Nov 1987).

Jaap Hoek, Optimized Renormalization Group Transformations, in Proceedings of the International Symposium on Field Theory on the Lattice, eds A. Billoire, R. Lacaze, A. Morel, O. Napoly and J. Zinn-Justin (Nucl. Phys. B, Proceedings supplements section).

[3] D. Barkai, K.J.M. Moriarty and C. Rebbi, Comp. Phys. Comm. 32 (1984) 1.

[4] Ph. de Forcrand, in Proceedings of a NATO Workshop on Lattice Gauge Theories: A Challenge in Large-Scale Computing, eds. B. Bunk, K.H. Mutter and K. Schilling, Plenum Press (New York) 1986.

TRIESTE CONFERENCE ON DIGITAL MICROELECTRONICS AND MICROPROCESSORS IN PARTICLE PHYSICS SUMMARY AND CONCLUDING REMARKS*

Thomas Nash
Advanced Computer Program
Fermi National Accelerator Laboratory
Batavia, Illinois 60510 USA

Abstract

This paper is a written version of the Concluding Remarks presented at the International Conference on the Impact of Digital Microelectronics and Microprocessors on Particle Physics. The Conference emphasized on-line data acquisition and triggering problems in high energy physics. Among the participants there was a clearly growing consensus that as these real time systems become larger they require more attention from the beginning to overall system coherence and manageability issues. We consider what this means for SSC/LHC era detectors. Given the interesting results on pixel silicon, neural networks, and parallel microprocessor based computers presented at Trieste, we speculate on some surprisingly simple, though still very radical, ideas on systems solutions for those huge detectors.

Trieste was the sixth location for a series of conferences focussing variously on the many aspects of on-line and off-line computing in experimental high energy physics. Starting at CERN, these conferences have been marked uniformly by the intense, fast changing nature of the subject matter they cover and the spectacular interest of their sites (Padua, Guanajuato, Amsterdam, Asilomar, Trieste, and next year, Oxford). Because Asilomar covered software issues in depth, the announced emphasis at Trieste was on hardware rather than software. To a large extent, the high energy physics community interprets this distinction between software and hardware as off-line *versus* on-line computing. Asilomar was populated by those working on off-line computing problems; Trieste's participants were primarily on-line and trigger specialists.

The difference between the style and concerns of these two groups, both essential to the success of HEP experiments, is extraordinary, almost that of two cultures with little communication. The on-line culture is willing to try anything, hardware or software, if it appears to have a chance of doing a required job. They place a strong emphasis on code efficiency and hardware cost effectiveness. The amount of effort required for a specific task and the overall coherence and manageability of experiment wide real time systems have not been at the forefront of their traditional concerns. Some individuals in this culture cannot understand, as one of them expressed it at Trieste, why all the hundreds and hundreds of MIPS

---

* This written version was prepared without access to the transparencies or papers prepared by other speakers at the conference. Although it thereby differs in detail and sequence from the actual concluding remarks given in Trieste, it expresses the same general conclusions.

worth of frequently idle on-line processors can't be applied to the off-line computing load.

Among those who develop off-line code, concerns are dominated by the huge scale of the software packages that must be prepared. Data and code management techniques are frequently used and there is considerable attention paid to the coherence of the overall skeleton of large programs. Ease of use and reliability of computers, peripherals, and operating systems takes precedence for many over processor cost effectiveness. When the computer cycles required for off-line processing inevitably reach finite budgetary limitations, cost effective processor systems developed by on-line types are, at best, tolerated.

At Trieste, despite the absence of most off-line software developers, many of their system level management concerns were beginning to make themselves felt among those responsible for large on-line data acquisition and processing systems. The ALEPH on-line electronics system is a prime example of this, even to the extent of using SASD tools in developing its on-line software. At OPAL and HERA's H1 on-line system management is being addressed through advanced human interfacing techniques taking advantage of Macintosh Hypercard tools.

Certainly, on-line systems coherence is far from being prevalent these days, particularly among smaller experiments attempting to retrofit modern processors into existing electronics. Conetti described the extraordinary variety of ways in which Fermilab's ACP Processors are being incorporated into experiments. Although widely used in standard configurations for off-line, there is no standard on-line implementation of these systems because there is nothing approaching a standard data acquisition system. In present day large experiments, the sociology of multi institution collaborations works against data acquisition system coherence. The typical scenario at design report time is to divide up the responsibilities among the various groups allowing each to define the approach to be used in specific subsystems. A few years later the problem of interfacing these subsystems becomes a hot topic. The situation is, to some extent, due to the way approval committees operate. Expertise is compartmentalized by subdetector and institution. The overall system is based on disparate subsystems, developed on the basis of correct, strongly held individual views, that don't mesh as a system.

Most experiments attempt to standardize at some level. Nonetheless, many still end up with systems that incorporate both Fastbus and VME or otherwise mix standards. Much of this results from a plague of what might be called *standards evangelism* in our business. Individuals who have developed some degree of personal expertise in something like VME or Transputers or UNIX or ACP Systems become strong advocates. At some level, this is an understandable tendency to protect an individual's intellectual investment in the expertise. However, the increasing complexity of detectors cries out for unbiased attitudes, a *secular humanist* approach, toward standards. We need to cool off the all too common electronics and computing standards religious wars which every laboratory has encountered.

The fact that it may prove possible to operate large present day experiments, handicapped as they are by the kind of institutionally mandated bottom up design described above, does not mean that this will continue to be possible for SSC/LHC era detectors. The order of magnitude increase in complexity of those behemoth detectors of the future simply *requires* a new and structured approach to on-line data acquisition and processor systems. It is not surprising, and very pleasing, how much agreement there was on this fundamental conclusion among participants at Trieste. We all agree that we need to get together and work out the rules for what might be called (borrowing the terminology from software) a *Structured Analysis Structured Design* (SASD) approach to data acquisition/trigger systems. We need to work out the rules, but here are some ideas that found easy agreement:

1. **Plan the overall system first**, at the same time as the physics is planned. Worry about protocols later.

2. Require **modularity** with a well defined protocol between modules.

3. The system should be as **homogeneous** as possible with a minimum of domain boundaries between different types of electronics. There is some feeling that there is at least one fundamental domain boundary between the type of protocols and electronics required for data acquisition and the type required and supported commercially for computing systems. At the very least, this boundary occurs at the point where data is recorded, but some of us feel it may be appropriate to use such commercial computer standards as far upstream as possible.

4. **Keep it simple stupid (KISS)!** Here this famous designer's rule implies such things as keeping control and data signals separate and minimizing sexy but hard to program hardware. Hardwired, specialized data driven, and esoteric parallel approaches must be reserved for where they are absolutely required, perhaps in lowest level triggers.

5. **SASD** for on-line software and electronic **CAD/CAE** design and simulation tools extended to the intermodule and system level.

6. The system must be **assemblable**. Related to modularity this is an issue of being able to bring large pieces of the system together without out having to redesign the data routing. Some radical ideas have been expressed at the Fermilab ACP regarding automatic data routing where data is sent to classes of available processors or memory rather than specific addresses. This approach for large data acquisition systems is similar to that used in specialized data driven trigger systems such as the Fermilab ECL-CAMAC (part of LeCroy's ECLine) and the Nevis data driven trigger processor.

Outlandish speculation is tolerated and encouraged in concluding remarks like these. With that license, let us see how we might apply the philosophy of these rules to designing an SSC data acquisition/trigger system that is simple and manageable, though of the enormous scale required. It is expected that SSC experiments will require something approaching a trigger reduction of 10^8 and will have to cope with such complications as several events in a crossing. A major portion of this trigger reduction will necessarily be in processors which use software written in high level languages because this is the only way to deal with such a level of trigger sophistication.

Three papers presented at this conference give a hint of what the future may bring. Irwin Gaines described the new generation of ACP Multiprocessors being developed at Fermilab with individual processors 10 times faster (we have since learned they will be 20 times faster) than the present generation. What he didn't say is that a DARPA funded project at Texas Instruments is underway to develop a Gallium Arsenide version of the same instruction set RISC processors as the ACP is using. They are expecting 200 MHz (about 100 VAX 11/780 power per chip set) by the end of next year. It is no longer unreasonable, therefore, to anticipate single board processor "nodes" available in 1995 with 100 VAX per node at the usual module cost of around $2500.

Spieler described his exciting work at LBL on pixel silicon detectors and the possibility of incorporating digitizing electronics directly on-silicon. Particularly interesting is the technique of bump bonding which allows connecting wafers full of integrated circuits that match the pixel spacing directly under the detector silicon. Listening to Bruce Denby's talk on neural network and cellular automaton applications to track and cluster finding, it doesn't take too much imagination to conceive of coupling VLSI implementations of these ideas (already being developed at Cal Tech and Bell Labs) to Spieler's pixel silicon.

A front end trigger based on VLSI could thus reasonably be expected to provide a trigger reduction of say $10^{2\pm1}$ leaving $10^{6\pm1}$ for higher level triggers. Can all of this remainder be handled with high level programmable processors along the lines of GaAs versions of ACP nodes? If so, we could avoid the difficulties of hardwired or other low level trigger techniques. How much processing power could we reasonably afford with 100 VAX power GaAs processing nodes? $5 million is hardly exorbitant for such a system on the scale of an SSC detector, and that would buy 200,000 VAX equivalents. The 2000 nodes could be arranged in some sensible tree structured, self routing architecture. Using a totally unjustified hand waving argument, we can scale up from CDF's planned 100 to 1 reduction using about 50 VAXes to 200,000, and we find that a reduction of 10^6 or so with such a system in 1995 doesn't seem at all beyond the realm of possibility.

It almost seems that all we need to get a hold of the extraordinary demands of SSC/LHC type detectors are strong R&D efforts in such areas as neural networks, multiprocessors, data acquisition systems, and silicon and other detectors. At Trieste it was preaching to the converted to conclude that perhaps what we need is a few less experiments in particle physics these days -- and a few more R&D projects .

THE VAX 11/785 DATA ACQUISITION FACILITY AT THE ENERGY RESEARCH LABORATORY, DHAHRAN, SAUDI ARABIA*

R.E. Abdel-Aal and H.A. AL-Juwair
Energy Research Laboratory, Research Institute, KFUPM, Dhahran, Saudi Arabia

1. OVERVIEW

The Energy Research Laboratory (ERL) has been recently established by the Research Institute of King Fahd University of Petroleum and Minerals (KFUPM). Major research facilities include a 350 KV accelerator for the production of 14 MeV neutrons and a 3 MV tandetron system. The ERL data acquisition and analysis system uses a VAX 11/785 cluster arrangement which allows increased reliability and facilitates future growth. An HSC50 controller supports all mass-storage devices (two RA81 disks and two tape units model TA78 and TU78). This reduces the demand on CPU time and main memory and frees the VAX UNIBUS for data acquisition. Two CAMAC branches are interfaced to the UNIBUS via MBD-11 drivers. Data is acquired using Lecroy 3511 fast CAMAC ADCs which have increased system throughput, reduced dead time, and simplified the interface hardware. The system runs the TUNL XSYS software.

2. PERFORMANCE

Compared with other XSYS installations, the ERL system offers a number of advantages leading to an overall reduction in dead time and in CPU time requirements. These include: faster CAMAC data acquisition, VAX sorting, and tape storage and reduced VAX QIO times. The MBD takes about 11+(n-1)6 usec to store an n-parameter event into the VAX memory. Singles spectra acquired at a pulser rate of 90 KHz consume only about 50% of CPU time with large buffers. The vax buffer switching times have been reduced by a factor of up to 1.7 at ERL, thus reducing the associated dead time. Dead time is about 2% for singles data at 50 KHz with large buffers. Taping at 6250 cpi does not add to this dead time.

* This work is supported by the Research Institute, KFUPM, Dhahran, Saudi Arabia.

FAST DATA TRANSFER PROCESSOR

By

M. ABDEL MEGED* and W. GASTI**

ABSTRACT

A special fast data transfer processor has been developed and integrated in an image multiprocessing system. The system is provided by an adequte and efficient mass storage capabilities (1024 x 1024 x 32 bit). Each processor in such system addresses its own original and final images during its bus operation. When a processor termintes its job, the occupied spaces in the storage will be free and may be used by another requesting processor. If the free spaces are not sufficient a data displacement or data compression is needed. This process will be done by the developed transfer processor instead of using a microprocessor to save time. Further capabilities has been considered, in the design of such processor, such as inverse zoom and interlacement operations to fit teledetection images. To cover all these capabilities the following algorithm is applied for both original and final images

$$A_c = A_o + (I \times 2^k) + [(M \times 2^k) \times J]$$

where Ac, Ao are the calculated and starting addresses; I, J are the current line and column (incremented by zoom factors); while M is the total number of lines and K is the interlacement factor.

The mass storage is seen by the processor as a tape of infinite length, where the images are arranged column by column. To reduce the hardware complexity the execution of the above algorithm is done in two sucessive phases using one adder and one multiplier. A pipe-line technique is applied to accelerate the address calculation. In such technique the address calculation for the Write cycle is done during the Read cycle and vise versa.

---

* Physics Department, Faculty of Science, Tanta University, Egypt.

** Assistant,in Electronic Applications Laboratory, STRASBOURG, FRANCE.

AN INTELLIGENT BUS ARBITRATION TIMING CONTROLLER

By

M. Abdel Meged*

ABSTRACT

In multiprocessing systems, many processors are connected together through general bus. Before a processor starts its transfer operation it must assure that the bus is free and will remain free from interference of other processors during its transaction. An Intelligent bus Arbitration Timing Controller (IATC) has been developed to avoid such bus conflict problems. An Arbitration cycle is initiated each 400 ns, if the bus is not busy, searching for true Arbitration Request (AR). When one or more (AR) lines are detected, the (IATC) will generate an Arbitration Grant for the processor of highest priority level.This (IATC) makes use of PROM'S implementation and combines the advanteges of two main arbitration techniques, named fixed priority leveles and circular priority ordering. The switching from one technique to the other is simply done by charging a configuration register.

The determination of the winning processor (or a group of processors) is carried out by decoding the contents of the PROM'S. The granted processor retains control of the bus until the end of its transfer operation, but a time-out is considered to inhibit the arbitration grant to avoid monpolization by a processor. During the transaction of a processor the arbitration cycle is synchronized by the Data Ready Signal (DRDY) in order to select the next winniq processor Such pipeline architecture is applied to minimize the time of arbitration. The choice of (DRDY) signal, instead of the address cycle, is found more convenient to represent the state of (AR) lines particularly for block transfer process.

---

* Physics Department, Faculty of Science, Tanta University, Egypt.

Communications for DELPHI online system

T. Adye, Rutherford and Appleton Laboratory, U.K.
F. Harris, Nuclear Physics Laboratory, University of Oxford, U.K.
P. Lorenz, University of Wuppertal, FRG.
I. Martinez, University of Santander, Spain

Abstract of Poster Session

The DELPHI online system is heterogeneous with differing communications media and computing system environments. Each detector, or group of 'small' detectors, is monitored by a Microvax(EC) with overall control residing in a large VAX(OPS). The main data acquisition tree is based on Fastbus with embedded 680x0/OS9 processors for control, trigger processing and formatting. The controls for gas, high voltage, etc., are based in 6809/FLEX processors embedded in G64.

Ethernet/Cheapernet links the VAX/Fastbus Masters/G64 processors. There is a requirement to unify as far as possible the software effort going into providing basic communications functions for message passing in this distributed system. The unification has come in the implementation of a standard programming interface for Transport Service (TS) via the CATS(Common Access to Transport Service), and the implementation of the Remote Procedure Call(RPC) system over CATS in the various environments.

The transport protocols used have been DECNET for inter/intra communications, and ISO TP4 for heterogeneous communications between the VAX systems and Fastbus and Slow Controls processors on Ethernet. The latter has used VOTS(VAX OSI TS) and the porting of the OSIAM package to the G64 and 680x0/OS9 environments. The porting for the Fastbus masters has just begun, together with the attendant implementation of RPC/CATS. Development work has begun on the setting up of CATS running over Fastbus itself (P Charpentier, Saclay).

CATS offers calls for making multiple, duplex connections between processes. These can be used for synchronous and asynchronous applications via 'blocked' and 'unblocked' calls for sending/receiving messages.

RPC offers transparent use of procedures in remote embedded processors, the system handling the problems of coding messages and interfacing to the transport mechanism. Although naturally a synchronous mechanism it can be used for asynchronous applications by the server calling the client back with a reverse procedure call to a 'server' procedure in the client.

The above system of RPC/CATS has been used successfully recently for the DELPHI Solenoid tests, and will be used for Slow Controls. It is also planned to use the system for communication involving the Fastbus masters for control functions.

This work has been accomplished in close coordination with DD, notably T.Berners-Lee, Y.Grandjean, G.Heiman and C.Piney. Current development work has begun for the 680x0/OS9, coordinated by DD.

A Special Purpose Processor for Ising Spin Systems and for Digital Image Processing.

Gaetano Roberto Aiello, Marco Budinich, Edoardo Milotti

Microprocessor Laboratory,
INFN Trieste and University of Trieste, Italy

The classical Metropolis or "heat-bath" algorithms are the most common choice for computing the ground state properties of Ising spin systems. If the model under study is described by a nearest-neighbour interaction Hamiltonian, the Metropolis and "heat-bath" algorithms have a "local" character, i.e. each step can be performed using variables which are "local" to each node in the Ising lattice (namely, the node spin variable, and the spins of the adjacent nodes). This "local" character can be exploited to give modified, highly parallel, algorithms.

What we have done is to build a dedicated processor to calculate with the maximum possible parallelism these kind of algorithms.

Our machine can be thought as a lattice of very simple (one bit) processing nodes: each node representing an Ising spin. Every processor can communicate with its adjacent neighbours and changes its state at each clock cycle according to very simple, predefined, rules. In this way our "processor lattice" mimics the evolution of an Ising lattice simulated on a traditional computer under the Metropolis or "heat-bath" algorithms with the difference that in our case all the spins are updated in parallel.

Each node is a simple one bit processor formed by a small look-up memory (which stores the transition probabilities between node states), a pseudo random number generator (of the feedback shift-register type), a comparator, and a flip-flop that stores the present node state.

Because of the "local" character of the interaction, the whole machine state can be updated in a checkerboard fashion, thereby achieving a high parallelism. Such a machine can be software-reconfigured to simulate lattices other than two-dimensional.

When we set out to build the machine, we understood that machine I/O and the pseudo random number generator were the main problems that we had to tackle, because they made the node structure considerably more complex than initially thought. Eventually we succeeded in shrinking each node into three commercial chips: two 8-bit-wide RAM's and one re-Programmable Gate Array.We have built a prototype machine with 12 nodes arranged into a two-dimensional lattice, with a total power of about 70 Mspin-flips/sec.

The machine could be considerably shrinked using VLSI techniques. Then, even with a (relatively) small lattice size, it could become an interesting image processing device, because of the equivalence between noise removal from a digitally stored image, and the problem of finding the ground state of an Ising spin system (with nearest-neighbour interaction and nonuniform external magnetic field).

REFERENCES.

B.M.McCoy and T.T.Wu: "The Two-dimensional Ising Model".

S.Kirkpatrick, C.D.Gelatt, M.P.Vecchi: "Optimization by Simulated Annealing", Science, **220**, p. 671 (1983).

P.Carnevali, L.Colletti, S.Patarnello: "Image Processing by Simulated Annealing", IBM J. Res. Develop., **29**, p. 569 (1985).

FAST TWO-DIMENSIONAL ELECTROMAGNETIC CLUSTER-FINDING IN THE NEW UA1 FIRST-LEVEL TRIGGER PROCESSOR

N.Bains[1], S.A.Baird[3], D.Campbell[3], M.Cawthraw[3], D.Charlton[1], J.Coughlan[3], E.Eisenhandler[2], N.Ellis[1], I.F.Fensome[2], P.Flynn[3], S.Galagedera[3], J.Garvey[1], G.Grayer[3], J.Gregory[1], R.Halsall[3], M.P.Jimack[1], P.Jovanovic[1], I.R.Kenyon[1], M.Landon[2], T.P.Shah[3], R.Stephens[3]

[1] University of Birmingham, Birmingham, England
[2] Queen Mary College, University of London, London, England
[3] Rutherford Appleton Laboratory, Chilton, Didcot, Oxfordshire, England

Experiments at hadron colliders have great difficulty in triggering on electrons and photons because of the high background due to jets. We describe a fast method for finding narrow showers in a calorimeter and requiring that they are isolated both laterally and in depth.

The new trigger processor[1)] uses electromagnetic cells of 20 cm x 20 cm in the upgraded UA1 calorimeter. At the first stage of the processor, digitised signals from e.m. and hadronic cells are converted into energy and transverse energy. Transverse energy in the e.m. cells is summed both vertically and horizontally in overlapping pairs, to allow for narrow electromagnetic showers that cross cell boundaries. These sums are then compared with selected thresholds to identify hit patterns. In addition, we look for low-level energy deposition in each e.m. cell, and for small amounts of total energy in the hadronic part of the calorimeter behind the e.m. cells.

The three kinds of comparator outputs are then fed to 32 cluster-finding modules. Each receives data from 40 cells under test and surrounding cells. The modules contain an array of programmable array-logic devices (PALs). Two types are used: one to recognise e.m. trigger candidates exceeding the required transverse energy threshold without any requirement on isolation, while the other requires isolated energy deposition in a maximum of four trigger cells.

The non-isolated PALs require simply that a horizontal or a vertical pair of cells has exceeded its transverse energy threshold; they recognize clusters of any size but 'reasonable' shape by using signals from four of the eight adjacent cells in veto to demand a 'corner'.

The isolated PALs are more novel. Their inputs are two horizontal and two vertical comparators on cell-pair E_t sums and 16 comparators on low-level deposition in individual cells. Using the horizontal, vertical and low-level comparators, they select every possible configuration of shower deposition in a group of one to four cells. The low-level comparators of cells surrounding this group are then used as vetoes, in the tightest possible configuration, to determine that the shower was isolated and therefore more characteristic of an e.m. shower than a hadronic jet. All this is done in one simple operation, comparing inputs with pre-programmed patterns in the PAL.

The two types of PAL are multiplexed to allow either option. Following each PAL, the hadronic energy behind the shower can be used in veto if desired to ensure that the longitudinal energy-deposition profile is also consistent with an e.m. shower. Finally, the number of clusters on each module is counted in a simple 40-bit adder tree using PROMs and adders.

The logic is cycled eight times per event. On each cycle both the PAL option and the use or non-use of the hadron veto can be independently selected. The PAL and hadron-veto section of the module takes about 60 ns and the adder tree about 75 ns. Latches at the input, before the adder stage and at the end allow pipelined operation so that data can be processed every 75 ns. The equivalent peak rate of each module is of the order of 10 000 mips. Full internal test facilities are provided.

1) N. Bains et al, 'The New UA1 First-Level Trigger Processor', contributed paper to this conference.

A FASTBUS DIGITAL READOUT MODULE FOR STREAMER TUBES

F. Beconcini [a], G.M. Bilei [b], R. Castaldi [a], U. Cazzola [a], R. Dell'Orso [b], P.G. Verdini [a]

[a] Dipartimento di Fisica dell'Università and Istituto Nazionale di Fisica Nucleare, Pisa, Italy

[b] Dipartimento di Fisica dell'Università and Istituto Nazionale di Fisica Nucleare, Perugia, Italy

ABSTRACT

A FASTBUS slave module has been built for the acquisition and preliminary online elaboration of digital data coming from limited streamer tube strip readout electronics. The module, based on the Motorola MC68020 microprocessor, performs fast readout with zero suppression, cluster searching and monitoring of the apparatus.

1. INTRODUCTION

A FASTBUS Digital Readout Module (WICDRM) has been designed to read the front-end electronics of streamer tube cathode strips. Signals coming from each strip are preamplified, discriminated, shaped and digitally stored into shift registers which will then be daisy-chained in order to have a continuous stream of data for the serial readout. Eight channels of preamplifiers, discriminators and shift registers are packaged in a single hybrid[1)] based on a four-channel custom chip developed by SGS. Four hybrids are usually packaged in a 32 channel board. An I/O card, mounted on the FASTBUS auxiliary backplane, transfers the data from eight different input daisy-chains into the FASTBUS acquisition module, converting between whatsoever transmission protocol may be chosen (for example, optical links or differential pairs) and the TTL levels used in the WICDRM.

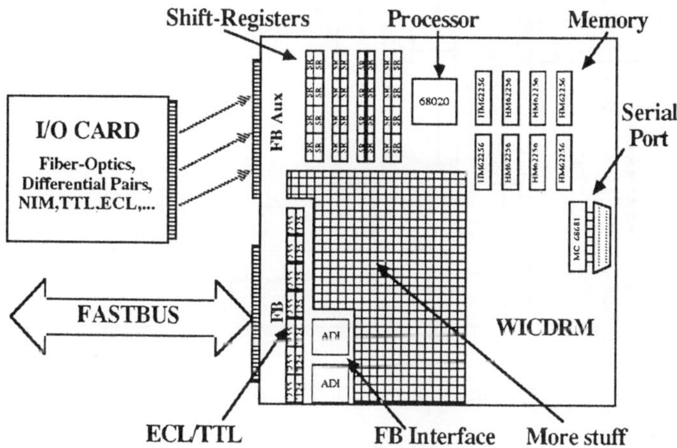

Fig.1. Architecture of the WICDRM

The data, digitally stored into the front-end shift registers, are daisy-chained and clocked with a (currently) 3.2 MHz rate into eight parallel input buffers inside the module, under the full control of the MC68020 microprocessor. During the buffer readout, one has the option of letting the processor write into RAM memory either those datawords which contain at least one cluster (defined as a set of one or more adjacent hit strips), together with the coordinates of these words within the given daisy-chain, or the full extent of the strip data.

Immediately after each first level trigger, all the digital data streams can be read and, if the trigger has in the meantime been accepted, the MC68020 will perform a second stage of elaboration, compacting the data, correcting it for possible mismappings and preparing it for further analysis by the FASTBUS Master Processor. Even if the trigger is not valid, the data may be processed locally in order to monitor the performance of the apparatus.

A processor interrupt by FASTBUS is implemented, as well as three external interrupts, and three different sources of Service Requests are used to flag the status of the current process to the FASTBUS master. For debugging purposes, an interrupt-driven serial port has also been included in the design.

2. THE FRONT END PROCESSOR

During the data acquisition cycle the WICDRM must perform the following tasks: fast readout and zero suppression; clustering and formatting of validated data; correcting for possible cable mismappings; and monitoring of the channel status (strip by strip hit count and possibly efficiency calibration with cosmic muon tracks).

During stand-alone operation the device can be used for extensive testing of the front-end electronics. The choice of the MC68020[2)] as CPU was mainly due to its high instruction execution speed, its full 32 bit address/data architecture, and the good software support. The microprocessor operates at its maximum clock speed of 16 MHz. The memories are static RAMs (32K x 8, 85 nsec access time) with no wait cycles. A block diagram of the front-end processor is shown in Fig. 2. The signals drawn on the left of the figure are directed to the front-end SGS electronics, those on the right are the internal connections with the FASTBUS interface.

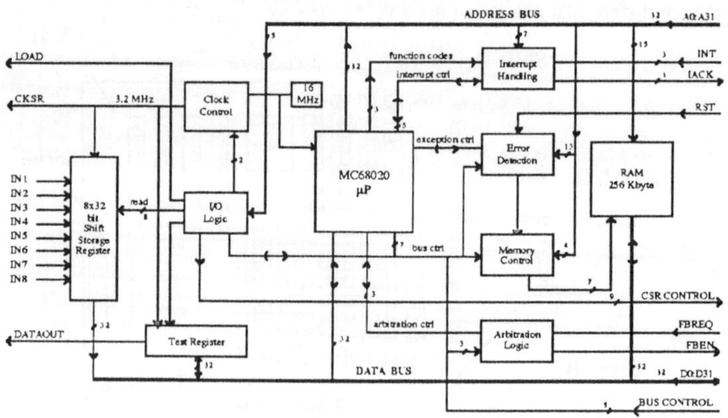

Fig.2. Block diagram of the front end processor

Upon reception of an interrupt, the processor can send a LOAD signal to the front-end electronics, latching the digital information. Subsequently the shift register clock (CKSR) is enabled and serial data from all eight input channels (IN1:IN8) is shifted into eight by 32-bit wide latches.

When the eight input buffers are full, the processor transfers the latched data into memory while optionally performing a preliminary zero suppression. The readout procedure continues until all the daisy-chained data has been transferred and processed.

If the trigger is accepted, the zero-suppressed data is further elaborated by the MC68020. The processor evaluates the size of each cluster of adjacent strips fired, and its location inside the apparatus. This information, properly coded, is read by the master and eventually transferred to mass storage.

The module has also a 32 bit Read-Write register, provided for the test of the front-end shift register chain. It is possible to load any 32-bit pattern and serially inject it at the beginning of each chain. Once this "a priori" known pattern has been shifted through the whole chain, the readback of the same bit configuration ensures that neither interruptions are present, nor are there malfunctioning units along the chain.

The memory is divided in two banks, one of which is dedicated to program and the other to data storage. For security purpose the program memory is interlocked to prevent write data cycle operations by the processor, by causing a BUS ERROR assertion if this happens.

The 256 KBytes of static RAM present on the board can be DMA-accessed by FASTBUS for fast Block-Transfer Read-Write operations.

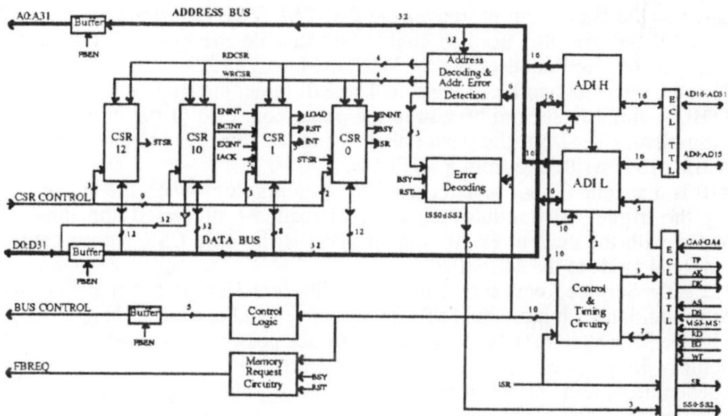

Fig.3. Block diagram of the FASTBUS interface

The arbitration logic handles the FASTBUS request (FBREQ) to access the memory by generating a Bus Request (BR) to the MC68020. The processor answers by asserting the Bus Grant (BG) to indicate that the bus will be available at the end of the current bus cycle, and enters an idle state. The BG generates a FASTBUS Enable (FBEN) signal that opens the Control, Data and Address bus buffers, giving FASTBUS access to the memory.

3. THE FASTBUS INTERFACE

The WICDRM is a single-width slave module that performs most of the standard operations specified by the FASTBUS protocol[3)], such as Broadcast Operations (either General, Class N or TP), Geographical Addressing in Data and Control spaces, Secondary Addressing (Read-Write), Random Data operations (Read-Write), Handshake Block Transfers (Read-Write) and Slave Status responses (implemented SS responses are 0, 1, 2, 3, 6 and 7). Parity is neither checked nor generated by the device.

The block diagram of the FASTBUS Interface is shown in Fig. 3. The bus-driving hardware is a mixture of 10K and 100K ECL logic. The 100K ECL-TTL bus transceivers are used only for the AD lines. 10K ECL logic is used on all the other information and timing lines.

The interface has been implemented using two ADI (Fujitsu MB114F307) chips[4]. The ADI Low (ADIL) handles the sixteen low-order lines (AD0 through AD15), while the ADI High (ADIH) handles the high-order lines (AD16 through AD31).

The two chips will perform slightly different functions: for instance only the ADIL chip takes care of class N Broadcast detection, and compares the five GA lines with the corresponding AD lines to perform the Geographical Address checking. Each chip contains a 16-bit wide Next Transfer Address register (NTA), a Data Latch, a 16-bit class N register (CSR #7, on the ADIL only), in addition to the logic needed to write and read these registers. The NTA can be modified or examined during the secondary address cycle, or incremented during a block transfer.

The Control and Timing Circuitry generates all the necessary signals for the control of the ADI chips (for example to load, read and increment the NTA, or to read and write CSR #7, to open and close data buffers etc.). In response to a FASTBUS cycle this circuitry also generates the handshake signals AK and DK according to FASTBUS protocol. In the WICDRM five Control and Status Registers are implemented: CSRs #0, #1, #7, #10 and #12. All the CSRs can be read or written by FASTBUS.

Bits 16 through 31 of CSR #0 indicate the Device Identification (ID). For the WICDRM an ID of 001B (hex) has been assigned. The MC68020 has write access to specific bits of this CSR such as the Busy flag; moreover the FASTBUS Service Request bit can be set. Other bits of CSR#0 are also used to enable and disable the external interrupts to the MC68020 and the Service Request from the WICDRM.

CSR #1 is the Interrupt and Reset register. Three different interrupts can be set or cleared from FASTBUS, and a Reset can be sent to the processor. Two of the three implemented interrupts can also be set from the front panel, while the third is automatically generated whenever CSR #10 is written to from FASTBUS.

CSR #10 is a special 32 bit register used for data tagging: when the current event is accepted by the trigger, all modules should be Broadcast addressed and the CSR #10 register loaded with the current event number. This is the only CSR whose contents are accessible in Read mode to the MC68020.

CSR #12 is the Service Request register. Three different SRs can be set and cleared from FASTBUS; in addition each individual SR can be enabled and disabled. These three SRs can also be set by the MC68020 to indicate the status if three possible processes.

The Memory Request circuitry generates a memory access request to the processor (FBREQ) depending on the status of the Busy and Reset flags. In particular the request is aborted if the device is busy or if the FASTBUS tries to access the program memory while the Reset signal is inactive.

In response to FASTBUS cycles the Error Decoding logic generates the appropriate Slave Status response (SS). A response SS = 1 indicates that the device is busy; SS = 2 means that the module cannot accept any more data during the block transfer; SS = 3 means that FASTBUS is trying to access the program memory while the Reset line is inactive; a Data Error - Accept (SS = 7) is generated if an invalid address is detected, and during the subsequent data cycle SS = 6 (Data Error - Reject) is asserted.

4. ACKNOWLEDGEMENTS

We want to express our thanks to B. Bertolucci, M. Breidenbach, R.S. Larsen, L. Paffrath and D.J. Sherden for the many valuable suggestions and encouragements received.

5. REFERENCES

1) "Test of the SGS D779-based readout electronics for the SLD limited streamer tube strips", F. Beconcini et al.; submitted to the 1987 IEEE Nuclear Science Symposium.

2) MC68020 32-bit Microprocessor User's Manual, Second Edition, Prentice-Hall, New York 1985.

3) IEEE Standard FASTBUS Modular High Speed Data Acquisition and Control System, ANSI/IEEE Std 960-1986.

4) R.Skegg and A.Daviel, IEEE Trans. Nucl. Sci. NS-32, n.1, 305 (1985).

The Energy Sum Processsor (E.S.P.)

(D. Bulfone and L. Lanceri - Microprocessor Laboratory and INFN, Trieste - Italy)

The E.S.P. (Energy Sum Processor) is a Fastbus module developed for the "Total Energy Trigger" of the DELPHI detector at CERN.

The "Total Energy Trigger" is produced by adding the values of the energy deposited in the calorimeters (Hadronic Calorimeter (HC), High Density Projection Chamber (HPC) and Forward ElectroMagnetic Calorimeter (FEMC)) and by comparing the result with thresholds.

The E.S.P. module will be used to process data coming from the FEMC: a disk-like fine grained lead glass calorimeter placed in the two end-caps of the DELPHI detector.

Using four dedicated "Energy Data Buses" (see figure below), the values of the energy deposited on the six sectors of each quadrant of the FEMC are sequentially presented, in parallel for each quadrant, at the inputs of the E.S.P. The module performs a fast digital sum obtaining the values of the energy deposited in each quadrant and in the whole end-cap. Such values are stored in the module and are accessible from Fastbus and from an external connector. About 2.5 μsec. are spent by the E.S.P. module to complete its operations.

Programmable threshold values can be pre-loaded in the E.S.P. module allowing local decision taking. The sequence of operations for the processing of the data is programmable and sequencer-like controlled by 24-bit patterns coming from a "Control Memory". Such a flexibility makes the E.S.P. independent of the particular segmentation of the detector considered; test sequences can also be executed.

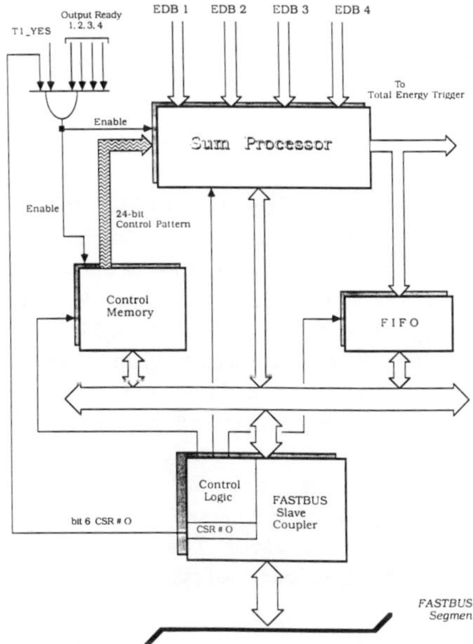

SOFTWARE SUPPORT FOR THE CERN HOST INTERFACE

D. Burckhart, T.J. Berners-Lee, C. Bizeau, R. Divia, P. Gallno, B. Heurley, K. Hollingworth, D. Jacobs, R.A. McLaren, H. Müller, C. Parkman, E. van de Bij, A. van Praag (CERN); A. Guglielmi, M. Rance (DEC); T. Almeida, P. Gomes (LIP); P. Alves (INESC)

ABSTRACT

A system of software modules has been designed for the CERN Host Interface (CHI) [1] project, a set of interfaces between VAX series computers and both FASTBUS [2] and VMEbus [3]. The CHI hardware allows the software to be split between its microprocessor and the VAX. Provision has been made for multi-user support on the host and communication software as well as library packages. Software tools assist the user in preparing his application software.

1. INTRODUCTION

In order to ensure good coordination of the hardware and software projects, a software system was designed in parallel with the hardware development of the CHI. By taking advantage of the common hardware architecture for the FASTBUS and VMEbus versions of the CHI, and by using modular software, a significant part of the support software can be shared for both versions. Remote Procedure Calls (RPC) [4] form the standard communication technique for system and application software and allow the use of library packages such as the NIM FASTBUS Standard Routines [5] locally on the processor or remotely on the host. Fast data transfers between the host and the instrumentation bus are supported making use of Direct Memory Access facilities designed into the CHI. An enhanced VAX/VMS driver reduces operating system overheads.

2. REQUIREMENTS

In conventional data acquisition for CAMAC [6] based experiments the driver formed part of the data acquisition kernel. Sequential readout lists had to be pre-prepared using a specialized language for FASTBUS in the CERN FASTBUS Interface (CFI) [6]. For the current generation of High Energy Physics experiments at LEP, the online software architecture had to be extended for new instrumentation buses which make heavy use of embedded processing for their multi-level trigger stages and for data reduction. The size and complexity of the readout architecture of those experiments require more flexible readout algorithms than are possible with readout lists. This can only be provided by the close coupling of a user programmable general processor to the instrumentation bus in use. On board intelligence provides the user with maximum flexibility for his particular event readout software. During the development phase of test or data acquisition programs a user-friendly and time-saving debugging environment is required, whereas at run time, fast execution and data transfer are important.

3. HOST TO INTERFACE COMMUNICATION

A software protocol run over the hardware link accomplishes control of the direction of the link, multiplexing of data between different tasks at either end, and flow control, so that data are never sent unless a receiving buffer exists. In collaboration with Digital Equipment Corporation the original VAX/VMS drivers has been enhanced to include concurrent access by many tasks. Send and receive transactions are handled in a single request in order to reduce operating system overhead for the communication protocol. The functionality provided to the user is that of an ISO standard transport service, allowing the creation and deletion of logical task-task connections. These connections are used by the RPC system, and are also directly available to application programs.

4. DEVELOPMENT ENVIRONMENT

Application software may be developed and debugged on the VAX, e.g. using CHI-resident routines remotely. At a later stage time-critical modules may be loaded into the CHI, and may then call the same library routines locally. At run time, when a program on one processor calls a subroutine on the other, parameters are passed to the subroutine, the subroutine is executed, and status and data are returned to the caller.

5. FASTBUS ACCESS

The NIM FASTBUS Standard routines have been implemented as a CHI-resident library which is based on the implementation of the Standard FASTBUS routines for the GPM [8] [9]. Use is made of the "key" address technique, i.e. the FASTBUS action of a single cycle routine consists of one MC68020 MOVE instruction to a FASTBUS key address. FASTBUS data may be transferred to and from CHI memory, or the "external buffer" mode of the Standard Routines may be used to transfer it directly to or from the VAX. Service Request and Interrupt Messages, accepted by the CHI FASTBUS port, result in the execution of the corresponding Standard service routines.

6. CONCLUSIONS

The CHI support software provides a flexible, user-friendly environment for application code running on the host or on the processor. Taking advantage of its built-in hardware features like DMA facilities and FASTBUS key addresses, high efficiency is achieved. Remote

Procedure Calls are used as a communication technique and an enhanced DEC Driver minimizes communication overheads.

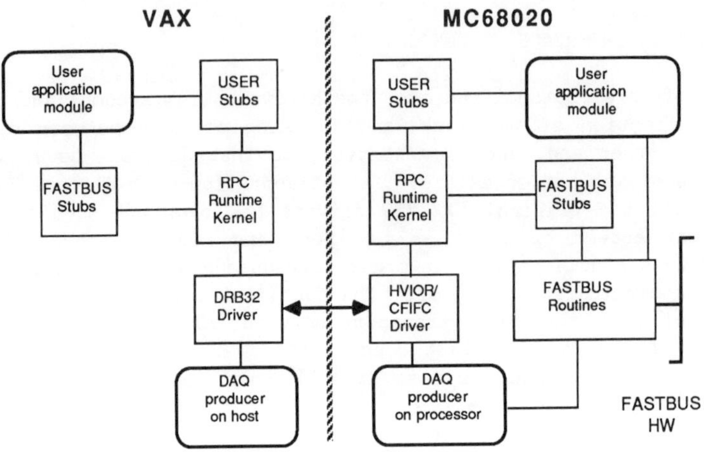

Fig. 1: CHI FASTBUS software overview

7. REFERENCES

[1] McLaren, R.A. et al., "The CERN Host Interface". Presented at the Nuclear Science Symposium, San Francisco, 21-23 October 1987

[2] ANSI/IEEE Std 960-1986, "IEEE Standard FASTBUS Modular High-Speed Data Acquisition and Control System"

[3] IEEE Standard 1014, "VMEbus: A Standard Specification for a Versatile Backplane Bus"

[4] Berners-Lee, T.J., "Experience with Remote Procedure Call in Data Acquisition and Control", Proceedings of the 5th Conference on Real-Time Computer Applications in Nuclear, Particle and Plasma Physics, San Francisco, May 1987

[5] U.S. NIM Committee, "FASTBUS Standard Routines", U.S. Department of Energy DOE/ER-0324, March 1987

[6] CAMAC, updated specification, Esone Committe, EUR8500, 1983

[7] Jacobs, D. et al., "CFI user guide", private communication, January 1985

[8] Müller, H., "A General Purpose Master (GPM) and Memory Module (DSM) for Online Applications", IEEE Transactions on Nuclear Science NS-33, no.1, February 1986

[9] Kozlowski, T. and Foreman, W.M., "An implementation of the new IEEE Standard Routines for FASTBUS". IEEE Transactions on Nuclear Science, NS-34, no.4, August 1987.

TRANSPUTER T800 PERFORMANCE IN A FORTRAN ENVIRONMENT

G.Cecchet,S.Centro,F.Marzano,M.Mattioli,S.Petrarca,D.Pascoli,L.Zanello
(INFN and Dep. of Physics,University of Padova,Pavia,Roma - Italy)

The performance of the transputer T800 has been measured using several Fortran programs:

BENCH #1 :This program simulates the time evolution of a themohydraulical modelization ("hydrocode") for a nuclear reactor. It is an iteratively structured simulation and has a behaviour similar to Montecarlo type algorithms. The 5000 lines, 40 subroutines, 200 kbytes (IBM370 code), Fortran program performs heavy double precision floating point computation and requires litte I/O. It has been used over a number of years to measure the performance of many different machines.
N.DoDUC, 'Fortran Central Processor Time BENCHMARK', July 1986.
(Program and IBM data courtesy of A. Fucci,CERN,Geneva,Switzerland))

Processor Engine	Execution time(s)	Relative Performance	Notes
VAX 11/780	1927	1.0	The reference !
MicroVax II	2434	.8	
VAX 8350	2112	.9	
VAX 8600	537	3.6	
VAX 8650	384	5.0	
VAX 8800	368	5.4	The fastest VAX
IBM 4361-L5	1410	1.4	
CERN/SLAC 3081E	328	5.9	
IBM 3090/200	70	27.5	The fastest IBM
MacII -16.6 MHz	4265	.5	Compiler Absoft 2.3
T800-20 MHz /5 cycle	1156	1.7	Compiler 3L/Inmos 1.1

BENCH #2 : The Whetstone benchmark (32 and 64 bit,Fortran version) :
R32=T800 / VAX11-780= 1.4 , R64=1.5 .
'FORTRAN + 5 cycle memory 'speed is 1.8 and 1.3 Megawhetstone/s .
'OCCAM-2+4 memory cycle (on chip memory disabled)' speed is 2.3 and 1.6 Mwhet/s.
'OCCAM-2 and memory on chip enabled' speed is 4.0 and 1.6 Mwhet/s .

BENCH #3 : GUIDE (Light collection simulation of a BGO crystal, G.Bellomi, L3 exp.)
R=T800-20 MHz / VAX11-780 = 1.3

BENCH #4 : BGO (Electromagnectic showers simulation in BGO crystal)
R=T800-20 MHz / VAX11-780 = 1.5

BENCH #5 : LEPTO52 (Lund Montecarlo for electroproduction)
R= T800-20 MHz / VAX11-780 = 1.5

Our conclusions are that in CPU intensive Fortran environment, the transputer T800-20Mh+DRAM 100 ns + Fortran v1.1 are equivalent to 1.5 VAX11/780 NOW (March 1988).Improvements are expected quite soon (better compiler,faster transputer and memory chips).

The figures we quote have been obtained using standard programs written for traditional mainframes; we believe that modifications exploiting the special capabilities of the transputer will give at least another factor 2-3 in price/performance , without resorting to problem dependent and sophisticated tecniques.A 4 megabyte program running in one processor + 4 Mbyte memory engine can "easily" be cut in 5 one Mbyte programs (20% overhead) to be run on 5 one processor+1Mbyte engines.

For newly designed large MonteCarlo programs, on the other hand, the effort of optimizing the algorithms on a large ensemble of transputers should provide large improvement factors.

A Fastbus acquisition system based on the Fastbus Intersegment Processor and its Event-oriented Memory (EDIP).

Ph.Charpentier (CERN)
G.Goujon, M.Gros, M.Mur, B.Paul, P.Siegrist (DPhPE, CEN Saclay)

Abstract

The Fastbus Intersegment Processor (FIP) is a 32-bit microprocessor based Fastbus Master serving as a node in multi-segment, multi-layer data acquisition systems Its architecture allows independant, parallel processing at each node in the system, and supports asynchronous operation of each layer via Event buffering and message communication.

An attached memory module, the Event Directory for Intersegment Processor, is a hardware-assisted multi-event buffer which maintains Event activity and history when used by independant source and sink masters.

Their use in the The Delphi TPC Data acquisition system is reviewed.

1 The Fastbus Intersegment Processor

The FIP is a single Fastbus module linking a Fastbus Cable Segment and a Fastbus Crate Segment. The FIP is controlled and read-out from the Cable port, and acts as a Master on the Crate Segment.

When used as a Front-End Processor, the MC68020-/68881 CPU system may receive the trigger interrupt, control and read Front-End Slaves, and prepare a multi-Event buffer for asynchronous read-out by the next layer Master through the Cable Segment port. Synchronisation is made by Cable to CPU Interrupts and CPU to Cable Service Requests.

A Dynamic interrupt level allocation allows an efficient real-time operation for the various tasks serving the Trigger, the SR from the Slaves, and the Cable layer communication interrupts. Fastbus block transfers on the Crate are implemented by a fast (up to 20 Mbytes/s) DMA sequencer.

Read block transfers on the Cable side are optimised by using a read-ahead mechanism.

In addition, the FIP provides a Simplex Segment Interconnect function, which allows direct access from the Cable to the Crate according to a programmed Route Table.

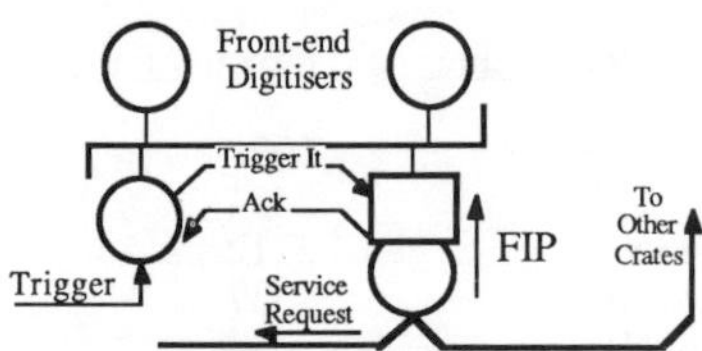

Fig1. The FIP as a Front-End Processor

2 The Event Directory for Intersegment Processor (EDIP)

At the higher acquisition layer, the FIP maintains an efficient multi-Event structure in the EDIP companion board.This dual-port memory maintains Event Status, location and history, and manages the free buffer space. Events may be written, read, or released in the circular buffer structure by simply giving their modulo-256 number.

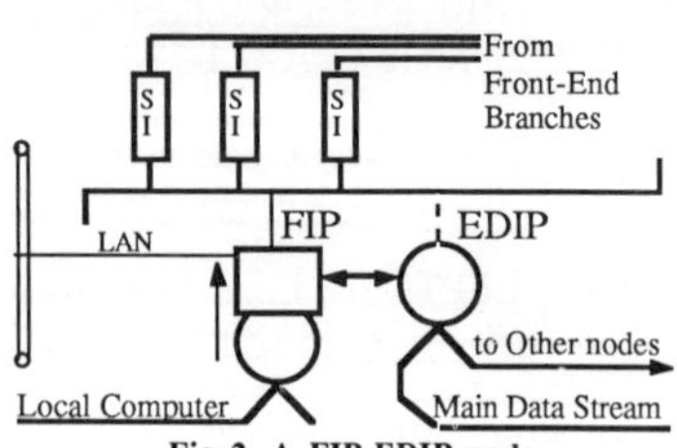

Fig 2. A FIP-EDIP node.

A FAST ZERO SUPPRESSION ALGORITHM FOR THE FORWARD ELECTROMAGNETIC CALORIMETER OF DELPHI IMPLEMENTED ON THE DSP 56000

D. Crosetto[1], *G. Rinaudo*[1], *A. Werbrouck*[2]

(1) Istituto Nazionale di Fisica Nucleare
Via P. Giuria 1. - 10125 TORINO

(2) Dipartimento di Informatica
C.so Svizzera 185. -10149 TORINO
ITALY

The Forward ElectroMagnetic Calorimeter of DELPHI consists of two endcaps of about 5000 counters each. Signals generated in the counters are read by phototriodes, sent to shaping amplifiers and digitized by 12 bit ADC's.

From Monte Carlo simulation, we expect that in a typical event about 5% of the channels are above threshold. However, in order to obtain a good resolution in the position reconstruction of the tracks, it is necessary to store not only the data of the counters above threshold but also the values of the adjacent counters. We expect therefore, for a good event, to have in each endcap about 200 data from the channels above threshold and about 1200 data from adjacent counters.

The data are stored by the front end electronics in a complicated order, which reflects the mechanical organization of the blocks in the modules. An efficient algorithm is therefore needed to read the data from the front end electronics organize them, apply the previous described intelligent zero suppression and store orderly the accepted data in a buffer without repetition. The time available for this analysis is a few msec, to allow sufficient time for the third level trigger to process the data and calculate clusters within 20 msec.

To perform this job, we developped a fast algorithm in assembly language of the Motorola DSP56000. The program was tested with a 56000 simulator on a personal computer.

The program is divided in three steps. In the first step the date are rearranged from the complex wiring order to a linearized matrix form: this step takes 1.5 msec. In the second step all elements above threshold are marked along with their nearest 8 neighbors in a linear matrix. In the third step all the marked elements are moved to the output buffer. Second and third steps need 5.3 msec.

The speed of the program profits from three peculiar features of DSP, which are particularly useful in problems like the present one: fast input-output, presence of internal RAM program memory as well as some powerful instructions (such as "CMPM", compare magnitude).

AN ACQUISITION SYSTEM BASED ON A NETWORK OF MICROVAX's RUNNING THE DEC VAXELN OPERATING SYSTEM

I. D'Antone, G. Mandrioli, P. Matteuzzi, and G. Sanzani
INFN Sez. di Bologna and Physics Department, University of Bologna, Italy

C. Battista, C. Bloise, A. F. Grillo, A. Marini, and F. Ronga
INFN Laboratori Nazionali di Frascati, Italy

G. Mancarella and A. Surdo
INFN Sez. di Lecce and Physics Department, University of Lecce, Italy

A. Giannasca, E. Lamanna, O. Palamara, S. Petrera and L. Petrillo
INFN Sez. di Roma and Physics Department, University of Roma "La Sapienza", Italy

ABSTRACT

We describe an acquisition system based on a network (Ethernet/DECNET) of MicroVAX's running under the VAXELN operating system. VAXELN is a Digital Equipment software product for the development of dedicated real time systems for VAX processors. A central VAX running under the VAX/VMS operating system is used as file server and as an interface with respect to the user's world. This acquisition system has been realized for the MACRO Experiment (Monopole, Astrophysics and Cosmic Ray Observatory) at the INFN Gran Sasso Laboratory.

1. MACRO Acquisition System.

The MACRO Experiment [(1)] has been designed for the study of cosmic rays and for the search of GUT monopoles , dark matter and neutrino stellar collapses. Briefly, the detector, installed underground at the Gran Sasso Laboratory, consists of six equal and almost independent entities (supermodules). Each supermodule has streamer tubes and liquid scintillators as active devices. The tracking is done by using streamer tubes.

The acquisition system has been designed taking into account the following requirements:

1) The system must be modular and must consist of a network of microcomputers to match the apparatus modularity. In this way the system size can be tailored to the real necessities.
2) The system must allow an easy access from remote locations to each computer or microcomputers in order to control the apparatus even from large distances (i.e. USA).
3) The system must be largely based on commercial products, both for software and for hardware, in order to be easily maintained.

Taking into account these requirements, we have chosen a system based on a MicroVAXII network (Ethernet/DECNET). The MicroVAX's are running a VAXELN system. VAXELN is a Digital Equipment software product for the development of dedicated, real time systems for VAX processors. A central VAX, running under VAX/VMS operating system, is used as file server and interface with respect to the user's world.The "CERN-Fisher CAMAC system crate", that uses the CES 2280 Qbus-CAMAC interfaces with the DMA controller 2281, has been chosen as the main hardware standard.

The general layout of the acquisition system is shown in Fig.1). Three MicroVAX's will control three groups of two supermodules each. Each MicroVAX is connected to two CAMAC parallel branches. A fourth MicroVAX will act as the supervisor for the neutrino events from a stellar collapse.

All the computers are connected via Ethernet under DECNET protocol. This choice does not imply any limitation on the data throughput that, in our case, is due the CPU used. We obtained, in absence of high software overload, a data throughput of more than 300 Kbytes/sec between two MicroVAX's in VAXELN, and more than 150 Kbytes/sec between a MicroVAX in VAXELN and the VAX8200. A big advantage of this network solution consists in the reduction of cables,due to the fact that is not necessary to have all the CAMAC crates in the same location.

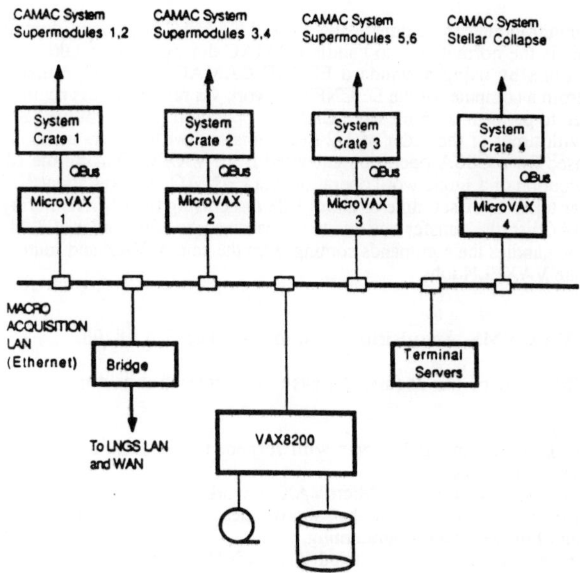

Fig.1 - General layout of the MACRO Data Acquisition System.

The tasks that each MicroVAX has to perform are: CAMAC readout, event filtering, data reduction and data forwarding to the VAX8200. Ancillary utilities have been implemented to help the control and the debugging of the apparatus: in fact during the acquisition an user from any DECNET node can:

1) require a copy of the raw data buffer from a given MicroVAX;
2) require the execution of CAMAC operations by using Remote Procedure Calls utilities for the CAMAC handling.

2. VAXELN System for the MICROVax's.

VAXELN has the following advantage respect to the VMS operating system:

1) optimized performances for real time operations due to the small overload of the operating system;
2) powerful and efficient message exchange facilities that are useful to build multi-jobs applications, even if the jobs are running onto different machines;
3) easiness to write device drivers and to work with peripherals. Device drivers can be written using high level languages (Epascal or C).

A machine running a VAXELN system is dedicated to a particular application and it is not possible to perform any operation that was not planned at the building time.
The VAXELN system for MACRO consists of six user jobs running with different priorities.The system includes four system jobs like the Ethernet driver. In each job many concurrent subprocesses can run with different priorities; in our system more than fifteen subprocesses run concurrently: the number of subprocesses depends on the number of activated DECNET connections.The software has been written in EPASCAL, the VAXELN PASCAL version.
The exchange of data between jobs is done using the VAXELN message facilities, useful in the handling of the event data.

The CAMAC input/output can be done using a list of CAMAC operations, contained in a file on VAX8200. This is the normal way to handle CAMAC during the data taking. Moreover, the access can be done by using a standard ESONE CAMAC library(2), either directly inside VAXELN, or from a computer of the DECNET network via remote access routines.
The time needed to perform a 16-bit word CAMAC operation (CSSA routine) is 74 microsec, to be compared with times of the order of 1-2 msec obtained with drivers under VMS. This time becomes 16 msec for a CSSA operation executed from VAX8200 using the MicroVAX as a server. The execution of a single word operation via CAMAC list takes about 30 microsec.
A DMA transfer takes an offset time of about 1.08 msec using the ESONE library and 0.8 msec using the CAMAC list; the transfer rate due to the hardware is of the order of 3.2 microsec/word.
A centralized job handles the commands coming from the central VAX and routes the commands to the appropriate VAXELN jobs.

3. THE VAX/VMS Acquisition System for the VAX8200.

The VMS part of the acquisition system performs, at this moment, the following functions:

1) Network server and I/O server with respect to the ELN systems running on the MicroVAX network;
2) router of the directives to the MicroVAX network;
3) router of the user directives to the subprocesses components of the VMS system;
4) handler of the general histogramming;
5) spooler of complete event data to other VAX/VMS computers in the network;
6) collector of the alarm conditions.

This part is constituted by several parallel subprocesses scheduled according to a prefixed scheme of priorities and synchronized via Event Flags.
Other multiple processes, substantially decoupled from the core of the system, are the interfaces toward the users (Consoles, Histogram Presenters).
The data sharing among the processes is performed via mailboxes and global sections, while the raw data flow among the different components of the system is arbitrated by the Model Buffer Manager(3).
The network service is performed by the use of the DECNET communications. The system implements synchronous DECNET transparent communications in the subprocess that routes the directives to a companion process of the ELN systems (send commands/wait for the completion) and asynchronous DECNET nontransparent communications in the subprocess that handles unsolicited data coming from the ELN systems (messages containing event data, alarms, etc.).
The asynchronous Net Server subprocess includes a procedure that builds the event data structure starting from the information (fragmented due to the transmission protocol) received from the network and originated from different MicroVAX's (more than one MicroVAX can be involved in the acquisition of the same event).

4. Overall Performances.

The performances has been studied using a random trigger generator.
The data flow related to a trigger is shown in Fig.2. The performances obtained with events involving only one MicroVAX are shown in Fig.3: the CAMAC list generated 10 DMA transfers and 80 word program transfers for a total of 13.07 Kbytes.
The data logging was done on disk. At 10% of dead time a throughput of about 40 Kbytes/sec has been obtained. This value corresponds to about 30 tapes/day (a tape is supposed to contain 120 Mbytes). The limitation is due to the VAX8200 CPU that is saturated for this data throughput. We think that the current Ethernet hardware could support a data throughput at least a factor three higher, if a CPU more powerful than the VAX8200 is used. However, this is not required for the MACRO Experiment.

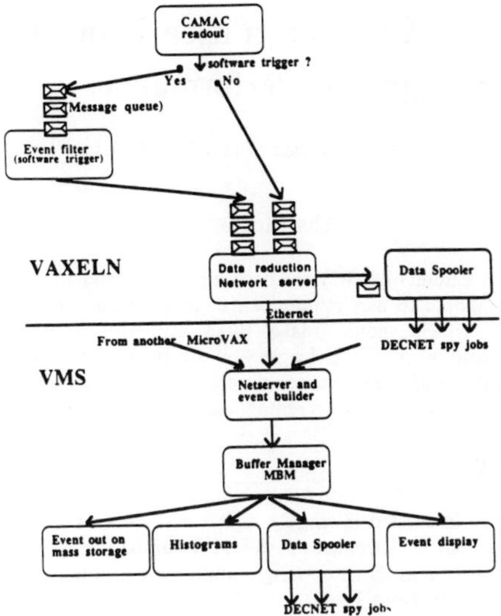

Fig.2 - Event flow from a MicroVAX to the VAX8200.

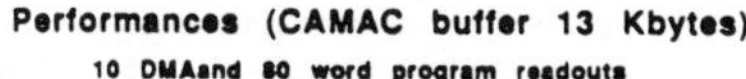

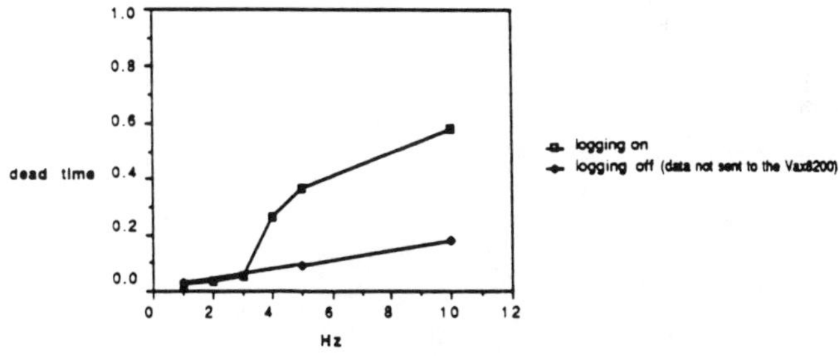

Fig.3 - Performances for random triggers.

REFERENCES

1. MACRO Collaboration, The MACRO Detector at the Gran Sasso Laboratory, NIM A264, 18 (1987)
2. Giannasca, A. et al., Internal note IFU 893, Physics Deptartment, Univ. of Roma "La Sapienza", 1987.
3. Vande Vyvre, P., The MODEL Buffer Manager, CERN Mini and Micro Computer Newsletter, No. 17, pag.13, December 1987.
"The MODEL Buffer Manager user's guide; MBM Vs. 2.007"; CERN, May 1987.

A Bipolar Cell Array for a nsec Time Digitizer; Design and Performance.

G. Delavallade [1)], J.J. Jaeger [2)] and J.P. Vanuxem [1)]

Abstract

Application Specific Integrated Circuits (ASIC) are becoming more and more popular, as they offer large scale integration exactly tailored to user requirements. In addition, semi-custom technologies are affordable at moderate cost, which should make them attractive in high energy physics for low-to-medium volume applications. An ECL Logic Cell Array has been designed to integrate the front end part of the LTD, [1] a high performance time digitizer developed for LEP detectors. The basic circuit is described and the selected array presented, after which results are given. [2]

1. THE BASIC CIRCUIT

The circuit which has been integrated is entirely digital and consists of the front-end part of the LTD (LEP Time Digitizer), a high performance, FASTBUS compatible, yet low cost time-to-digital converter offering 2 ns resolution, 32 μs dynamic range, multi-hit and multi-event buffering capability, as well as data formatting and fast read-out speeds.

Fig. 1 shows the basic implementation for one channel which uses the standard ECL 10K series of SSI and MSI integrated circuits. Both the 125 MHz CLOCK and TEST inputs are common to all the 48 channels implemented in the LTD. All inputs, including the 48 channel input (IN) signals are differential, thus ensuring a good noise immunity by the inherent common mode rejection of the differential receivers. The heart of the circuit is a 2 ns time interpolator, which determines in which of 4 possible time bins of 2 ns a hit occured on the wire during a clock period of 8 ns.

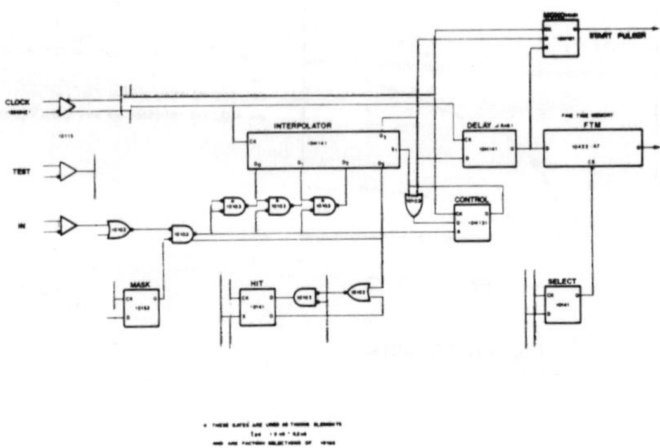

Figure 1: Front End Circuit of the LTD (one channel)

---

1) CERN, Division EP, Geneva, Switzerland
2) IN2P3, Collège de France, Paris, France

The motives for integrating this circuit are numerous. The first and principal one is simply.....SPACE! The basic circuit was redesigned with this in mind, and a special effort was made to try and optimize the logic and incorporate 4 channels in the chip, so that 48 channels could be accommodated on a FASTBUS card. Rapid calculation then showed that a semi-custom bipolar technology offering chips with a complexity of about 1000 equivalent gates would do the job, providing that some drastic requirements (such as flip-flops toggling at 250 MHz) could be met. Minimizing the power consumption was also a major concern in the choice of the technology.

2. CHOICE OF ASIC TECHNOLOGY AND TYPE OF ARRAY

The ACE (Advanced Customized ECL) family of semi-custom ICs from Philips consists of seven different prediffused circuits with a complexity ranging from 200 to 3000 equivalent gates. The main advantages of the family are low propagation delays, low power consumption and the availability of numerous logic functions in a powerful library. Two versions of the family exist: the "Low Power" and "Turbo". The Low Power version has an equivalent gate delay of 350 ps at 0.8 mW/gate, thus producing a factor of merit (i.e. the product of the equivalent gate delay by the power consumption per equivalent gate) of 0.28 pJ, one of the best figures available on the ASIC market to-day.

Within the ACE Low Power family the ACE 900L Logic Array was chosen as it fits well with the requirements of the circuit, offering 900 equivalent gates, one of the lowest possible power consumptions in its category and the best price/performance ratio. The array contains in its centre area 36 major sites, each one being composed of 38 transistors and 16 resistors. On the periphery of the chip are 58 I/O sites, which are used to interface the circuit with TTL or ECL (10 KH or 100 K) external signals, and 22 minor sites, which may also contain some logic. The present design uses all 36 major sites, all 58 I/O sites (and pins !) and 12 minor sites. The dimensions of the chip are about 6mm x 5 mm, and the chip is housed in a PGA−64 package.

3. RESULTS AND CONCLUSION

With such a semi-custom LSI circuit, a gain of a factor 30 in component count and 5 in printed circuit board space (10 if one excludes the external delay elements from this comparison) has been achieved for the front end part of the LTD. A second major benefit is the reduction of power consumption by more than a factor 5; the typical power consumption of the chip itself (without including the external pull−downs) is 1.3 Watt. Other advantages include an invaluable reliability improvement which is the direct result of the drastic reduction of the number of components, and an overall (component and assembly) cost reduction which increases with quantities ordered, but is already effective for orders as low as 1000 pieces.

The development time for this circuit was 6 weeks, including schematic capture, simulations, routing and test pattern generation, and prototypes were made available within 12 weeks after the development phase.

In conclusion the development of this ASIC has allowed a new time-to-digital conversion technique to be successfully brought to life. Because of its numerous advantages, it is expected that such semi-custom technologies will play a major role in the development of front end electronics for the read-out of future LEP (and post-LEP!) detectors.

References

[1] The LTD: A FASTBUS Time Digitizer for LEP Detectors, G. Delavallade and J.P. Vanuxem, NIM (1985) 596−604, also EP−Electronics Note 85−06 and DELPHI Note 87−68 ELEC 28.

[2] Application of Bipolar Cell Array Technology to the Development of a Time Digitizer, G. Delavallade, J.J. Jaeger, J.P. Vanuxem, CERN EP−Electronics Note 87−03, also DELPHI Note 87−69 ELEC 29.

A HIGHLY PARALLEL ALGORITHM FOR TRACK FINDING

Mauro Dell'Orso

Dipartimento di Fisica - Università di Pisa
Piazza Torricelli 2 - 56100 Pisa - Italy

Luciano Ristori

INFN Sezione di Pisa
Via Vecchia Livornese 582a - 56010 S. Piero a Grado
Pisa - Italy

ABSTRACT

We describe a very fast algorithm for track finding, which is applicable to a whole class of detectors like drift chambers, silicon microstrip detectors etc. The algorithm uses a *pattern bank* stored in a large memory and organized into a tree structure.

INTRODUCTION

The analysis of data collected by modern High Energy Physics experiments often requires a lot of computing power. One of the most demanding tasks is usually track reconstruction. In this paper we describe a very fast algorithm for track finding, which is applicable to a whole class of detectors like drift chambers, silicon microstrip detectors etc.

For sake of simplicity we will assume that our detector consists of a number of parallel layers, each layer being segmented into a number of *bins*. When charged particles cross the detector they *hit* one bin per layer. As shown by the authors [1], with this abstraction the problem of track finding is reduced to a search for hit patterns *matching* with the *event* (the set of hits in the detector). The hit patterns, computed once for all, are stored in a large memory called *pattern bank*.

In this paper we want to show how the pattern bank can be organized into a tree structure to speed up the pattern matching process.

SUCCESSIVE APPROXIMATIONS

The basic idea is to follow a successive approximation strategy and apply our pattern matching algorithm to the same event seen with increasing spatial resolution. Lower spatial resolution is simulated by logically ORing adjacent bins.

Fig. 1 shows an event with four tracks crossing four parallel layers. From top to bottom the spatial resolution of the detector is improved by a factor two every step. The image is confused at the beginning and becomes clearer as the resolution improves. In the following we discuss the case of straight tracks, but the same considerations apply to curved tracks in a magnetic field.

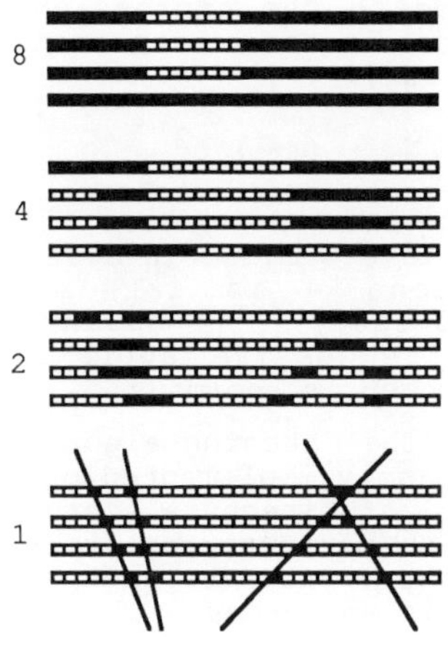

Fig. 1

Fig. 2 shows how a single track is seen when each detector plane is considered as being only two bins.

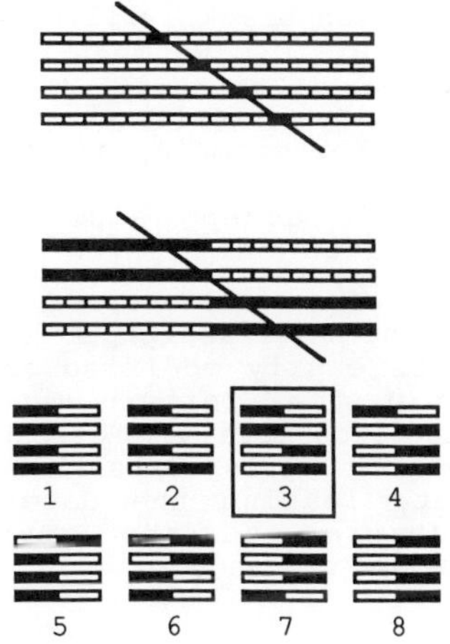

Fig. 2

In this case the total number of patterns compatible with a straight line is eight. Pattern number 3 is the one that matches. Since we have one track candidate at this level of spatial resolution, we now double the number of distinguishable bins in each plane and proceed to match the four patterns shown in fig.3.

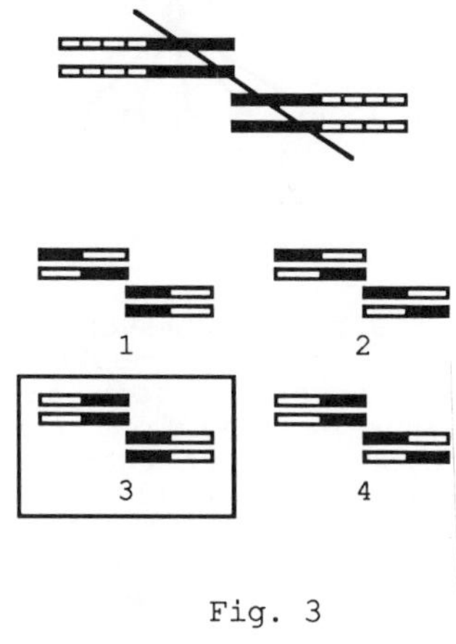

Fig. 3

Pattern number 3 in fig.2 is said to *generate* the four *sub-patterns* in fig.3. Since we still have one track candidate we go on halving the bin size. This process is iterated until we either reach the actual resolution of the detector (success) or we are left with no track candidate (failure).

TREE SEARCH

The pattern bank can thus be arranged in a tree structure as shown in fig.4. Increasing depth corresponds to increasing spatial resolution. Each node represents one pattern and it is linked to all the subpatterns it generates when the spatial resolution is improved by a factor two.

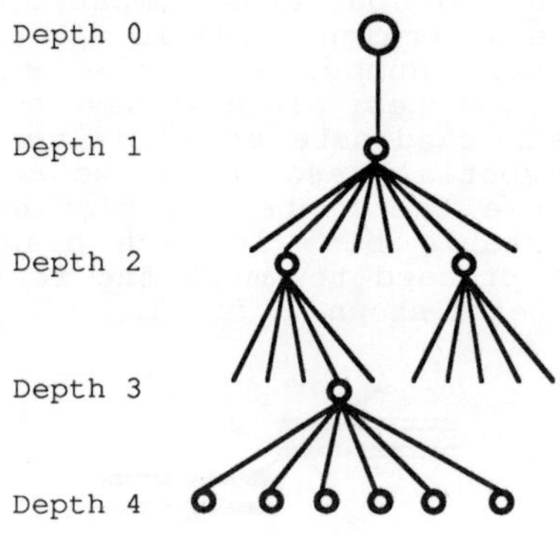

Fig.4

The pattern matching process can be implemented as a tree search. We scan all the patterns hanging from one node and every pattern that matches correctly with the current event is considered a track candidate and enables the search at the next deeper level in the tree. A track is found whenever this search reaches the bottom of the tree.

This tree-search is obviously much faster than a purely sequential search. The average number of patterns one has to examine to find a track is given by:

$$N_m = k*N_l \qquad (1)$$

where N_l is the total number of depth levels in the tree and k is the average number of patterns hanging from a single node. k ranges typically from 4 to 8 depending on the particular geometrical arrangement. We then have:

$$N_m = k*\log_2(n) \qquad (2)$$

where n is the number of bins per detector layer and k does not depend significantly on n. Expression (2) is to be compared to the expression which holds for a purely sequential search [1]:

$$N_m = (m-1)*n^2 \qquad (3)$$

where m is the number of detector layers Since in most applications n is rather large (100÷1000), the advantage of the successive approximation approach is enormous.

The matching algorithm can be easily implemented on a parallel architecture because different patterns can be compared to the event data independently and in any order.

MISSING HITS

In real experimental situations each plane detects particles with an efficiency which is less than one. This means that there is a finite probability that some of the hits, in a given track, will be missing. Therefore we must be prepared to accept cases where we have only a partial pattern match. For example, if we have a detector with eight layers, we might accept also tracks that match only seven or six hits.

The tree-search algorithm may be easily modified to accept partial pattern matches: we need to modify only the way each pattern is compared to the event, leaving the data base structure and the visiting strategy unchanged.

MONTE-CARLO SIMULATION

We simulated a simple detector, consisting of 4 layers with 128 bins per layer and

100% efficiency, to evaluate the performances of the tree search algorithm.

One measure of the algorithm speed is the number of *steps* carried out through the tree, that is the actual number of hit patterns examined to find all the tracks. The distribution of the number of steps for our simple detector is shown in fig. 5. Each simulated event consists of one randomly generated track crossing the four detector layers. The fluctuations in the channel contents are generated by the actual tree structure. The mean value of the number of steps is 54.

It is interesting to study the performance of the algorithm on events with many tracks. The table below shows the mean number of steps per track as a function of the number of tracks in the event.

tracks	steps/track
1	54
2	60
3	68
4	75
5	82
6	89
7	97
8	104
9	111
10	118

The number of steps increases, but only by a factor 2 in going from 1 to 10 tracks per event.

The detector noise, that is the presence of hits not associated to any real particle, has a big impact on the algorithm performances. The next table shows the mean number of steps, for single track events, as a function of the detector noise.

noise	steps
0%	54
1%	82
2%	135
3%	212
4%	315
5%	446

The first column in the table contains the probability for each detector bin to be hit by the noise. The number of steps rapidly increases with the noise because the algorithm visits fake branches activated by noise hits at low resolution.

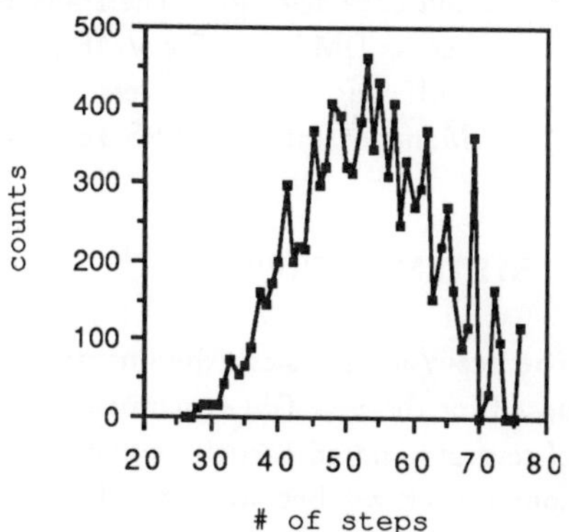

Fig. 5

References:

1 M.Dell'Orso, L.Ristori. "VLSI STRUCTURES FOR TRACK FINDING". Contribution to this conference.

A CAMAC CRATE CONTROLLER AS A VSB DEVICE

J. Hoffmann, H.Sohlbach

GSI, Gesellschaft für Schwerionenforschung mbH
Planckstr. 1, D-6100 Darmstadt
Fed. Rep. of Germany

ABSTRACT

A fast interface between the VME subsystem bus VSB together with a CAMAC crate controller was designed and built. Up to 15 CAMAC crates can be connected to a single VME processor board. The transfer rate is up to 4 MBytes/s. Differential line drivers allow a distance of up to 50 m between the VME processor and the last CAMAC crate. Independent transfers can take place between different VME processors and their connected CAMAC crates in parallel at the same time. In a first version of the CAMAC controller only programmed CAMAC NAF operations are provided. An extended version is currently under development. It will incorporate a micro-processor allowing flexible readout procedures including NAF list processing, pattern selection, Q-check, and data selection. The transfer will be performed with direct memory access DMA into the VME processor's local memory without using the VME bus. A special version to connect VSB to a FASTBUS Segment Manager/Interface LRS-1821 is also under development.

1. SYSTEM OVERVIEW

The heavy ion physics experiments at energies of 100 to 1.500 MeV/nucleon as proposed for the new GSI synchrotron facility SIS will produce events with multiplicities of several hundred particles and primary data rates of about 10^4 events/s. Under such conditions it will become very difficult to set selective hardware triggers. Therefore the information of some 10^5 particles/s have to be processed for calibration, reconstruction of particle tracks, and sorting to get software trigger conditions.
These requirements led us to the necessity of designing and building a VME-based data acquisition system with a high grade of parallelism. The data digitized in CAMAC and FASTBUS ADCs and TDCs will be transferred into VME processors in parallel, one to 15 CAMAC crates or in the future one FASTBUS segment per VME processor. These processors can be located in one to 15 VME crates linked by a crate interconnect bus (e.g. the CES VMV bus). They will collect, calibrate, sort, and select the data to build

a subevent buffer. The subevents of all processors of one VME crate will be collected by one processor, the subcrate manager. The event builder, one specific processor in the VME master crate, will collect the subevents of its own crate and all subevents of the other VME crates to build the total event buffer. This complete event might be analyzed further by other VME processors to generate particle tracks and software trigger conditions. It will also be transferred to a VAX host computer via a parallel direct memory access (DMA) interface. As VME frontend processors we selected the CES FIC-8230[1)] boards with a Motorola 68020 processor chip and a DMA controller for both the VME and the VSB bus. Figure 1 shows an overview of the VME system.

2. THE VSB-CAMAC CONNECTION

In the past a connection between a VME processor and CAMAC or FASTBUS was realized in several institutes and by the industry using special VME modules (CAMAC branch driver, I/O modules). The VME processor was used to read in the data in a software loop, occupying not only the VME bus but also the processor itself.
We are using the new VME Subsystem Bus VSB[2)] to connect up to 15 CAMAC crates or in the future one FASTBUS segment directly to the VME processor board, not touching the VME bus for the transfer. The VME processor board CES FIC-8230 additionally provides a direct memory access between the processor's local memory and the VSB as well as the VME bus.
To drive up to 50 m cable between the VME processor and the CAMAC or FASTBUS crate differential line drivers are used. Since the bus transfers are handshaked, larger distances are possible if reduced transfer rates are acceptable. The adaptor board (single Euroboard size) is connected at the rear of the VME crate at the processor's location. This allows up to 10 VME processors in the first VME crate and up to 15 in the following ones. An optical link is under design.

3. THE CAMAC CONTROLLER

The advanced version currently under development will read out the digitized data by loadable lists specifying the CAMAC module station N, the subaddress A, the function F, and a repetition factor. The NAF-list will be loaded by the VME processor into an internal memory of the controller. The CAMAC controller then waits for a LAM triggering the CAMAC readout controlled by the NAF list. The data are stored in an internal 4kByte memory in the CAMAC controller. After the readout the CAMAC controller asserts a VSB interrupt signal to trigger the VSB master, the FIC-8230. This processor then initializes the DMA in its DMA controller which is then transferring the data from the CAMAC controller buffer into the VME processor's memory. Additionally single programmed CAMAC operations can be initiated at any time by the VME processor.
For several experiments conditions for the readout may additionally be necessary, e.g.

Q-response, lower level discrimination, or correlations. All these functions can be handled best by a micro-processor. We decided to use a Motorola 68030 processor with 25 MHz to be built into the CAMAC controller. All internal registers of the CAMAC controller will have read/write access from the VME processor via the VSB. Figure 2 shows the block diagram of the CAMAC-VSB interface.
The controller follows the A1 specification, i.e. there can be no auxiliary crate controller in the CAMAC crate. The mapping of the CAMAC addresses into the VSB address space (32 bit) is variable and is defined by a rocker switch on the VSB driver board. The lower 20 address bits are used for CAMAC addressing according to the following scheme (Bits 0 and 1 are always set to 0 for four byte addressing):

Bit 2 - Bit 5 : A (subaddress)
Bit 6 - Bit 10 : F (function)
Bit 11 - Bit 15 : N (station number)
Bit 16 - Bit 19 : C (crate number)

If C=0 is addressed, the corresponding NAF information will be sent to all crates on the bus (broadcast). The CAMAC controller status register provides among others bits for the Q- and X-lines, the list process enable, the VSB interrupt enable, and the single CAMAC execution.
The crate controller has a block mode capability which is also supported by the DMA controller of the VME processor. Due to the internal buffer size of the DMA controller the maximum size of a block is four 32-bit words.
Each transferred data word has 32 bits with 16 or 24 bit CAMAC data and in addition address information. This will make the later handling of the event data easy even if modules are skipped during the readout.

4. CURRENT STATUS AND FUTURE PLANS

The first version of the CAMAC controller without the micro-processor is in production. To read out CAMAC data with this controller the VME processor has to execute single I/O instructions. When a LAM is set on the CAMAC dataway, the controller asserts the VSB interrupt signal to trigger the VSB master, the FIC-8230.
Beside the development and the production of the new micro-processor based CAMAC crate controller, an interface between VSB and FASTBUS in planned. We will use the LeCroy Segment Manager/Interface LRS-1821 to connect its ECL-bus with VSB.

5. REFERENCES

[1] FIC-8230 User's Manual, CES SA., Geneva (1987)
[2] The Parallel Subsystem Bus of the IEC 821 Bus, Rev.C (1986)

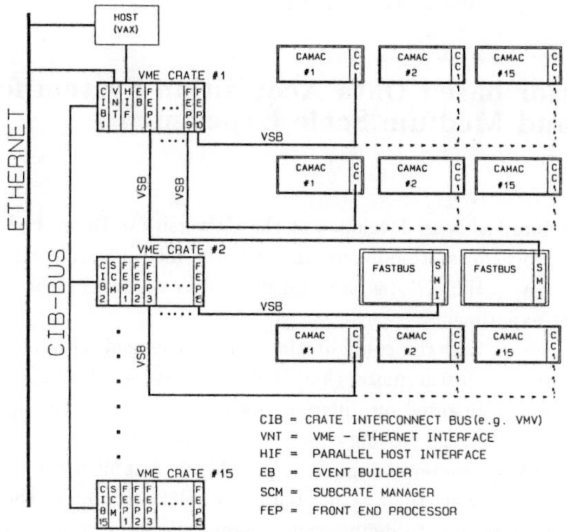

Figure 1: VME system overview

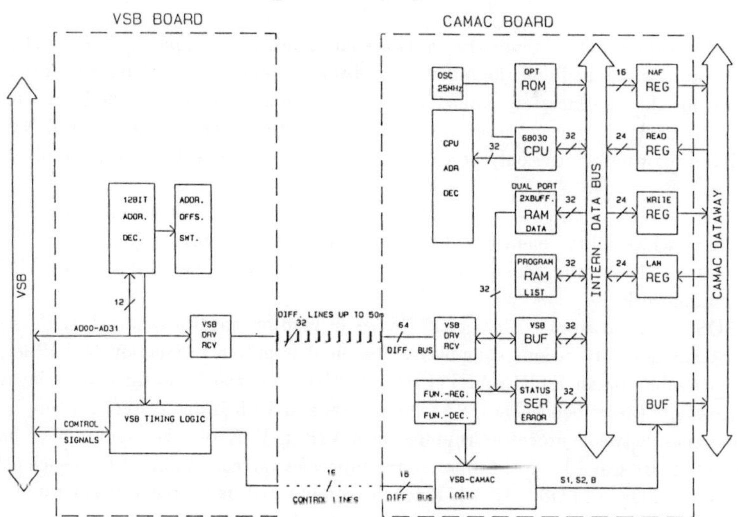

Figure 2: Block diagram of the CAMAC - VSB interface

K.Honscheid, J.Manns, P.Schütz
Physikalisches Institut der Universität Bonn
5300 Bonn 1, Nußallee 12
WEST GERMANY

A Multiprocessor based Data Acquisition System for Small and Medium Scale Experiments

The SOS data acquisition system has been developed at the University of Bonn, Fed. Rep. of Germany as a versatile multiprocessor system for use in small and medium scale experiments. It is designed for data rates up to 100 KBytes/sec and due to its modular structure it can easily be adapted to different experiments.

Modularity is used here in a sense, that the configuration can be changed without having to modify any part of the software and furthermore, that SOS is not restricted to a special kind of processor but can easily be implemented on different computers. In SOS this is achieved by using a hardware independent message passing concept called 'Logical Names'. With these Logical Names, which are simple character strings, it is possible to establish communication links between processes without knowing their actual hardware relationship, i.e. the path. A process wanting to accept messages, has to define such a name. When a connection to this process is requested, SOS translates the Logical Name into physical addresses hidden to the user.

A second feature of SOS extends the hardware independence even more. After reset or power up the system is scanned and the different hardware components are detected. Special driver programs for the interfaces are loaded automatically and from then on the hardware can be accessed by using the normal inter-process communication channels, i.e. Logical Names. So the driver programs are the only processes which have to know details of the hardware modules.

In addition, this modularity is supported by the hardware, since it is arranged in a simple bus structure to which all the elements of the data acquisition system are connected. This structure is called VIPnet and consists physically of VMEbus backplanes and a CHEAPERNET local area network.

Since SOS processes work with Logical Names instead of 'hardware paths', the system can easily be extended to various computers. Required is only a connection to VIPnet, i.e. a VMEbus module or an ETHERNET/CHEAPERNET interface. So far SOS is installed on VIP processors (described below), VAX computers and VERSAdos systems. There is no logical difference between processes running on a VIP, a VAX or wheresoever SOS has been implemented. By using Logical Names all the processes can communicate with each other. So the main advantage of SOS is that it hides the complicated multiprocessor environment from the user. He has to deal only with a simple multitasking system.

Some hardware modules have been specially designed for SOS. The most important module is the VIP VMEbus microprocessor board which is equipped with a dual-ported memory section, a mailbox interrupt, a local bus at the P2 connector and an ETHERNET/CHEAPERNET local area network interface and is therefore optimized for a multiprocessor environment. Interfaces to special hardware, e.g. CAMAC, are available.

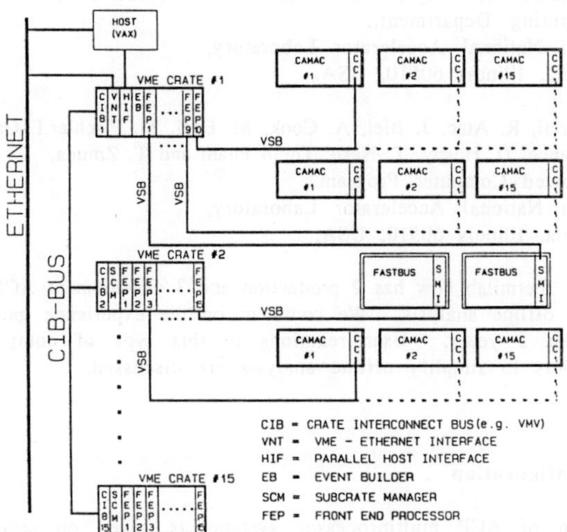

Figure 1: VME system overview

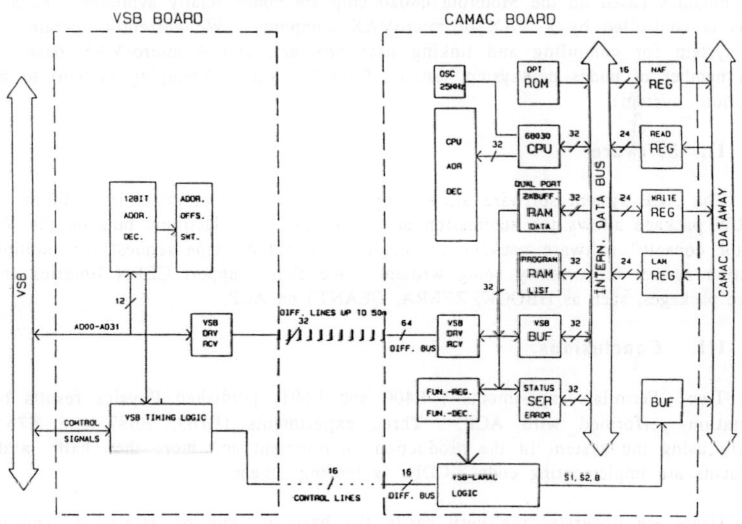

Figure 2: Block diagram of the CAMAC - VSB interface

Experience with the ACP multiprocessor systems in the Fermilab Computing Center.

C. Kaliher, A. Kreymer, P. Lebrun and A. Valderrama,
Computing Department,.
Fermi National Accelerator Laboratory,
Batavia, Illinois 60510, USA.

H. Areti, R. Atac, J. Biel, A. Cook, M. Edel, M. Fischler,I. Gaines,
R. Hance, D. Husby, T. Nash, Thinh Pham and T. Zmuda,
Advanced Computer Program.
Fermi National Accelerator Laboratory,
Batavia, Illinois 60510, USA.

Abstract . Fermilab now has 2 production and 3 development ACP systems in operation for large scale offline analysis. We comment on the experience gained from these systems over the past 2 years. User reactions to this type of computing are summarised and software tools to simplify offline analysis are discussed.

I. Hardware configuration .

The first generation of ACP multiprocessor systems is based on single board computers running FORTRAN 77 programs at speeds approching that of a VAX 11/780. VME modules based on the Motorola 68020 chip are commercially available. Each of these systems is controlled by a VAX or microVAX computer. We currently operate 1 VAX-based system for compiling and linking user program, and 4 micro-VAX based systems, with a number of nodes per system ranging from 5 (small debugging system) to 65 (full production system).

II. Software

The ACP system software allows implementation of user code, in addition, the ACPRUN package allows job submission and retrievals from electronic bins on the VAX. A "operator console" software package to monitor and satisfy tape request for multiple production system is currently being written.. We also support CERN libraries and software packages, such as HBOOK, ZEBRA, GEANT3 on ACP.

III. Conclusions.

Two Fermilab experiments (E400 and E691) published Physics results based on computation performed with ACP. Three experiments (E769, E687 and E731) are currently using the system in the production environment and more then half a dozen of experiments are implementing code. CDF is leading them.

Users are understanding very easily the basic priciple of paralelism, and decide rather quickly on how to proceed. Problems occurs when (i) the node code is not writrten in F77, (ii) the COMMON block layout is messy with a large number of small data transfer must be performed and (iii) if debugging on the node is substantial. This is because the ABSOFT M68020 F77 linker, compiler and interactive debugger are rather limited and were not written for a "supercomputer" environment.

THE MEGA EXPERIMENT AND THE USE OF MICROPROCESSORS FOR DATA ACQUISITION AND EVENT SELECTION

Thomas Kozlowski, James F. Amann, Martin D. Cooper,
Will M. Foreman, Gary E. Hogan, Franklin J. Naivar,
Michael A. Oothoudt, Wayne Smith
Los Alamos National Laboratory*
Los Alamos, N.M. 87545

Peter S. Cooper
Fermi National Accelerator Laboratory+
Batavia, Il. 60510

E. Barrie Hughes, Charles C. H. Jui, John N. Otis
W. W. Hansen Laboratories of Physics
Stanford University**
Stanford Ca. 94305

Randy Fisk, Donald Koetke, Robert Manweiler
Valparaiso University++
Valparaiso, In. 46383

Kevin I. Hahn, John K. Markey
Yale University***
New Haven, Ct. 06511

----

* Supported by U.S. DOE Contract W-7405-ENG-36.
+ Operated by URA, under contract with the U.S. DOE.
** Supported by U.S. NSF Grants PHY 84-11168, 87-13411.
++ Supported by U.S. DOE Grant DE-AC02-84ER40130.
***Supported by U.S. DOE Grant DE-FG02-87ER40313.

1 EXPERIMENT OVERVIEW

1.1 Goals

The MEGA experiment at the Los Alamos Meson Physics Facility (LAMPF) will attempt to measure the branching ratio (not yet observed) for the decay of the muon to an electron and a gamma ray with a sensitivity of about 10**-13. To achieve this sensitivity the experiment must observe a large number of Michel decays (3x10**7 per second for 1.2x10**7 seconds), and as a consequence deal with high front-end rates. The LAMPF duty factor of approximately 6 percent results in even higher instantaneous rates (5x10**8 decays per second). A goal of the experiment is to limit the amount of data tapes to 2000 (6250 bpi). It follows from these constraints that a great deal of event rejection must be done on-line, hence the use of an array of microprocessors for this purpose. Because of manpower and funding constraints it was decided to make as much use as possible of existing and commercial hardware and software.

1.2 Detector

Since the muons will stop in a thin target, the desired signal will consist of a 52.8 MeV electron and a 52.8 MeV photon travelling in opposite directions from the decay point. The MEGA dectector is optimized for the detection and measurement of the energy and the direction of the electron and the photon from candidate events. It consists of an "electron-arm" and a "photon-arm" inside a superconducting solenoid.

The electron-arm consists of eight cylindrical wire chambers to determine the helical trajectory of the electron, and scintillator arrays to provide electron timing information. The "photon-arm" consists of four converter and pair-spectrometer layers. Each layer consists of lead photon converter, scintillators, and wire chambers. The energy and conversion point of the photon are measured. In addition, the photon-arm provides a fast hardware trigger indicating detection of a single high energy photon consistent with a muon decay to an electron and a gamma. About 2400 triggers are expected per second.

1.3 Data Acquisition

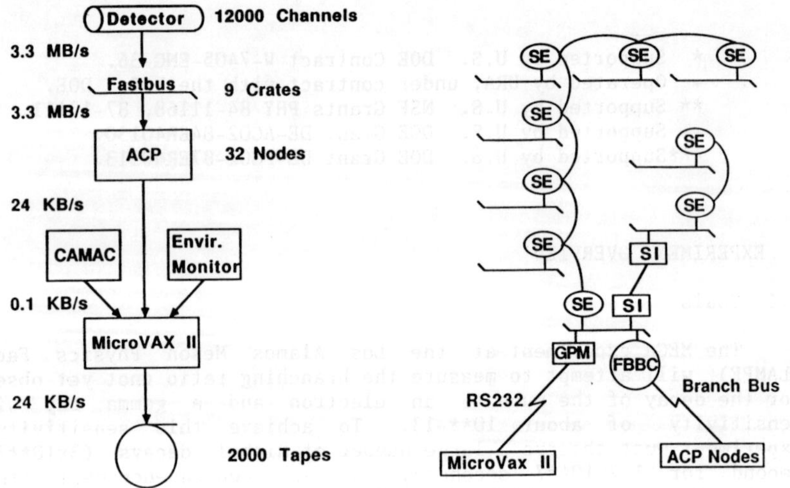

Figure 1. Data acquisition overview. Figure 2. FASTBUS system.

The detector includes approximately 12000 analog and digital channels. A hardware trigger strobes event data from the front-end electronics into 150 or more multiple-event buffered FASTBUS ADCs, TDCs and latches. A single event is expected to be about 1400 bytes long. About 20 events will be acquired during each LAMPF 500 microsecond beam spill. Because of the short time between events during a spill, all events in a single spill are stored in buffers in the FASTBUS modules. FASTBUS is used because its bus speed matches the MEGA requirements of a 3.3 Mbytes/second average data rate (2400 triggers/second at 1400 bytes per event), and because of availability of commercial hardware.

At the end of the beam spill all data is moved into the array of microprocessors for further processing. In order that the FASTBUS modules are able to accept data from the next beam spill the data must be read out in the 8 millisecond interval between beam spills. The microprocessor array is the Fermilab ACP system (Advanced Computer Program) [1]. This system was chosen because it is strongly supported at Fermilab and the hardware is available commercially. The microprocessors reject most events, sending only about 1 out of 200 as good decay candidates to the host computer for logging. The rates for the various stages of the data flow through the system are summarized in Figure 1. A Digital Equipment Corporation (DEC) MicroVax is used for experiment control and data logging.

2 FASTBUS SYSTEM

2.1 Configuration

There is a single FASTBUS master, the General Purpose Master (GPM) developed at CERN [2] and available commercially [3]. It is a single board FASTBUS module incorporating a Motorola 68000 processor, 512KB of RAM, 64-128KB of memory ported to FASTBUS, two RS232 ports, and a parallel port (not used by this application). For program development and support much use has been made of the cross software tools available from CERN (MONICA monitor, assembler, compilers, linker and loader) [4]. Remote Procedure Call (RPC) software from CERN [5] has been used to implement a VAX-GPM link over the RS232 lines. The 1 KB/s rate available over the RS232 lines is adequate for the application. The connection to the ACP system is made through the Fermilab designed FBBC (FASTBUS Branch Bus Controller) [6].

Nine crates of FASTBUS are required. The network is diagrammed in figure 2. There is not sufficient time available between beam spills to do separate block transfers for all 150 FASTBUS modules. A solution ("MEGAblock") has been provided by the manufacturer of the MEGA FASTBUS modules [7] that makes use of the FASTBUS "daisy-chain" lines. This concept allows any number of a subset of modules in a crate to be read out by a single block transfer from the initial module in the chain.

2.2 Data Flow

The data acquisition sequence is as follows: On the receipt of an "end-of-beam-spill" front panel trigger, the FASTBUS master executes about 10 block transfer reads, moving data from the FASTBUS modules to its local memory. It then issues a write to the FBBC to map to a particular VME crate and node memory address, followed by a single block write to the FBBC to transfer all the data from a single spill to a ACP processor node.

In the time available between data transfers, the FASTBUS master polls the ACP nodes to find available nodes (not busy) for future data transfers, and updates "clocks" in all ACP nodes (the nodes have no hardware clocks) that are used by the reconstruction algorithm to terminate excessively long reconstructions. The total time must be less than about 8 ms. Based on both measurements and some estimates the time presently required is less than 7 milliseconds.

3 MICROPROCESSOR ARRAY

3.1 Configuration

In the "standard" Fermilab off-line system multiple crates of ACP processors are connected via the Fermilab designed Branch Bus, which in turn is interfaced to the host VAX via a DEC DR11 class device. Since the Branch Bus is a single master system, the FASTBUS system interfaces to the ACP system via a second Branch Bus for which the FBBC is the master. The present plan is to implement a system equivalent to 32 nodes of M68020 processors (two crates). More nodes may be added if more computing power is needed.

The standard off-line ACP system software is used with no modifications. The VAX side of the system runs as if the ACP system is running in an off-line data analysis mode.

3.2 Microprocessor Event Reconstruction

The FASTBUS data from a beam spill are unpacked in the ACP to form event candidates. The data permit rapid identification of the photon trigger time, the photon conversion point in the photon-arm, and electron scintillator times.

Software in the ACP must apply restrictions to these events to accept only 1 out of every 200 events for taping and later analysis. To develop and test software for this purpose, a Monte Carlo program was written to produce both Michel electrons and electron-photon events from muon decays at rest. For each electron scintillator hit, the time difference between the photon trigger and an electron scintillator hit is used to calculate the helical orbit of a 52.8 MeV electron that began with a muon decay in the thin stopping target and had a photon oppositely directed which converted at the specified coordinates in the photon arm. Once an orbit has been calculated for the chosen electron scintillator, anode and cathode hits are sought in the vicinity of the each track crossing a wire chamber. Failure to find these hits for each crossing is sufficient reason to reject the event. The full complement of hits found is then used to refine the reconstructed track trajectory and to obtain a good measure of the reconstructed energy. Finally, the direction of the photon is required to be tangent to the electron orbit at the point of decay. Each hit scintillator is analyzed to determine if it passes all the above cuts until one electron scintillator conforms to an electron track that meets all of these criteria. The on-line analysis of this event is then stopped and the event is written out to tape.

The Monte Carlo studies indicate a high acceptance for signal tracks (88%), an efficient rejection rate for background events (99%), and an average time for on-line event analysis in the ACP of 62 ms which is within a factor of 5 of the design value. We expect to achieve the design value with further optimization and algorithm improvement.

4 STATUS

All basic components of the FASTBUS, ACP and VAX data acquisition system have been implemented and tested at some level for a development run that took place in November 1987. A development run is scheduled for the end of the summer of 1988 that is expected to include one electron-arm chamber and most of a complete photon-arm layer. Actual data-taking runs are expected to begin 1989.

References

1. I. Gaines, et al., "Use of the Fermilab Advanced Computer Program Multi-Microprocessor as an On-Line Trigger Processor", IEEE Transactions on Nuclear Science, Vol. NS-32, pp. 1397-1404, August 1985.

2. H. Muller, "On-Line FASTBUS Processor for LEP", Proceedings of the Conference on Computing in High Energy Physics, Amsterdam, Netherlands, 1985.

3. Creative Electronics Systems, 70 route du Pont-Butin, 1213 Petit-Lancy 1, Switzerland, and Dr. B. Struck, D-2000 Tangstedt, Backerbarg 6, Federal Republic of Germany.

4. J. D. Blake, Use of Microprocessor Cross Software under VAX VMS, CERN Internal Report, CERN/DD/SW-LM/T9, Geneva, Switzerland.

5. T. J. Berners-Lee and A. Pastore, RPC User Manual, CERN DD-Division, OC-Group Internal Report, Geneva, Switzerland, April 27, 1987.

6. B. Flaugher, et al., "Integration of the ACP Microprocessor Farm with the CDF FASTBUS Data Acquisition System", IEEE Transactions on Nuclear Science, Vol. NS-34, pp. 865-869, August 1987.

7. Phillips Scientific, 305 Island Road, Mahwah, NJ 07430, U.S.A. "MEGAblock" is a registered trademark of Phillips Scientific.

Planned Online Event Reconstruction by 370E Emulators for the OPAL Experiment

L Levinson, R Yaari
Physics Department, Weizmann Institute, Rehovot, Israel

The OPAL experiment at LEP will have 5 370E emulators (4 IBM 168 units of CPU power equivalent) in the online event flow to reconstruct events. The emulators are supported in a VME environment. They work in a multi-processing mode, i.e. each emulator processes one event from start to finish, with their I/O requests handled by record servers in both VME and in the online VAX. In this system, unlike in other attached processor systems, the "user's" code is not split between a host and an attached processor. The offline reconstruction program is unchanged (including I/O statements) and resides entirely, and only, in the online emulators. The host is only a record server and down-loader. The OPAL standard offline reconstruction program, written, and frequently changed, by emulator non-experts, will be run on this system. It requires 6MB of memory. With 0.2Hz of multi-hadron events, there will 20 IBM 168 seconds of CPU time available per event.

ARCHITECTURE OF THE OPAL RECONSTRUCTION STAGE

The Reconstruction Stage in the OPAL DAQ hardware consists of a VME crate with a high speed optical link downward to the Software Filter Stage, fast parallel links upward to one or more VAX computers, a VME interface for each 370E, and a Motorola 68020 processor system running a UNIX-like real-time multi-tasking system (OS9). Selected events ($\sim$150K Bytes) are transferred from the Software Filter Stage directly into an available 370E. The reconstructed events are transferred via DMA into an 8MB VME/VSB buffer to await transfer to the VAX. The optical link, DMA, emulator interfaces, and VAX links all transfer data at from 4 to 6MB/sec. If an event was not selected for reconstruction, or if all emulators are busy, it is transferred directly to the output buffer.

RELEVANT CHARACTERISTICS OF THE 370E EMULATOR

The 370E emulator runs a standard IBM load module (i.e. after linking). The IBM 370 instruction set is emulated, including Super-Visor Calls, program exceptions and I/O interrupts. The complete IBM VS FORTRAN language is supported, including sequential and direct access I/O. Hbook, Zebra, Cernlib, etc. and the IBM VS Fortran library all run without change. Results are bit-for-bit identical to IBM. The 370E mini-operating system performs the needed IBM system services and mediates with a host that emulates the I/O services.

THE PROGRAM ENVIRONMENT

Separate record servers will handle events, μ-processor file I/O, VAX file I/O, and error messages to the OPAL online message system. Histograms can be dynamically requested and are written to a data base (Zebra RZ) for display by the OPAL human interface.

A FAST ELECTRONIC TRIGGER AND PREANALYSIS SYSTEM FOR THE CRYSTAL BALL AT LNS

C. Maiolino, P. Finocchiaro, E. Migneco and P. Piattelli

Istituto Nazionale di Fisica Nucleare, Laboratorio Nazionale del Sud
Viale A.Doria (angolo Via S.Sofia), 95125 Catania
Dipartimento di Fisica dell'Università di Catania
Corso Italia 57, 95129 Catania
ITALY

A trigger and preanalysis system, all set up using about sixty CAMAC modules connected via a fast bus in ECL logic, for the Crystal Ball at LNS is described.

The Crystal Ball multidetector is built up by 420 Barium Fluoride crystals [1]. Such a crystal emits two distinct light components, a fast and a slow one, the relative amount of which is strongly dependent on the type, charge and mass of the radiation detected [2]. From each crystal four signals will be taken: a time signal, the fast component signal (F) and two signals for the slow component (E and E') with different attenuation to expand the dynamic range.

All these signals will be digitized using four independent chains of FERA ADCs, whose readout will be enterely done via four ECL auxiliary buses, in zero suppression mode. The data are immediately transferred in four data stack modules to derandomize the input data flow to the trigger and to free the front end of the acquisition. Then the four data flows are synchronized, and the detector identification number is computed and added to each group of data. The next step is the particle-γ discrimination, performed by evaluating the F to E ratio, followed by the neutron identification, performed using the time of flight information. At this stage the event is completely reconstructed and all the calculations on multiplicity and sum energy can be carried out. The results of all the operations performed are added to the event for subsequent analysis. Finally the last section provides a validation on the event, eventually followed by its broadcasting to the computer system via an ECL differential bus.

REFERENCES

1) E.Migneco, Proceedings of the Beijing International Symposium on Physics at Tandem, Beijing China May 26-30 1986, 559, World Scientific Publishing.

2) C.Agodi, R.Alba, G.Bellia, R.Coniglione, A.Del Zoppo, C.Maiolino, E.Migneco, P.Piattelli, P.Sapienza and Yan Chen, Nucl. Instr. and Meth. A, in print.

A Simple Microprocessor for FASTBUS Slave Modules

W. T. Meyer and M. S. Gorbics
Ames Laboratory
Ames, Iowa 50011
USA

ABSTRACT

We describe a simple but powerful implementation of a microprocessor for use as an embedded processing engine in FASTBUS modules. This design uses a 16 MHz 68000 processor and 128 kilobytes of static random access memory, yet it is economical enough for use on boards which are produced in large quantities. The circuit contains 11 chips, draws less than 4 watts of power and costs less than $75 in parts. Software can be written using any assembler or compiler/linker which can create Motorola S-code or an absolute binary image, and several tools exist to aid software development.

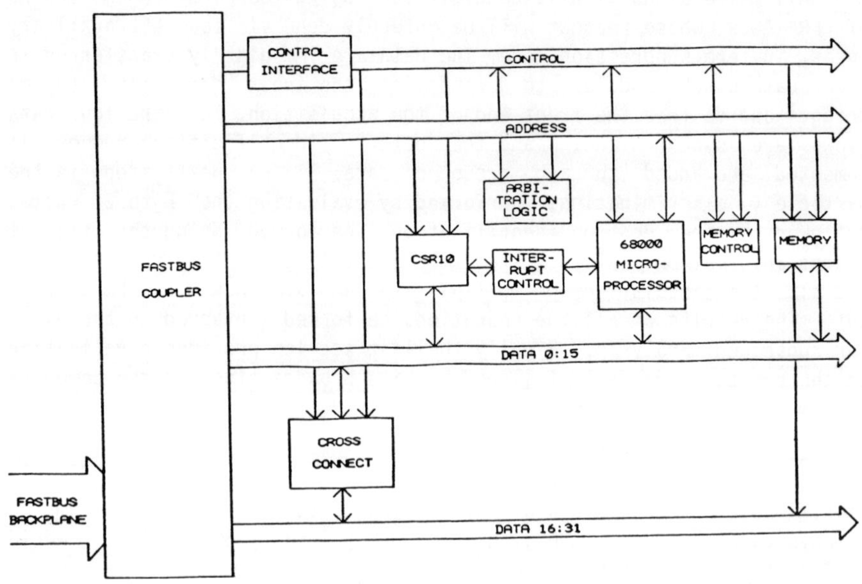

LEPICS
Parallel Processing in a High Energy Physics Computer Center

Richard P. Mount
California Institute of Technology
Pasadena, CA 91125, U.S.A.

LEPICS, the L3 Parallel Integrated Computing System, is under development to meet the CERN-site computing needs of the L3 collaboration. LEPICS will integrate the conventional mainframe environment, essential for many aspects of physics analysis, with cheap attached processors suitable for Monte Carlo, reconstruction, and perhaps some summary tape processing. The core of the system, an IBM 3090 mainframe, has already been installed at CERN, and part of the planned complement of attached processors has now been connected. In the period up to mid 1989, software will be developed which will make the power of an almost arbitrary number of attached processors available transparently to LEPICS users. This software will involve both a standard structure for L3 event processing jobs, and enhancements to the VM/CMS system to allow dynamic scheduling of attached processor resources which is essential if parallel processing is to be effective for a large user community.

The figure below outlines how a Manager and Interface Virtual Machines will make parallel processing available to the users' Task Virtual Machines. The Currently planned resource management functions are:

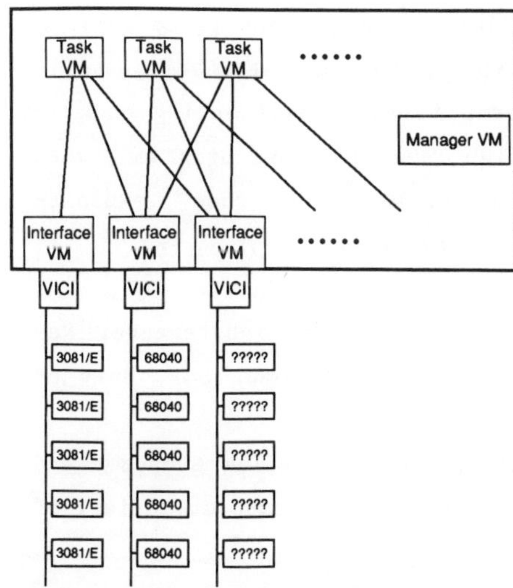

1. Schedule batch jobs taking into account declared multi-processor use.
2. Allocate resources dynamically based on relative task proirity and throughput considerations.
   - offer more processors to some executing tasks.
   - request release of processors by other executing tasks.
   - force release of processors if necessary.
3. Encourage throughput (e.g. by penalties for letting attached processors lie idle).

DATAFLOW COMPUTING TECHNIQUE FOR FAST PROCESSING

S.SANYAL
COMPUTER SYSTEMS & COMMUNICATIONS GROUP
TATA INSTITUTE OF FUNDAMENTAL RESEARCH
BOMBAY-400005, INDIA.

1.Introduction

In the present day researches in Physics and other advanced research areas high speed computation is an absolute must . In Particle Physics , in the areas of optimization of the lattice of bending and focussing magnets and the drift space between them , computers with very high speed are heavily used. Modelling of the Quantum Chromo Dynamics, Theory of Strong Interactions of Hardons involve high speed computing. Achieving these with very fast computers turns out to be prohibitively expensive. A possible cheaper alternative could be Dataflow Computing Technique.

2.Dataflow Computing Technique

This concept uses flow of data (operand) as a means of process scheduling rather than the standard Von Neumann computer architecture where process execution is sequential. In dataflow, speed of computation is only governed by the availability of data [1]. If a program is so structured that huge amount of data is available in parallel, all operations can be done in parallel in independent processors. A dataflow computer designed with multiple processing elements , available cheaply now a days in VLSI form, is capable of acting on multiple data elements in parallel. A high level language program , when compiled , will produce dataflow computer executable code , exploiting the inherent parallelism of the program [2].

3.References

[1] Arvind & Gostelow Kim P. : Dataflow Computer Architecture, Research & Goals, Tech Rept 113. Dept of Info. & Comp. Sc., Univ of California, Irvine,(1978).

[2] Sanyal S. & Mukherjee S. : Simulation of a Dataflow Computer and Some Results, Comp. Sc. & Informatics, Journal of Computer Society of India, 16, 9-19, (1986).

IC DESIGN USING STANDARD CELL AND MACRO CELL

F.Slorach
Rutherford Appleton Laboratory
Chilton,Oxon,England OX11 OQX

Introduction

As the number of methods of integrated circuit design increases, so it becomes vital to ensure that a circuit is implemented on the technology to which it is most suited.

Mixing Standard Cells with blocks of Full Custom layout can provide a route which strikes a good balance of all the important criteria in chip design.

The Reasons for Mixing Standard and Macro cells

One of the functions of the Microelectronics Group at RAL is to provide support for the scientific community, and in this role is often required to produce very specialised integrated circuits in relatively short time spans.

The Group required a method of integration and a CAD system which would meet these needs, as well as providing a route to low-cost and low-volume production.

Full Custom was inappropriate for reasons of time, cost, and the risk of failure in the first iteration. The specialist requirements in sections of the designs precluded the use of Gate Arrays and of the conventional Standard Cell approach.

Implementing these specialised sections of the design as full custom blocks (Macrocells), and building the remainder of the circuit from a cell library (Standard Cells) offered a good solution.

The SDA[1] suite of software was obtained on evaluation from ES2[2] and the Histogrammer design was carried out to explore the Macro cell plus Standard cell method. The SDA software is an integrated set of CAD programs, including analogue [3] and digital [4] simulators, automatic Place and Route, and tools to assist in the design of custom layout.

The Histogrammer has been fabricated and tested and has proven the viability of this approach to Integrated Circuit design. The second design (Coincidence Array) was commisioned for the OPAL Trigger Processor and has been sent for fabrication.

The CAD software reduces the design time for the circuits, and fabrication by direct-write technology cuts out mask-making. Thus the time taken for the whole process of integrating complex digital circuits has been cut significantly.

Mixing of Standard cells and Macro cells has been shown to meet the Group's requirements, and has given designers the means to produce working digital chips more quickly.

---

[1]Silicon Design Automation, Santa Clara, USA

[2]European Silicon Structures

[3]SPICE

[4]SILOS (property of SimuCad)

INTERMARRIAGE IN PERSONAL COMPUTING STANDARDS

B.G. Taylor

EP Division, CERN, 1211 Geneva 23, Switzerland

Abstract

This presentation highlights selected aspects of hardware and software standardisation relevant to the personal computers now widely applied in particle physics experiments. It shows how personal computer bus standards can be married to PC-independent bus standards, and illustrates how the present range of de facto user-interface standards has evolved by cross-fertilization during operating system development.

1 Hardware

Since the introduction of CAMAC over 20 years ago, nonproprietary instrumentation standards have played an important role at international research laboratories. The use of such standards has made it possible to interconnect electronic equipment of varied origin, and has greatly facilitated the work of integrating large systems whose elements have been developed by research collaborators in dispersed institutes.

With the more recent development of the VME/VSB bus family, experimenters have been provided with a high-performance successor to CAMAC which is now supported by over 100 industrial suppliers. The high degree of parallelism possible in a distributed configuration of VME multiprocessors, with independent VSB access to their private resources and buffer memory, minimizes dead-time in the read-out of a large detector facility and allows sophisticated triggering and filtering systems to be implemented.

Modern personal computers such as the Apple Macintosh family can be used as stand-alone software development workstations for such systems and their graphics-oriented user-interface has proved well-suited to control and monitoring tasks during data-taking. MacVEE systems make it possible to marry the internal bus of these personal computers with external international-standard bus systems to provide direct memory-mapped access to them from the PC microprocessor.

For example, the open-architecture 68020-based Macintosh II is provided with six Apple NuBus slots. It can be interfaced to multi-crate VMEbus and CAMAC systems by MICRON (MacVEE Interface Card Resident On NuBus) as shown in Fig. 1. The Apple NuBus is a time-quantized asynchronous computer-backplane bus structure, with geographical addressing and multiplexed 32-bit address and data lines. MICRON supports 8, 16 and 32-bit operations on the NuBus, and automatically performs the required VMEbus operations so that the interface is transparent to the user.

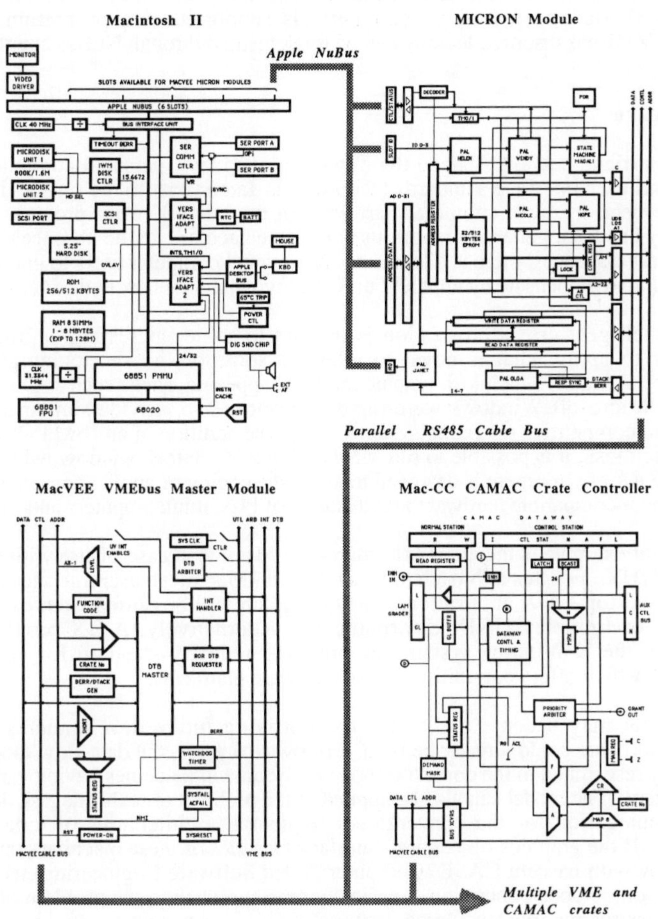

Fig. 1 Integrating a personal computer with international bus standards

The large addressing range of the 68020 allows 128 Mbytes of contiguous superslot address space to be dedicated to the external system and directly accessed by the microprocessor in 32-bit mode. In its normal slot space, the MICRON module itself can also accommodate up to 512 Kbytes of EPROM, a small part of which contains slot declaration data. All of the address non-alignment functionality of the 68020 can be supported through to VMEbus slaves, and data caching (which may be introduced in future 68030-based Macintosh family computers) is supported as a programmable option. Restricted VMEbus resource-locking can be implemented through NuBus attention cycles.

2 Software

Intermarriage has also shaped the evolution of today's personal computer operating systems and user-interface standards (see Fig. 2). In contrast to the outstanding progress which has been made in computer hardware in the last 20 years, progress in software technology has been generally laborious and unspectacular. In the absence of many truly revolutionary advances, there is a tendency for personal computer users to gravitate towards systems which compensate by apparent universality what they lack in distinctive originality.

But intelligent cross-fertilization is leading to some interesting approaches which endeavour to apply what is novel to alleviate some of the shortcomings of what is established. After the Macintosh graphic interface appeared, shells such as Digital Research GEM and Microsoft Windows were rapidly developed to provide similar user-interface functions to competitive operating systems. With the addition of an IBM PC emulator card for the Macintosh, it is possible to run MS-DOS in a Macintosh window, while IBM's own projected SAA environment is designed to allow the coding of applications that are portable between the incompatible hardware architectures of PCs, minicomputers and mainframes.

A recent example is the attempt to marry the Macintosh ease-of-use with the power of UNIX in A/UX. For this approach to be successful, Macintosh user interfaces will have to be developed for A/UX applications, most of which will be ported from existing AT&T System V or Berkeley UNIX environments. Alternatively, A/UX programmers can develop on the X-Window system, an emerging industry standard for network-based windowing which offers portability among hardware platforms.

But what may become one of the most striking future developments in personal computer software could already be born. For over 10 years, the data-flow model has been extensively researched in the context of non von Neumann machines having various degrees of parallelism. The model can also be applied to the problem of real-time task definition and intercommunication for the conventional sequential architectures of today's personal computers. If the graphics-object user-interface facilities of these machines can be married successfully with nascent CASE (Computer Aided Software Engineering) technology for automatic source-code generation, a radically new approach to the problem of application software construction (the term 'programming' seems inappropriate) could be created.

Physics users of personal computers have quite particular needs oriented to real-time data input and software development flexibility, combined with requirements for the recording of quite large data volumes and powerful tools for interactive output data manipulation and display. Preliminary work using MacVEE systems with Macintosh II at CERN and in the USA suggests that the intermarriage of CASE techniques with international instrumentation standards and conventional scientific libraries holds considerable promise for the future.

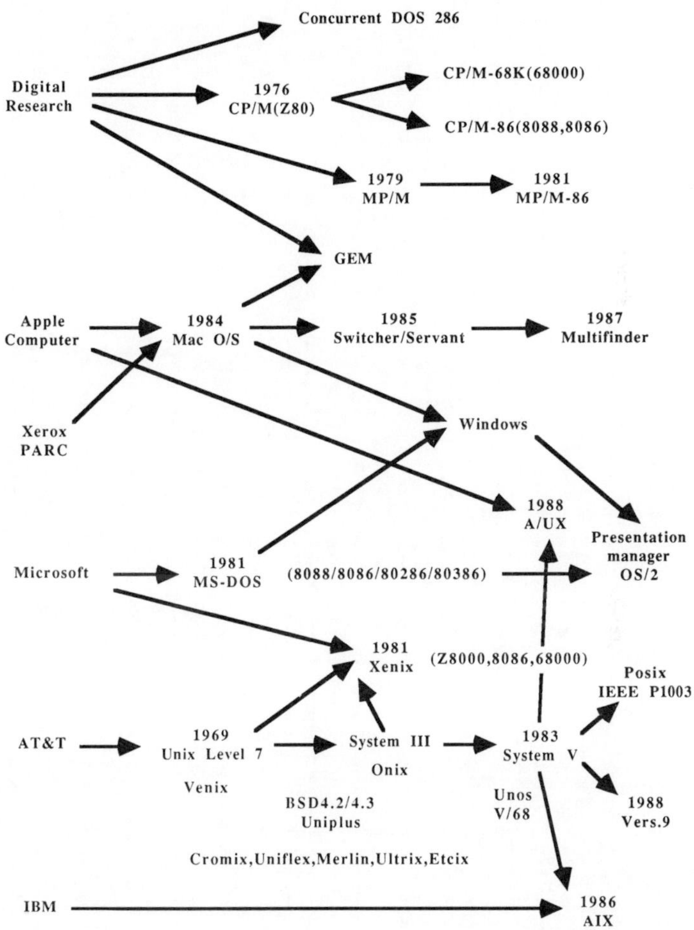

Fig. 2 Cross-fertilization in personal computer operating system development

List of partecipants

ABBES Mohamed	College de France - Laboratoire de Physique Corpusculaire Paris FRANCE
ABDEL MEGED Mahmoud M.	University of Qatar Doha QATAR
AIELLO Roberto	Dipartimento di Fisica dell' Università and INFN Trieste ITALY
AKIMEUKO Serguei	I.V. Kurchatov Institute of Atomic Energy Moscow U.S.S.R.
ALI Kamil Al-Sheikh	Scientific Research Council - Solar Energy Research Centre Baghdad IRAQ
ALMEIDA Teresa	L.I.P. Lisbon PORTUGAL
AMENDOLIA S.R.	INFN - Sezione di Pisa Pisa ITALY
AVETISJAN Albert	Yerevan Physics Institute Yerevan U.S.S.R.
BARREIRA Gaspar Pereira	L.I.P. Lisbon PORTUGAL
BARRELET Etienne	Universitè Paris VI et VII - LPHNE Paris FRANCE
BATTAIOTTO Pedro Eduardo	ICTP-INFN Microprocessor Laboratory Trieste ITALY
BECONCINI Fabio	INFN - Sezione di Pisa Pisa ITALY
BEKER Harold	CERN - Ep Division Geneva SWITZERLAND
BERNAUDIN Philippe	Universitè Paris XI - (Paris-Sud) - LAL Orsay FRANCE
BERTOCCHI Luciano	International Centre For Theoretical Physics Trieste ITALY
BERTRAND Jean-Luc	Universitè Paris XI - (Paris-Sud) - LAL Orsay FRANCE
BIBOK Gyoergy	Hungarian Academy of Sciences - Inst. Nucl. Res. Debrecen HUNGARY
BIFONE Angelo	Università degli Studi "La Sapienza" Roma ITALY
BINGEFORS Nils	University of Uppsala - Institute of Physics Uppsala SWEDEN
BUDINICH Marco	Dipartimento di Fisica dell' Università and INFN Trieste ITALY
BULFONE Daniele	Dipartimento di Fisica dell' Università and INFN Trieste ITALY

BURCKHART Doris	CERN - DD Division Geneva SWITZERLAND
BUSHNIN Juri	Moscow Institute For High Energy Physics Moscow U.S.S.R.
BUYTAERT Jan	Universitè Libre de Bruxelles - Institut de Physique Brussels BELGIUM
CABRAS Giuseppe	Università degli Studi di Udine and INFN Udine ITALY
CAFARELLA Lili	Università degli Studi "La Sapienza" Roma ITALY
CAMPBELL Alan	Deutsches Electronen Synchrotron (Desy) Hamburg GERMAN FEDERAL REPUBLIC
CARTER Jeremy	Royal Holloway & Bedford New College Egham UNITED KINGDOM
CASTELLI Edoardo	Dipartimento di Fisica dell' Università and INFN Trieste ITALY
CASTRO Andrea	Università degli Studi di Padova Padova ITALY
CECCHET Giorgio	INFN - Sezione di Pavia Pavia ITALY
CENTRO Sandro	Università degli Studi di Padova and INFN Padova ITALY
CHARPENTIER Philippe	CERN - EP Division Geneva SWITZERLAND
CHIARINI Armando	Università di Bologna - Laboratorio di Ingegneria Nucleare Bologna ITALY
CHOLLET Frederique	Laboratoire de Physique Des Particules (Lapp) Annecy Le Vieux FRANCE
CHOROWICZ Michel	CEN - Saclay Gif Sur Yvette FRANCE
CHRISTOFORI Pier Paolo	Università degli Studi "La Sapienza" Roma ITALY
COLAVITA Alberto Antonio	ICTP-INFN Microprocessor Laboratory Trieste ITALY
COMIN Mauro	Area di Ricerca - Sincrotrone Trieste ITALY
CONETTI Sergio	Mcgill University Montreal CANADA
COSTANTINI Flavio	INFN - Sezione di Pisa Pisa ITALY
COUTURES Christian	CEN - Saclay Gif Sur Yvette FRANCE
CROSETTO Dario	Istituto di Fisica dell' Università and INFN Torino ITALY

CUTTS Dave	Brown University Providence UNITED STATES OF AMERICA
D'ANTONE Ignazio	INFN - Sezione di Bologna Bologna ITALY
DARBO Giovanni	Università degli Studi di Genova and INFN Genova ITALY
DEGRE' Andrè	CERN - EP Division Geneva SWITZERLAND
DELL'ORSO Mauro	Università di Pisa Pisa ITALY
DENBY Bruce	Universitè Paris XI - (Paris-Sud) - LAL Orsay FRANCE
DI PIRRO Giampiero	INFN - Laboratori Nazionali di Frascati Frascati ITALY
DIOS Zoltan	Hungarian Academy of Sciences - Inst. Nucl. Res. Debrecen HUNGARY
DIVIA' Roberto	CERN - DD Division Geneva SWITZERLAND
DOBINSON Robert William	CERN - EP Division Geneva SWITZERLAND
DUCORPS Antoine	Universitè Paris XI - (Paris-Sud) - LAL Orsay FRANCE
EINSWEILER Kevin	CERN - EP Division Geneva SWITZERLAND
EISENHANDLER Eric	Queen Mary College London UNITED KINGDOM
ENDER Christoph	Universitat Heidelberg - Physikalishces Institute Heidelberg GERMAN FEDERAL REPUBLIC
FAIVRE Jean-Claude	CEN-Saclay - Service Phys. Nucl. Moyenne Energie Gif Sur Yvette FRANCE
FALCIANO Speranza	INFN - Sezione di Roma Roma ITALY
FANET Herve	CEN-Saclay - Service Phys. Nucl. Moyenne Energie Gif Sur Yvette FRANCE
FERNANDEZ Jorge Eduardo I.	Università di Bologna - Lab. di Ingegneria Nucleare Bologna ITALY
FINOCCHIARO Paolo	INFN - Laboratorio Nazionale del Sud Catania ITALY
FORMENTI Fabio	CERN - EP Division Geneva SWITZERLAND
FOSTER Brian	University of Bristol - H.H.Wills Physics Laboratory Bristol UNITED KINGDOM
FOURNIER Guy	CEN Saclay - Service Phys. Nucl. -Haute Energie Gif Sur Yvette FRANCE

FRENCH Marcus Julian — Rutherford Appleton Laboratory, Didcot UNITED KINGDOM

GAINES Irwin — Fermi National Accelerator Laboratory, Batavia UNITED STATES OF AMERICA

GAL Janos — Hungarian Academy of Sciences - Inst. Nucl. Res., Debrecen HUNGARY

GALEOTTI Stefano — INFN - Sezione di Pisa, Pisa ITALY

GAMBA Diego — Università di Torino - Sezione INFN, Torino ITALY

GASPAR Clara — CERN - EP Division, Geneva SWITZERLAND

GAVILLET Philippe — CERN - EF Division, Geneva SWITZERLAND

GIANNETTI Paola — ICTP-INFN Microprocessor Laboratory, Trieste ITALY

GLANZMAN Thomas — Stanford University - Stanford Linear Accelerator Center, Stanford UNITED STATES OF AMERICA

GOMES Paulo Filipe Do Carm — L.I.P., Lisbon PORTUGAL

GONCALVES Manuel — L.I.P., Lisbon PORTUGAL

GRAYER Geoffrey — Rutherford and Appleton Laboratories, Didcot UNITED KINGDOM

HARRIS John F. — Kernforschungsanlage Julich Gmbh, Julich GERMAN FEDERAL REPUBLIC

HEATH Greg — University of Oxford, Oxford UNITED KINGDOM

HEGYESI Gyula — Hungarian Academy of Sciences - Inst. Nucl. Res., Debrecen HUNGARY

HEIJNE Henricus — CERN - EF Division, Geneva SWITZERLAND

HINTON Robert — CERN - DD Division, Geneva SWITZERLAND

HOEK Jaap — Rutherford and Appleton Laboratories, Didcot UNITED KINGDOM

HOLLINGWORTH Kennet — CERN, Geneva SWITZERLAND

HONSCHEID Klaus — Universitat Bonn - Physikalisches Institut, Bonn GERMAN FEDERAL REPUBLIC

HRISOHO Aleksandar — Universitè de Paris XI (Paris-Sud) - LAL, Orsay FRANCE

HUET Michel — CEN Saclay - Service Phys. Nucl. - Haute Energie, Gif Sur Yvette FRANCE

INCANDELA Joseph R. — CERN - EP Division
Geneva SWITZERLAND

INNOCENTI P.G. — CERN - SPS Division
Geneva SWITZERLAND

JAROSLAWSKI Stanislav — Rutherford and Appleton Laboratories
Didcot UNITED KINGDOM

JOHANSSON A. — University of Stockholm
Stockholm SWEDEN

JOST Beat — CERN - Ep Division
Geneva SWITZERLAND

KACK Soren — University of Lund
Lund SWEDEN

KEH Stefan — CERN - EP Division
Geneva SWITZERLAND

KLOK Peter F. — Katholieke Universiteit Nijmegen
Nijmegen NETHERLANDS

KOZLOWSKI Thomas — Los Alamos National Laboratory
Los Alamos UNITED STATES OF AMERICA

KURT Adnan — Bogazici Universitesi
Istanbul TURKEY

LAMANNA Ernesto — Università degli Studi "La Sapienza"
Roma ITALY

LANCERI Livio — INFN - Sezione di Trieste
Trieste ITALY

LE COQ Jacques — Laboratoire de Physique Des Particules (Lapp)
Annecy Le Vieux FRANCE

LE DU Patrick — CEN - Saclay
Gif Sur Yvette FRANCE

LEBRUN Paul — Fermi National Accelerator Laboratory
Batavia UNITED STATES OF AMERICA

LEVINSON Lorne — Weizmann Institute of Sciences
Rehovot ISRAEL

LI Lin — Institut Fuer Theoretische Physik E.T.H.-Hoenggerberg
Zurich SWITZERLAND

LIPPI Ivano — INFN Sezione di Padova
Padova ITALY

LJUSLIN Chris — CERN - DD Division
Geneva SWITZERLAND

LOKEN James Gilbert — University of Oxford Nuclear Physics Laboratory
Oxford UNITED KINGDOM

LOMBARDO Qaria Paola — INFN - Sezione di Pisa
Pisa ITALY

LONDEI Alessandro — Università degli Studi "La Sapienza"
Roma ITALY

LONE Sigurd	CERN - DD Division Geneva SWITZERLAND
MACAVERO Egidio	Laben Milano ITALY
MAGGIORA Angelo	Università di Torino - Sezione INFN Torino ITALY
MAIOLINO Concettina	INFN - Laboratorio Nazionale del Sud Catania ITALY
MANDRIOLI Gianni	INFN - Sezione di Roma Roma ITALY
MANNS Jochen	Universitat Bonn Physikalisches Institut Bonn GERMAN FEDERAL REPUBLIC
MARCHIORO Alessandro	CERN - EP Division Geneva SWITZERLAND
MARON Gaetano	INFN-Laboratori Nazionali di Legnaro Legnaro ITALY
MARROCHESI P.S.	CERN - EP Division Geneva SWITZERLAND
MARTIN Jean-Pierre	CERN - EP Division Geneva SWITZERLAND
MARZANO Francesco	INFN - Sezione di Roma Roma ITALY
MATTEUZZI Pietro	INFN - Sezione di Roma Roma ITALY
MAY David	Inmos Ltd. Bristol UNITED KINGDOM
MCLAREN Robert	CERN - DD Division Geneva SWITZERLAND
McPHERSON George Michael	Rutherford and Appleton Laboratories Didcot UNITED KINGDOM
MEIJERS Franciscus	CERN - EP Division Geneva SWITZERLAND
MENCIK Maurice	Universitè Paris XI - (Paris-Sud) - LAL Orsay FRANCE
MERTENS Volker	CERN - EP Division Geneva SWITZERLAND
MEYER Walter Thomas	CERN - EP Division Geneva SWITZERLAND
MIOTTO Alessandro	CERN - EF Division Geneva SWITZERLAND
MJORNMARK Ulf	University of Lund Lund SWEDEN
MORANDO Maurizio	Università degli Studi di Padova Padova ITALY

MOUNT Richard	CERN - EP Division Geneva SWITZERLAND
NAGPAL Tarlok	Bishop University Lennoxville CANADA
NAMJOSHI Rohit	Cornell University - Cleo Collaboration/Wilson Laboratory Ithaca UNITED STATES OF AMERICA
NASH Tom	Fermi National Accelerator Laboratory Batavia UNITED STATES OF AMERICA
NOTZ Dieter	Deutsches Electronen Synchrotron (Desy) Hamburg GERMAN FEDERAL REPUBLIC
PAAL Andras	Hungarian Academy of Sciences - Inst. Nucl. Res. Debrecen HUNGARY
PANZIERI Daniele	Università di Torino and Sezione INFN Torino ITALY
PARISI Giorgio	Università degli Studi "La Sapienza" Roma ITALY
PASCOLI Donatella	Università degli Studi di Padova Padova ITALY
PEGORARO Matteo	Università degli Studi di Padova Padova ITALY
PERRIER Jacques	Universite de Genève - Ecole de Physique/Dpnc Geneva SWITZERLAND
PETROLO Emilio	INFN - Sezione di Roma Roma ITALY
PIATTELLI Paolo	INFN - Laboratorio Nazionale del Sud Catania ITALY
PITKE Madhukar	Tata Institute of Fundamental Research Bombay INDIA
PONTECORVO Ludovico	Università degli Studi "La Sapienza" Roma ITALY
POSE Rudolf Arthur	Akademie der Wissenschaften der DDR Berlin GERMAN DEMOCRATIC REPUBLIC
PROFESSI Jorge	INFN - Laboratori Nazionali di Legnaro Padova ITALY
QUINTON Stephen	Rutherford and Appleton Laboratories Didcot UNITED KINGDOM
REIS Mario	CERN Geneva SWITZERLAND
RICHTER Mathias	Gesellschaft Fur Schwerionenorschung Mbh Darmstadt GERMAN FEDERAL REPUBLIC
RINAUDO Giuseppina	Istituto di Fisica dell' Università and INFN Torino ITALY
RISTORI Luciano	INFN - Sezione di Pisa Pisa ITALY

RONGA Francesco	INFN - Laboratori Nazionali di Frascati Frascati ITALY
ROUGER Michael	CEN - Saclay - Service Phys. Nucl. - Moyenne Energie Gif Sur Yvette FRANCE
SALINA Gaetano	INFN - Sezione di Roma Roma ITALY
SCARLATELLA Michele	Stanford University - Stanford Linear Accelerator Center Stanford UNITED STATES OF AMERICA
SCHALK Terry	University of California - Institute of Particle Physics Santa Cruz UNITED STATES OF AMERICA
SEVER Franc	CERN - EP Division Geneva SWITZERLAND
SFORZI Giuliano	Informatica Friuli Venezia Giulia Trieste ITALY
SHIELD Peter	University of Oxford Oxford UNITED KINGDOM
SICKMULLER Harold	Universitat Heidelberg - Physikalishces Institute Heidelberg GERMAN FEDERAL REPUBLIC
SILVA Carlos	CERN Geneva SWITZERLAND
SLORACH Fergus	Rutherford and Appleton Laboratories Didcot UNITED KINGDOM
SPIELER Helmuth	Lawrence Berkeley Laboratory Berkeley UNITED STATES OF AMERICA
SPRENG Wolfgang	Gesellschaft Fur Schwerionenorschung Mbh Darmstadt GERMAN FEDERAL REPUBLIC
STAPNES Steinar	CERN - EP Division Geneva SWITZERLAND
STEFANINI Giorgio	CERN - EP Division Geneva SWITZERLAND
SZABO Istuan	Hungarian Academy of Sciences - Inst. Nucl. Res. Debrecen HUNGARY
TAYLOR Bruce G.	CERN - EP Division Geneva SWITZERLAND
THALER Jon	University of Illinois Urbana UNITED STATES OF AMERICA
TLACZALA Wieslaw	CERN - EP Division Geneva SWITZERLAND
TOFFOLI Tommaso	MIT Laboratory for Computer Science Cambridge UNITED STATES OF AMERICA
TOSELLO Flavio	Università di Torino - Sezione INFN Torino ITALY
TRASATTI Luciano	INFN - Laboratori Nazionali di Frascati Frascati ITALY

TRIPICCIONE Raffaele — INFN - Sezione di Pisa
Pisa ITALY

VANUXEM Jean Pierre — CERN - EP Division
Geneva SWITZERLAND

VERDINI Piero Giorgio — INFN - Sezione di Pisa
Pisa ITALY

VERKERK Catharinus — CERN - DD Division
Geneva SWITZERLAND

VERMEULEN Joseph Cornelis — NIKHEF
Amsterdam NETHERLANDS

VISSCHERS Jan Lammert — NIKHEF
Amsterdam NETHERLANDS

VON RUDEN Wolfgang — CERN - EF Division
Geneva SWITZERLAND

WANG Xiao Jing — INFN - Laboratori Nazionali di Legnaro
Padova ITALY

WESLEY TANASKOVIC Ines — United Nations University
Tokyo JAPAN

WILLIAMS David — CERN - DD Division
Geneva SWITZERLAND

YAARI Rafi — Weizmann Institute of Sciences
Rehovot ISRAEL

ZANELLO Lucia — Università degli Studi "La Sapienza" and INFN
Roma ITALY

ZOFFOLI Matteo — Università degli Studi "La Sapienza"
Roma ITALY

ZOGRAFOS Konstantinos — CERN - PS Division
Geneva SWITZERLAND

AUTHOR INDEX